AF410827

Internal Combustion Engines Improving Performance, Fuel Economy and Emissions

Internal Combustion Engines Improving Performance, Fuel Economy and Emissions

Special Issue Editors

Federico Millo
Lucio Postrioti

MDPI • Basel • Beijing • Wuhan • Barcelona • Belgrade • Manchester • Tokyo • Cluj • Tianjin

Special Issue Editors
Federico Millo
Corso Duca degli Abruzzi 24
Italy

Lucio Postrioti
Università degli Studi di Perugia
Italy

Editorial Office
MDPI
St. Alban-Anlage 66
4052 Basel, Switzerland

This is a reprint of articles from the Special Issue published online in the open access journal *Energies* (ISSN 1996-1073) (available at: https://www.mdpi.com/journal/energies/special_issues/ ICE_IP_FEE).

For citation purposes, cite each article independently as indicated on the article page online and as indicated below:

LastName, A.A.; LastName, B.B.; LastName, C.C. Article Title. *Journal Name* **Year**, *Article Number*, Page Range.

ISBN 978-3-03936-168-7 (Hbk)
ISBN 978-3-03936-169-4 (PDF)

Contents

About the Special Issue Editors

Federico Millo is a full-time professor of automotive internal combustion engines at Politecnico di Torino, Italy, where he also received his master's degree in mechanical engineering in 1989, before joining the faculty as a research assistant in 1991. He has published over 150 articles based on his research activity, focused on internal combustion engines and hybrid powertrains. He has been a principal investigator for a number of research projects with major OEMs, such as General Motors, FCA, Honda and Ferrari, and the coordinator for several research projects funded by national and regional Italian public institutions. He was the recipient of the 2011 Honda Initiation Grant Europe, and in 2016, he was recognized as an SAE Fellow for contributions to a wide range of IC engine issues and the energy management of hybrids. He was the first Italian from academia to be elevated to the role of Fellow.

Lucio Postrioti received his master's degree in mechanical engineering from the Università di Perugia in 1994 and his PhD in energy systems from Politecnico di Bari in 1999. In 1998, he joined the Università di Perugia Engineering Faculty, where has been Associate Professor of Mechantronics and Internal Combustion Engines since 2005. He has published more than 70 papers focused on his research activity, specifically about the experimental analysis of fuel injection systems and internal combustion engines. He has been a scientific director in several joint research programs with automotive OEMs (Ducati, General Motors, Lamborghini, Continental, Magneti Marell) and research programs funded by National Italian Institutions. He is the co-founder of two academic spin-off companies involved in the development of testing devices and serves as an R&D testing partner for several major automotive OEM companies.

Preface to "Internal Combustion Engines Improving Performance, Fuel Economy and Emissions"

Internal combustion engines are facing unprecedented challenges to reduce their adverse environmental impact in terms of pollutant and greenhouse emissions, while continuing the improvement of their performance. To achieve the ambitious goals of meeting US Tier3 and post-Euro 6 emissions standards within the extremely demanding Real Driving Emissions test protocols, while simultaneously reaching the challenging post-2020 CO2 emissions targets, the automotive industry is going to deploy an unparalleled mix of technological developments. These developments will range from the use of alternative fuels, advanced fuel injection and combustion technologies, to aftertreatment and powertrain electrification. This Special Issue aims, therefore, to encourage both academic and industrial researchers to present their latest findings concerning technologies which enable pollutant emissions reduction and fuel economy and performance improvements for internal combustion engines. This will provide readers with a comprehensive, unbiased, and scientifically sound overview of the most recent research and technological developments in this field.

Federico Millo, Lucio Postrioti
Special Issue Editors

Article

Development and Assessment of an Integrated 1D-3D CFD Codes Coupling Methodology for Diesel Engine Combustion Simulation and Optimization

Federico Millo [1,*], Andrea Piano [1], Benedetta Peiretti Paradisi [1], Mario Rocco Marzano [1], Andrea Bianco [2] and Francesco C. Pesce [3]

[1] Energy Department, Politecnico di Torino, 10129 Torino, Italy; andrea.piano@polito.it (A.P.); benedetta.peiretti@polito.it (B.P.P.); mario.marzano@polito.it (M.R.M.)
[2] POWERTECH Engineering, 10127 Torino, Italy; a.bianco@pwt-eng.com
[3] General Motors Global Propulsion Systems, 10129 Torino, Italy; francesco_concetto.pesce@gm.com
[*] Correspondence: federico.millo@polito.it

Received: 24 February 2020; Accepted: 30 March 2020; Published: 1 April 2020

Abstract: In this paper, an integrated and automated methodology for the coupling between 1D- and 3D-CFD simulation codes is presented, which has been developed to support the design and calibration of new diesel engines. The aim of the proposed methodology is to couple 1D engine models, which may be available in the early stage engine development phases, with 3D predictive combustion simulations, in order to obtain reliable estimates of engine performance and emissions for newly designed automotive diesel engines. The coupling procedure features simulations performed in 1D-CFD by means of GT-SUITE and in 3D-CFD by means of Converge, executed within a specifically designed calculation methodology. An assessment of the coupling procedure has been performed by comparing its results with experimental data acquired on an automotive diesel engine, considering different working points, including both part load and full load conditions. Different multiple injection schedules have been evaluated for part-load operation, including pre and post injections. The proposed methodology, featuring detailed 3D chemistry modeling, was proven to be capable assessing pollutant formation properly, specifically to estimate NOx concentrations. Soot formation trends were also well-matched for most of the explored working points. The proposed procedure can therefore be considered as a suitable methodology to support the design and calibration of new diesel engines, due to its ability to provide reliable engine performance and emissions estimations from the early stage of a new engine development.

Keywords: diesel engines; numerical simulation; pollutant emissions prediction; computational fluid dynamics

---

1. Introduction

Diesel engine performance and emissions are strongly dependent on the fuel spray injection, in-cylinder mixture formation, and combustion processes. The highly demanding legislative and environmental targets mandatorily require a clear understanding of the interaction between the fuel spray, the in-cylinder swirling flow field, and the piston bowl, to characterize the combustion efficiency and the pollutant formation phenomena correctly. Therefore, reliable Computational Fluid Dynamics (CFD) simulations are of paramount importance to integrate experimental studies for an efficient optimization and design of new Diesel combustion systems. Furthermore, with the exponential increase of the computational power of Central Processing Units (CPUs), extremely detailed physical and chemical models can be run with an acceptable computational time, becoming fundamental in the first stages of the design and development phase. On one hand, properly calibrated 1D-CFD models can be considered as adequate to reproduce the engine behavior accurately with a minimum computational effort. However, the development of new engine designs can condition the engine behavior significantly (e.g., a new piston bowl shape affects combustion) and requires a new calibration with experimental data in order to guarantee a satisfactory level of accuracy. On the other hand, the use of a 3D-CFD numerical model with its strong predictive capability can be helpful in simulating new engine designs to avoid an expensive and time-consuming experimental campaign. It is worth pointing out that a 0D/1D multizone combustion model could provide satisfying results with the aim to optimize the fuel injection strategy, as Piano et al. show in [1]. In their study, a detailed 1D model of the fuel injector coupled with a 1D model of the engine is employed to minimize Brake Specific Fuel Consumption (BSFC) and Combustion Noise (CN) without exceeding the Brake Specific NOx (BSNOx) baseline value. However, 3D-CFD detailed combustion analysis becomes essential when a substantial reshaping of the combustion chamber is introduced or when the post injection potential in soot emissions reduction is evaluated [2]. In addition, the two different numerical approaches could be considered complementary; in fact, co-simulations exploit the strongest points of both approaches, minimizing their drawbacks. For example, in engine simulation, an interesting option could be the development of a 1D/3D-CFD coupled methodology: 3D approach to simulate the complex components and flow processes in the combustion chamber, and 1D approach to solve the intake and exhaust systems gas flow. Following this procedure the possibility to integrate fast running 3D-CFD simulations into a 1D model is proposed in some commercial codes, where the two simulations run in parallel exchanging flow information (i.e., fluid pressure, velocity, temperature, and composition) at every timestep at specified boundaries. This approach is quite common and straightforward and could be pursued for detailed modeling of components (e.g., manifolds, compressors, turbines, valves, etc.), where the flow contains significant 3D effects, or 3D flow results are of specific interest. Several authors have chosen this option to optimize the geometry of complex engine components such as intake manifolds [3–5] or innovative cooling systems [6]. The purpose of this study is to propose an alternative possibility of 1D/3D coupling: 3D-CFD simulations are implemented to perform a detailed combustion analysis, using the 1D models to provide time-varying boundary conditions (BCs), a reliable injection rate profile, and to estimate brake specific quantities and other global engine parameters. To obtain high-fidelity results from a 3D-CFD in-cylinder simulation, it is necessary to impose realistic pressure and temperature traces at the intake and exhaust ports, as well as the initial thermodynamic conditions and species mass fractions inside the cylinder region. The integration of a 1D complete engine model to generate boundary and initial conditions is a well-established procedure, adopted for different applications ranging from dual fuel combustion analysis [7,8], to the combustion characterization in a two-stroke opposed piston engine [9], to the detection of knock tendency in spark ignition engines [10,11], as well as to the study of alternative combustion modes in Diesel engines [12]. However, while 1D-CFD codes could be implemented to support several types of analysis as intake/exhaust system optimization, turbo matching or global engine performance assessment, they are characterized by a limited predictive capability as far as the combustion process is concerned, especially in case of complex injection schedules (e.g., multiple injections, post injection for soot oxidation). Differently, a stand-alone 3D-CFD simulation could be suitable for detailed combustion analysis, but it is generally limited to the description of in-cylinder phenomena. The aim of the present work is therefore the development of an integrated and automated procedure able to achieve a realistic Diesel engine combustion simulation, as suitable support to predict engine performance parameters and emissions estimations in the early-stage design phase of a new engine, when no experimental data are available. Pursuing this target, the coupling of 1D models with the 3D-CFD simulation is fundamental to provide reliable and accurate combustion data to the "global" 1D-CFD model and not only to characterize the 3D simulation with proper BCs.

In the first part of the present work, the simulation methodology is described in detail, dwelling on the different setups, the CFD models chosen and the coupling technique between the 1D and the 3D setups. The essential elements of the described approach are a reliable spray calibration methodology and a combustion model coupled with a detailed chemistry scheme, to correctly characterize the emissions formation phenomena. A reliable spray modeling is fundamental to predict the fuel spray formation process and how it affects the combustion processes and should be accurately validated, as can be seen in other researches [13,14]. Moreover, one of the main novelties integrated into the simulation setup is the presence of a detailed 1D model of the injector, extensively described in [15,16], to provide a realistic injection rate also in presence of complex injection schedule, which is one of the fundamental inputs of a reliable 3D-CFD analysis. The 1D environment is also employed as a primary workflow for the whole engine design to obtain realistic time-dependent BCs. Finally, the presence of a 3D-CFD results post-processing is integrated, not only to guarantee consistency with respect to experimental data in terms of in-cylinder pressure and heat release rate comparison, but also to evaluate global engine parameters such as brake specific quantities, even in the preliminary engine design phase. All the steps of the numerical analysis are connected in an automated procedure thanks to the use of in-house developed Linux scripts, which allows exchanging data between the 1D-CFD software and the 3D-CFD one.

The entire setup is validated against experimental data on three working points of the engine map representatives of a typical type approval driving cycle, also considering complex injection strategies with pilot and post injections. The possibility to investigate the mixing improvement, combustion, and pollutant emissions formation when using different injection strategies becomes very attractive using 3D-CFD [17,18], and an example of this type of optimization is illustrated in the second part of the work. At one working point, a further assessment is carried out, varying the post injection strategy in order to minimize the soot emissions, being an interesting application of the present methodology. Other future applications can be the capability to study the influence of different fuel injection system designs on the spray targeting, as Leach et al. have done in [19] or an appropriate description of the combustion process involving complex piston bowl geometries [20,21]. Indeed, since the in-cylinder air and fuel motion control the combustion process, affecting the NOx and soot formation, the present multidimensional model could be a powerful tool to support the development phase of new piston bowl shapes improving air and fuel mixing to achieve a better spatial distribution of the fuel, also for complex, multi-pulse injection schedules [1], which are nowadays of paramount importance to achieve the ultra-low engine-out emissions levels necessary to fulfill the legislation limits.

2. Methodology

The engine selected for this study is a 4-cylinder Diesel turbocharged engine for light-duty applications, featuring a common rail injection system and a high-pressure Exhaust Gas Recirculation (EGR) loop. Table 1 provides more information about the engine.

Table 1. Main design characteristics of the test engine.

Cylinders	4
Bore × Stroke	79.7 mm × 80.1 mm
Displacement	1.6 L
Compression ratio	16:1
Turbocharger	Single-Stage with Variable Geometry Turbine (VGT)
Fuel injection system	Common rail Max Rail Pressure 2000 bar
Rated power and torque	100 kW @ 4000 rpm 320 Nm @ 2000 rpm

The test engine features a re-entrant type of bowl design, shown in Figure 1 and it is equipped with an 8-holes solenoid injector.

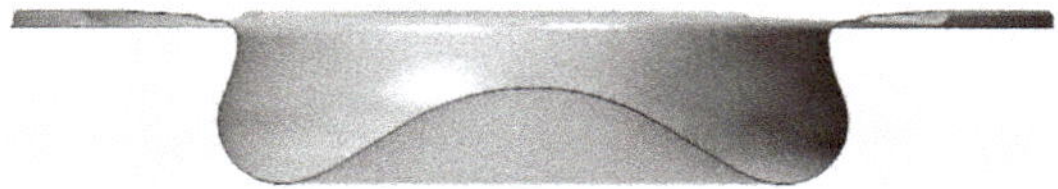

Figure 1. Re-entrant bowl design.

In order to validate the simulation setup, three different engine working points were explored, two of which at part load and one corresponding to the rated power, as depicted in Figure 2 on the engine operating map and listed in Table 2. The two part load points can be considered as representatives of operating conditions during the execution of type approval driving cycles, where the choice of the third working point was made to test the reliability of the numerical model also in high speed-high load conditions.

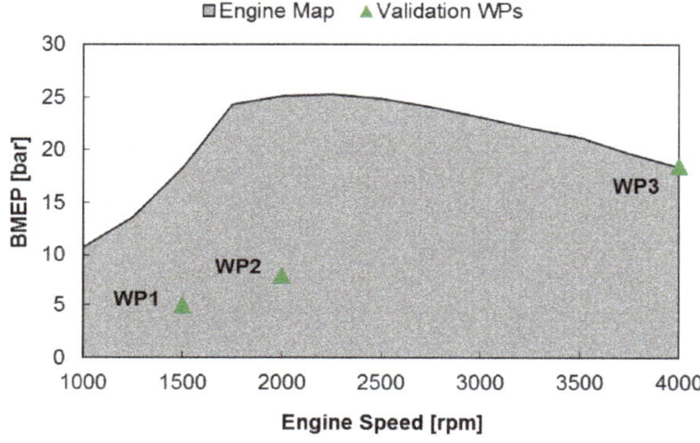

Figure 2. Engine selected WPs (Working Points) on the engine map.

Table 2. Selected engine WPs.

Speed (rpm)	BMEP (bar)
1500	5.0
2000	8.0
4000	18.5

In the definition of a proper simulation setup for a compression ignition engine, the two fundamental needs are a well-calibrated spray modeling and an accurate combustion modeling. An automated coupling of 1D and 3D models was developed to address these requirements, and the adopted methodology is summarized in the block diagram of Figure 3. The 1D model of the entire engine runs separately respect to the two steps of the 3D simulations, to properly characterize the in-cylinder multidimensional simulation with suitable time dependent BCs in terms of pressure, temperature and species concentration. The entire engine cycle is simulated by means of a 1D complete engine model; in this work, a commercially available software GT-SUITE was selected and the validation results have been already presented in [22,23]. Results from this preliminary phase are used to initialize the first step of the 3D numerical analysis: a so-called full cylinder "cold flow" simulation. A cold flow analysis starts during the exhaust stroke up to the intake valve closure (IVC) to capture the thermodynamic conditions and the charge motion during the gas exchange process. Then, the combustion process is simulated considering only one sector of the cylinder centered on a single spray axis of the injector. The injector rate comes from a 1D injector model, extensively calibrated and validated in [15,16], in which the only input is the control current signal, in terms of Energizing Times (ET) and Dwell Times (DT). Finally, the results from the 3D combustion simulation are post-processed by means of a 1D post-process tool available in GT-SUITE environment, the Cylinder Pressure Only Analysis (CPOA). The CPOA is a stand-alone calculation starting only from the measured cylinder pressure, the engine geometry and basic operating cycle average results (such as volumetric efficiency and fuel injected mass). The boundary conditions needed from the CPOA and the pressure trace along the cycle are taken from the results of the 3D sector simulation. In this way, the simulated in-cylinder pressure is analyzed using the same solution methodology as the initial 1D engine model, ensuring perfect consistency between the 3D and the 1D approach. Moreover, the final output of the present methodology is the description of the engine performance in terms of global engine combustion parameters and emissions estimations. All the interconnections between the 1D engine model, the injector model, and the 3D two-stages simulation are performed automatically using in-house developed Linux scripts. The presence of automated interconnections between the different models allows the opportunity to implement the present methodology iteratively in a two-way coupling procedure: at the end of the first 3D-CFD simulation the obtained combustion data should be provided again to the 1D-CFD detailed engine model, setting up a new computational loop since the chosen convergence criteria would be satisfied. Even if in the present paper the iterative procedure is not proposed, it could be considered for future works.

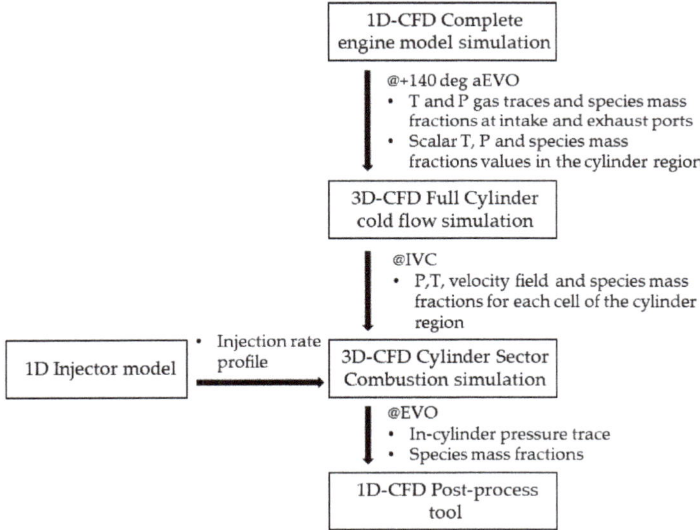

Figure 3. Block diagram schematizing the proposed methodology.

2.1. D-CFD Simulation Setup

The complete engine model used in the first phase of 1D simulations is a detailed model of the entire engine, including all the subsystems as the turbocharger, the EGR circuit, all the pipes and volumes from the intake to the exhaust system. In this way, the model is able to reproduce the engine's behavior accurately in all the different working points. The outputs obtained from the 1D model are imposed as boundary conditions in the 3D-CFD simulations. In detail, the temperature and pressure gas traces as a function of the crank angle and the value of the species mass fractions will be imposed at the intake and exhaust ports at the start of the subsequent 3D-CFD simulation, as well as the initial thermodynamics conditions and species mass fractions in the different regions in which the 3D cylinder model is divided. Finally, the results of the 3D sector combustion simulation are post-processed in a 1D model, to ensure consistency between the experimental data and the outcomes of the 3D-CFD simulation.

2.2. D-CFD Injector Simulation

An accurate and realistic injection profile is a fundamental input of the combustion 3D-CFD simulation. In this case, only a limited set of injection rates was available and to extend the availability of the model, there was the necessity to build a reliable injection rate profiles library. Therefore, a detailed 1D model of the solenoid injector was built in GT-SUITE [15,16]. The present injector model is able to calculate the injection rate starting from the Electronic Control Unit (ECU) parameters in terms of ET, DT and Rail Pressure.

2.3. D-CFD Simulation Setup

The 3D-CFD simulations are carried out using a commercially available software CONVERGE CFD, in two different steps. At first, a full cylinder cold flow simulation starts during the exhaust stroke until the IVC, to model the gas exchange process. This stage of the model aims to evaluate the thermodynamic conditions inside the cylinder, the gas exchange process, and the charge motion up to the end of the compression stroke, accounting for the interaction of the moving geometry with the fluid dynamics. As mentioned, the full cylinder simulation is automatically initialized using as time-varying boundary conditions the 1D-CFD complete engine model results. Then the 3D solution in terms of temperature, pressure, species concentration, and velocity field is mapped in all the cells of the domain at the IVC and imposed to a sector mesh to start the combustion simulation. The combustion simulation is performed only on a portion of the cylinder, on a sector of 45 degrees considering the presence of an 8-holes injector, as can be seen in Figure 4.

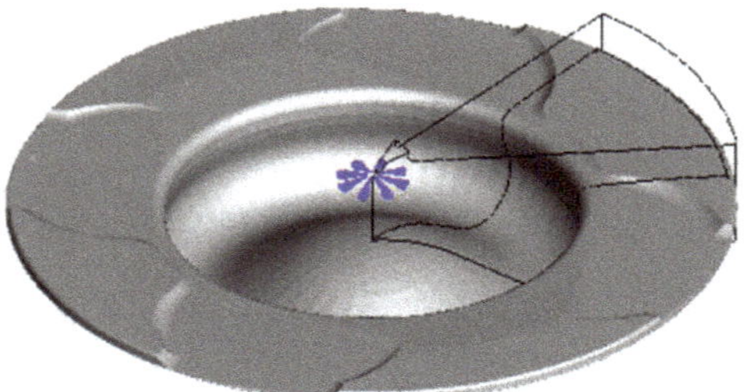

Figure 4. Selected cylinder sector from the entire combustion bowl geometry.

Simulating the combustion process only on a reduced volume of the entire cylinder geometry allows reducing the computational time significantly. This well-established simulation technique assumes that the swirl motion has a predominant effect, and therefore, the combustion process can be assumed as axial-symmetric [24]. A flat profile of the piston head is chosen hypothesizing that the presence of the valve seats does not affect the main outcomes of the combustion process considerably, as demonstrated by Bergin et al. in [25]. Since different sectors can give different results due to non-uniform charge distribution, the proper choice of the portion of the cylinder to be considered in the combustion simulation is performed comparing the results of different sector simulations with a combustion simulation carried out on the entire full cylinder geometry.

The base grid dimension in all the simulations is fixed at 0.5 mm, reaching the minimum size of 0.25 mm due to the two grid refinement techniques available in the 3D-CFD environment. Thanks to the high refinement of the base grid, a fixed embedding of the first level is considered adequate and is placed near the nozzle to depict the spray phenomena correctly. Apart from the near-nozzle region, it is quite challenging to choose a priori where a refinement of the grid is necessary. In this case, the

Adaptive Mesh refinement (AMR) method is applied in all the sector volume, where high velocity and temperature gradients grow up and the grid is scaled according to the same abovementioned rules of the fixed embedding, as reported in [26].

As far as turbulence is concerned, the Reynolds-averaged Navier–Stokes (RANS) based Re-Normalization Group (RNG) k-ε model [27] is adopted. Considering the spray model, the atomization and the breakup of the droplets are calculated by means of calibrated Kelvin Helmholtz and Rayleigh Taylor (KH-RT) model [28]. The "blob" injection method of Reitz and Diwakar is used [28], in which parcels of liquid with a characteristic size equal to the effective nozzle diameter are injected into the computational domain. The O'Rourke model is chosen to simulate the turbulent dispersion of the spray parcels, distributing the parcels evenly throughout the cone of the injector [26]. The No Time Counter method (NTC) [29] is used as a collision model with the addition of a collision mesh and the dynamic droplet drag [30] to account for the possibility of variations in the drop shape. As far as the spray-wall interaction is concerned, the O'Rourke wall film model is adopted. Finally, the fuel evaporation is described by means of the Frossling evaporation model [31], which converts the evaporated liquid fuel mass into the specified source species. The validation of the spray model was done considering the breakup constants of the KH-RT model, the discharge coefficient and the spray angle values as calibration parameters, in comparison with penetration curves experimental data coming from constant volume vessel tests performed at the University of Perugia [32,33]. These data were available only for a reference injection, whose characteristics are shown in Table 3 (see also Figure 5 for a sketch of the mentioned injection schedule).

Table 3. Main characteristics of the reference injection data.

Ref. Case–8-Holes	Solenoid Common Rail Injector
Vessel Pressure (bar)	11.28
Vessel Temperature (°C)	20
Rail Pressure (bar)	400
Energizing Time P2 (ms) ET-P2	0.215
Dwell Time P2 (ms) DT-P2	0.81
Energizing Time P1 (ms) ET-P1	0.21
Dwell Time P1 (ms) DT-P1	0.41
Energizing Time Main (ms) ET-Main	0.32

Figure 5, top, shows the experimental injector current and injection rate profile used for the spray model calibration; the injection schedule presents two pilot injections and one main injection. To calibrate the breakup model, the constant volume vessel was reproduced in the 3D-CFD software and the injection rate was simulated by means of the 1D injector model. Figure 5, bottom, displays the results in terms of spray penetration of the three injections, comparing the numerical results with the experiments. A good agreement was found for the two pilot injections, while some differences can be seen in the case of the main injection. In this case, the discrepancies between the experimental and the simulated results could be addressed to the momentum transfer from the liquid jet to the air, and the corresponding possible local variations in the air density inside the test vessel in case of large injection pulses. For the purpose of this study, these results were considered acceptable according to the size of the baseline bowl (bowl radius approximately equal to 25 mm) and to the limited liquid spray penetration in real engine operating conditions. Further investigations could be carried out to more clearly understand the root causes of these discrepancies.

Concerning the combustion model, the SAGE detailed chemistry solver is implemented with the Skeletal Zeuch mechanism, the reduced version of the complete Zeuch mechanism, enhanced by the inclusion of soot reactions from Mauss's work [34]. It features 121 species, including Poly-cyclic

Aromatic Hydrocarbons (PAHs); therefore, it is possible to use the Particulate Mimic (PM) model [35], based on the method of moments, to predict the cell-averaged soot mass and number density. Since B10 fuel (10% biodiesel and 90% petrodiesel blend) was used for the experimental activity, in the numerical analysis the properties of the fuel are set equal to the ones of B10, as reported in Table 4.

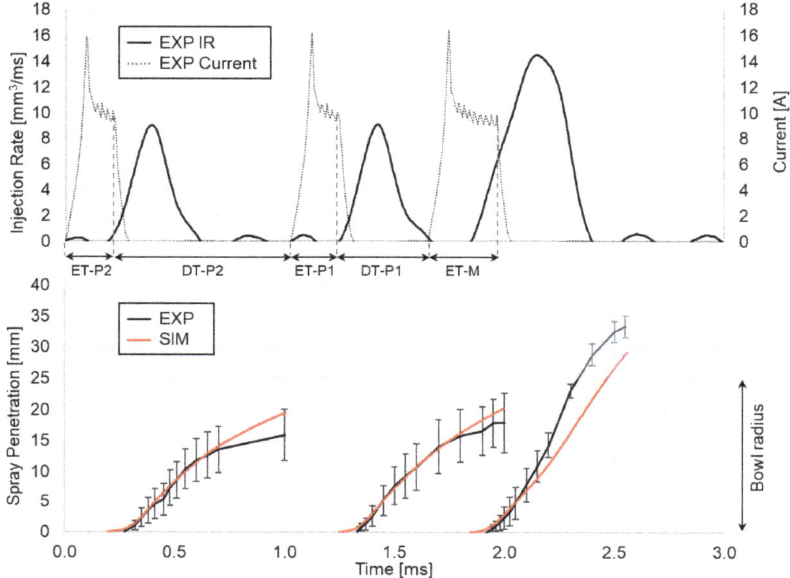

Figure 5. Top: Experimental injection current (dotted black) and hydraulic injection schedule (solid black). Bottom: Numerical spray penetration (red) compared with the experimental data (black) obtained in conditions shown in Table 3.

Table 4. Main characteristics of the B10 Diesel blend [36].

Characteristic	Method	Value
Density (at 15°)	EN ISO 3675	0.833 (kg/dm³)
Cetane Number	EN ISO 5165	54.2
Sulphur content	EN ISO 20,846	6.3 (mg/kg)
Lower heating value	ASTM D 240	42.32 (MJ/kg)
Higher heating value	ASTM D 240	45.20 (MJ/kg)
Stoichiometric air/fuel ratio	-	14.45

3. Results and Discussion

The discussed methodology was validated on three working points (see Figure 2 and Table 2) in terms of in-cylinder pressure and heat release rate, in comparison with the available experimental results. Firstly, the validation of the sector mesh simulation approach is presented in Figure 6, where the in-cylinder pressure obtained in the case of a full cylinder combustion simulation is compared with the results of the selected sector combustion analysis for the 4000 rpm, 18.5 bar of Brake Mean Effective Pressure (BMEP) WP. The comparison can be considered as acceptable since the sector simulation is able to reproduce correctly the entire combustion process allowing a noticeable advantage in terms of computational time saving, as highlighted in Figure 7.

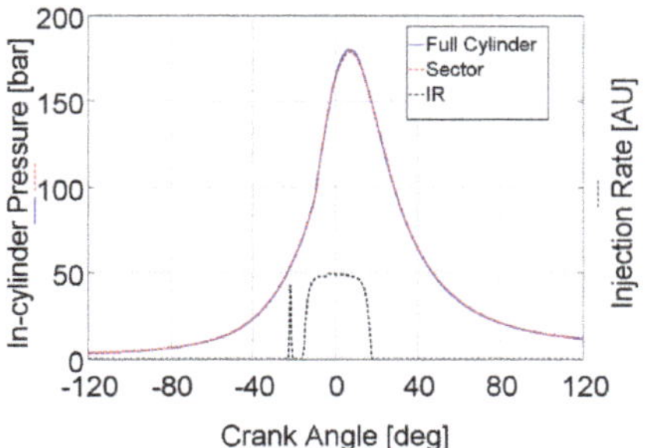

Figure 6. In-cylinder pressure comparison between full-cylinder geometry (solid blue) and sector mesh approach (dashed red), injection rate (dashed black). WP: 4000 rpm, 18.5 bar BMEP.

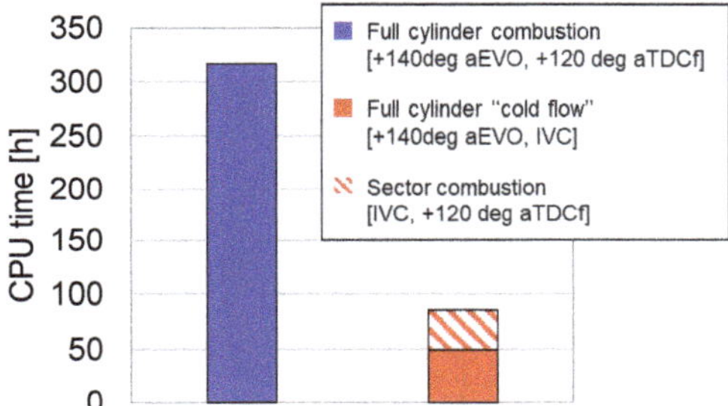

Figure 7. Computational time comparison of the full cylinder (blue) and the sector mesh approaches (red–white), which needs a preliminary full cylinder cold flow analysis (red). The CPU time is referred to simulations distributed on 24 cores, Intel Xeon E5–2680 v3 2.50 GHz processor.

Regarding the validation of the coupled 1D/3D approach, Figure 8 displays the comparison of experiments and numerical results for the working point 1500 rpm × 5.0 bar BMEP, where the injection profile from the 1D detailed injector model is highlighted in black dashed line. A good agreement is obtained, both in terms of in-cylinder pressure and heat release rate. The combustion timing is correctly captured by the numerical model for pilots, main and post injections.

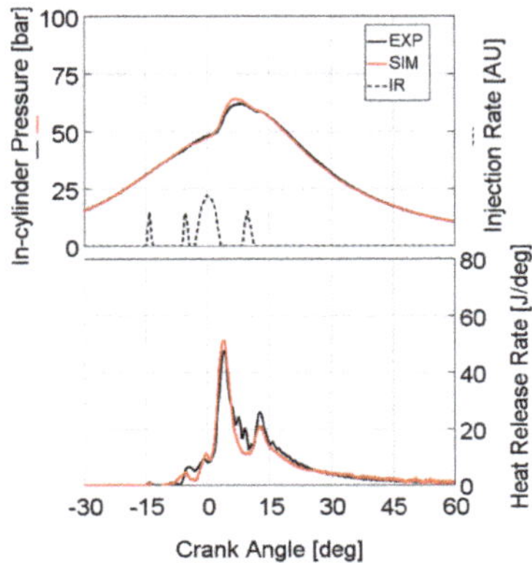

Figure 8. Top: simulated (red) vs. experimental (black) in-cylinder pressure, injection rate (dashed black). Bottom: simulated (red) vs. experimental (black) Heat Release Rate. WP: 1500 rpm × 5.0 bar BMEP.

The in-cylinder pressure and heat release rate compared with the experimental data are shown in Figure 9 for the second WP (2000 rpm × 8.0 bar BMEP). The combustion duration and the ignition delay are both correctly captured by the numerical simulation.

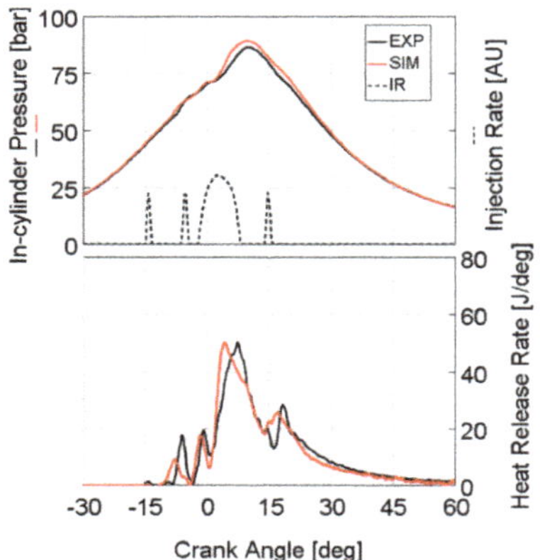

Figure 9. Top: simulated (red) vs. experimental (black) in-cylinder pressure, injection rate (dashed black). Bottom: simulated (red) vs. experimental (black) Heat Release Rate. WP: 2000 rpm × 8.0 bar BMEP.

The capability of capturing the combustion process accurately is also confirmed in the last test engine working point, 4000 rpm × 18.5 bar BMEP, as shown in Figure 10, which features one pilot and main injection with high rail pressure, generally showing an excellent agreement between experiments and modeling results.

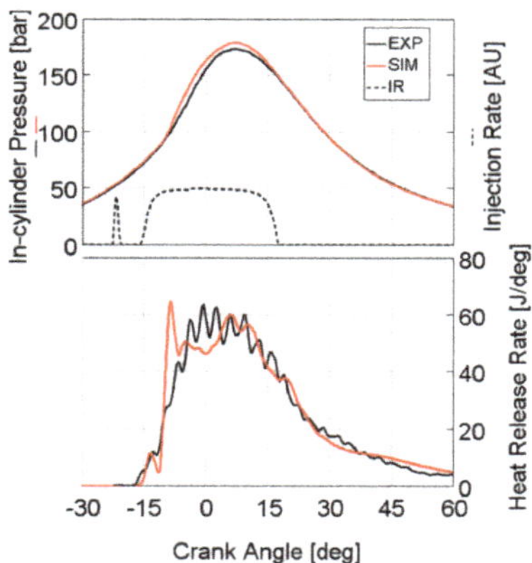

Figure 10. Top: simulated (red) vs. experimental (black) in-cylinder pressure, injection rate (dashed black). Bottom: simulated (red) vs. experimental (black) Heat Release Rate. WP: 4000 rpm × 18.5 bar BMEP.

The excellent predictive capabilities of the proposed methodology could also be highlighted in Figure 11, in which the comparison between experiments and simulation results of the main combustion parameters is depicted. More specifically, the 3D-CFD sector simulation is able to predict Peak Pressure, Crank Angle (CA) at Peak Pressure, 10%–90% Combustion Duration and MFB50% with a high accuracy level. To further quantify the level of reliability of the simulation results, Table 5 shows the absolute error of the combustion parameters presented in Figure 11, with respect to the experimental data. For example, in terms of Peak Pressure, the error remains lower than +/−2 bar and can be considered as acceptable, as well as for the other combustion parameters.

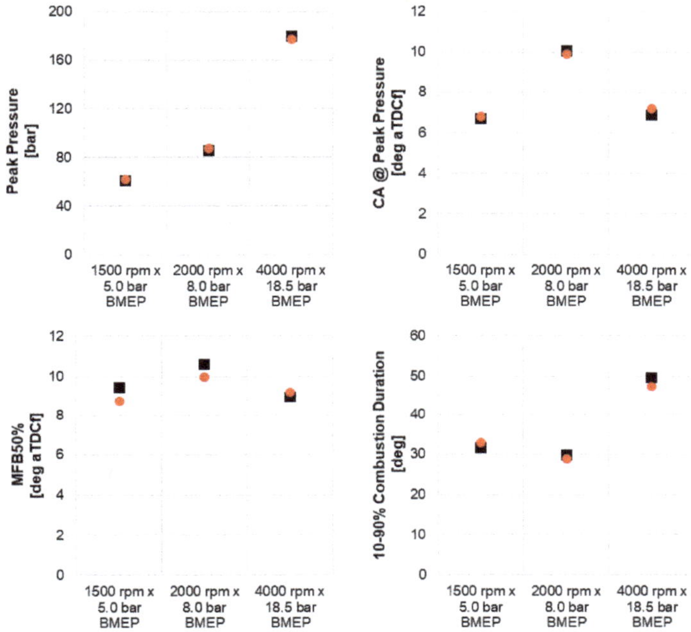

Figure 11. Peak pressure (**top–left**), Crank Angle at Peak Pressure (**top–right**), 10–90% Combustion Duration (**bottom–right**), and MFB50% (**bottom–left**) comparison between experiments (black square) and simulation results (red circle).

Table 5. Absolute errors of the simulation results with respect to the experimental data for the main combustion parameters presented in Figure 11.

Engine WP	Peak Pressure	CA @ Peak Pressure	MFB 50%	10–90% Comb Duration
-	Error (bar)	Error (deg)	Error (deg)	Error (deg)
1500 rpm × 5.0 bar BMEP	+1.20	+0.11	−0.68	+1.29
2000 rpm × 8.0 bar BMEP	+2.08	−0.12	−0.62	−0.99
4000 rpm × 18.5 bar BMEP	−1.97	+0.30	+0.25	−2.28

Regarding the results in terms of pollutant emissions, Figure 12 shows normalized NOx emissions simulation results in comparison with experimental data. A very good agreement is noticeable in terms of trend across the three tested WPs. Concerning the soot values, Figure 13 presents normalized soot emissions comparing the experimental data with both empirical (Hiroyasu [37], blue bar) and detailed soot (PM, red bar) models. It can be highlighted that the empirical Hiroyasu model leads to a lower accuracy in soot prediction with respect to the detailed PM model. In fact, it is not able to capture the trend across the three WPs in terms of soot emissions, with a not negligible overestimation of the soot produced at the WP 1500 rpm × 5.0 bar BMEP. On the contrary, the PM model, supported by the detailed chemistry solver, depicts correctly the experimental trend. For these reasons, the detailed PM model was considered for the numerical soot prediction.

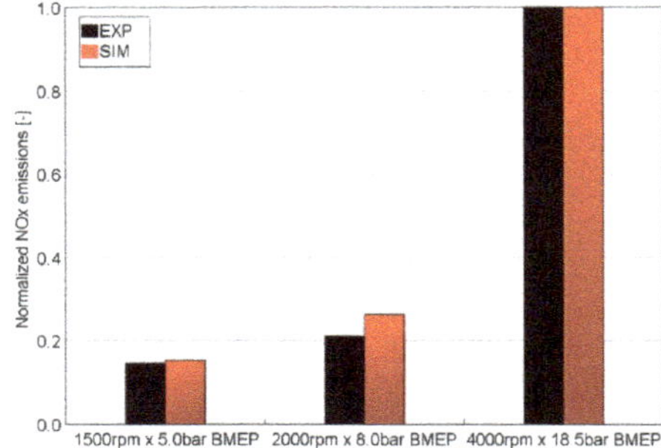

Figure 12. NOx emissions normalized respect to the corresponding value for the 4000 rpm × 18.5 bar BMEP WP, experimental data (black) and simulated (red).

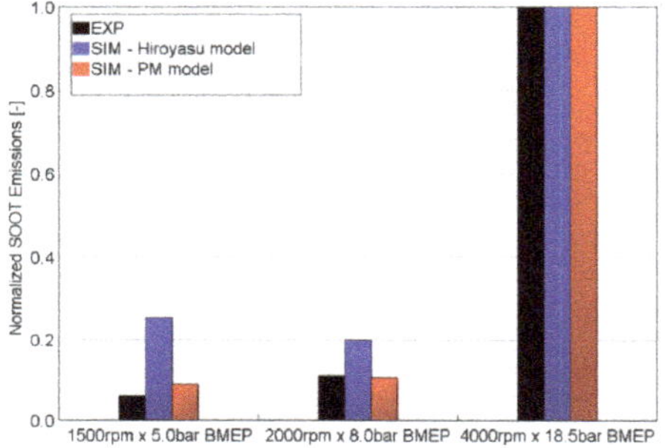

Figure 13. Soot emissions normalized respect to the corresponding value for the 4000 rpm × 18.5 bar BMEP WP. Experiments (black) compared with numerical results obtained by means of the Hiroyasu empirical model (blue) and detailed PM model (red).

Thanks to the above described satisfactory validation of the proposed methodology, it can be considered as a valuable tool to scrutinize and optimize different combustion chamber and injector nozzle designs, also for complex multiple injection strategies.

As a possible example of application of the proposed methodology, the effect of different injection parameters on post injection efficacy in terms of soot reduction has been investigated and reported in this work. Four different injection schedules have been analyzed on a single working point (1500 rpm × 5.0 bar BMEP) varying either the timing or the post injected quantity, as shown in Figure 14. The Start of Injection (SOI) of the main and of the two pilot injections is kept constant. Moreover, the amount of fuel injected in the main injection is adjusted so to keep the engine BMEP at a constant level; this means that, at constant rail pressure, the main injection duration is reduced once the post injected quantity is increased, causing a longer post event. The injector detailed model is used to obtain the different injection profiles. As shown in Figure 14, the different injection schedules tested include a sweep of DT (600, 1000, 1400 µs) and a sweep of ET (170 and 210 µs).

The in-cylinder pressure and the simulated heat release rate are shown in Figure 15 considering the DT sweep. The predictivity of the numerical model can be considered as acceptable, although the model seems to slightly underestimate the heat release of the main injection for all the three cases, while slightly overestimating the heat release of the post injection for the post injections with higher DTs (case c and d). Moreover, the model seems to be unable to correctly predict the heat release of the post injection with the shorter DT (case a), suggesting that the effects of pulse to pulse interactions in terms of air and fuel mixing are not fully captured.

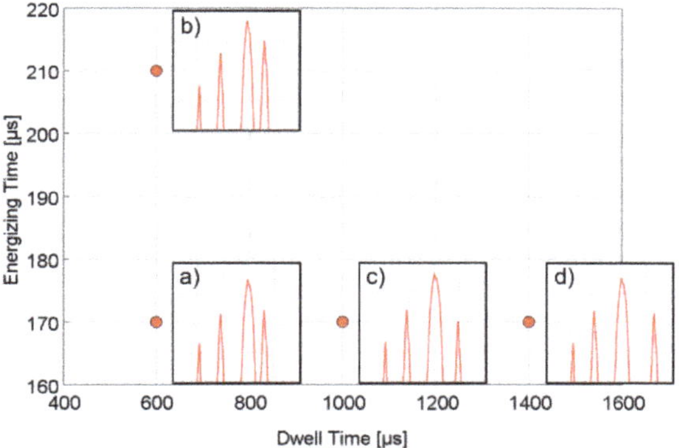

Figure 14. Injection schedules varying the ET and the DT of post injection, WP: 1500 rpm × 5.0 bar BMEP.

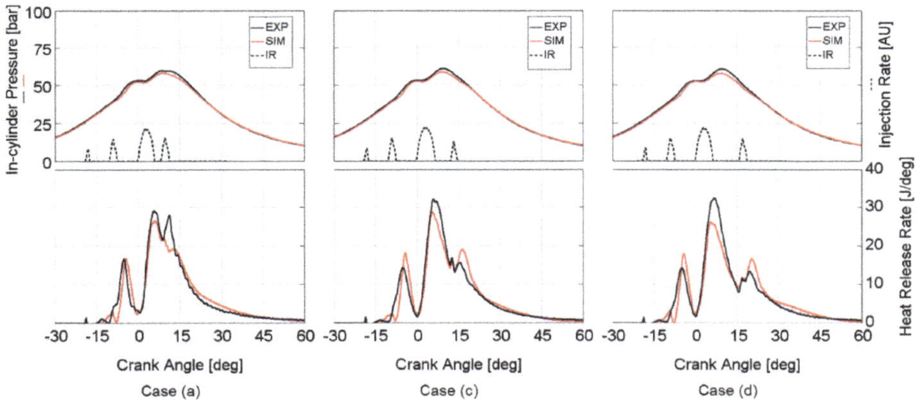

Figure 15. Effect of post injection DT–Top: simulated (red) vs. experimental (black) in-cylinder pressure, injection rate (dashed black). Bottom: simulated (red) vs. experimental (black) Heat Release Rate. WP: 1500 rpm × 5.0 bar BMEP–Case (a), (c) and (d).

In Figure 16, the results obtained with a constant DT (600 µs) and different ETs of the post injection (case a vs. case b) are presented. Although the same discrepancies previously highlighted between the experimental and simulated heat release rate of the post injection could be pointed out, the overall evolution of the combustion process seems to be reasonably captured by the model, and further analysis, which will be described in the following paragraphs, will confirm its predicting capabilities.

The soot emissions results for the four post injection strategies analyzed are shown in Figure 17. Although the discrepancy between experimental and simulation results is quite evident in absolute terms, the simulation is capable to correctly capture the main trends among the different post injection strategies, highlighting that:

- the shorter post injection event is more effective in terms of soot emission reduction (soot emissions of case a are less than a half of soot emissions of case b);
- the influence of the post injection timing is definitely less important than the effect of the post injection quantity (soot emissions of case a, c and d, corresponding to different DTs, are comparable to each other, while are all significantly lower than soot emissions of case b, corresponding to a higher post injection quantity).

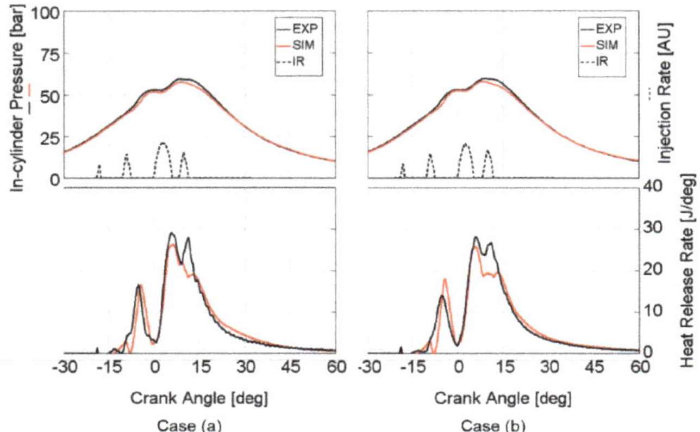

Figure 16. Effect of post injection ET–Top: simulated (red) vs. experimental (black) in-cylinder pressure, injection rate (dashed black). Bottom: simulated (red) vs. experimental (black) Heat Release Rate. WP: 1500 rpm × 5.0 bar BMEP–Case (a) and (b).

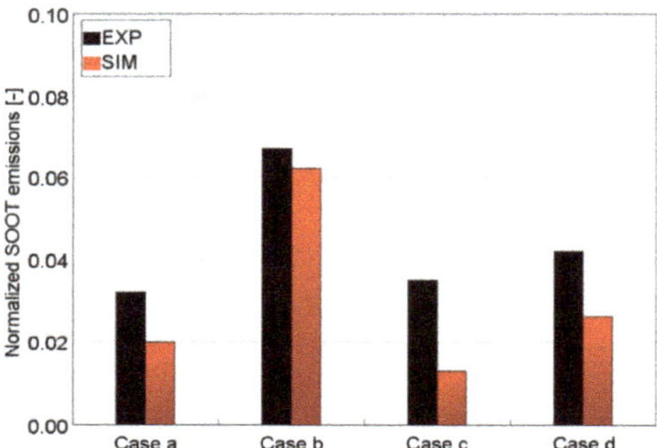

Figure 17. Normalized soot emissions for the injection schedules reported in Figure 14, experimental data (black) and simulated (red), WP: 1500 rpm × 5.0 bar BMEP.

In order to more clearly understand the phenomenon, two different post injection schedules were selected and analyzed, with constant Dwell Time (DT = 600 µs) and different Energizing Times of the post injection event (ET = 170 and 210 µs, respectively), corresponding to case (a) and case (b) of the test matrix of Figure 14. The numerical results related to the injection rate (top), the heat release rate (center) and the time-varying total soot mass (bottom) of the chosen injection schedules are shown in Figure 18.

Firstly, it can be highlighted that the two curves of the total soot mass start to diverge during the post injection event, and the case with the higher post injected quantity (case b) presents the lower peak of total soot mass, probably due to the shorter main injection event. At the same time, the oxidation process appears to be more effective and faster for the case with the lower post injected quantity (case a). More in detail, the mechanisms of soot formation and oxidation have been investigated at four different crank angle degrees highlighted in Figure 18, analyzing the contour plots of the temperature, of the Turbulent Kinetic Energy (TKE), and of the soot mass distribution as reported in Figures 19–22, respectively. The selected plane of the combustion chamber where the contour plots are presented is placed on the spray axis, corresponding to the middle of the sector geometry used to perform the 3D combustion simulation, as shown in Figure 19.

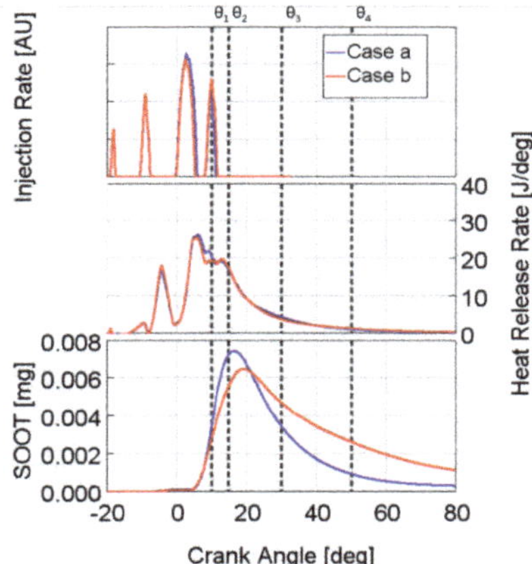

Figure 18. Numerical results for the injection rate (top), heat release rate (center) and the total soot mass (bottom), red lines for case (a) (ET = 170 µs) and blue lines for case (b) (ET = 210 µs) of the post injection test matrix of Figure 14.

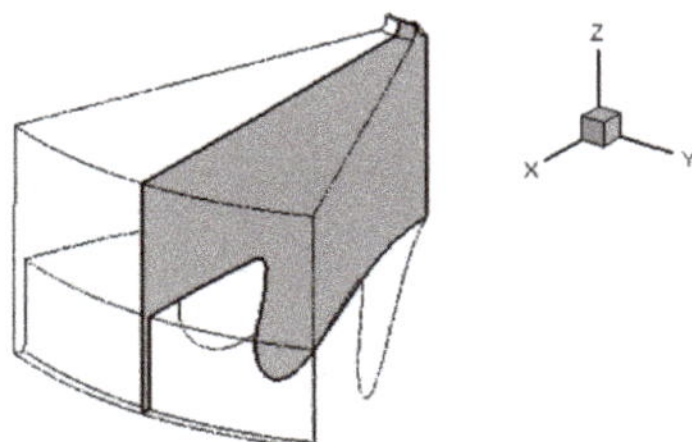

Figure 19. Plane of the sector geometry selected to represent the contour plots of Figures 20–22.

- **At θ_1 = 10 deg aTDCf** –2 deg CA after the start of the post injection events As shown in Figure 18, during the post injection events the soot mass curves of the two different injection schedules start to diverge, with a higher soot production in the case with shorter post injection (case a), which corresponds to a longer main injection. However, at this stage of the combustion process, the two cases are still showing almost identical soot masses, as confirmed by the contour plots of Figure 22, which do not yet show significant differences, although a higher temperature distribution in the spray area for case a) can be noticed in Figure 20, due to the more prolonged main injection (and, of course, combustion) event.

- **At θ_2=15 deg aTDCf**–3 deg CA after the end of the 210 µs post injection At this stage, looking at Figure 22, it is evident that the post injection with a shorter duration (case a) presents a higher production of soot within the bowl, while it shows a lower quantity of soot around the bowl rim. From the analysis of Figures 20 and 21, it is quite evident that case b), due to the longer post injection, shows higher temperatures and higher turbulence intensities in the fuel spray jet and in the bowl rim regions, thus leading to a higher soot mass concentration in these regions. Nevertheless, the total quantity of soot at this CA remains higher for case a), being the soot production within the bowl clearly predominant.

- **At θ_3=30 deg aTDCf**–18 deg CA after the end of the 210 µs post injection Going ahead in the combustion process, the soot concentration for case (b) is significantly higher than for case (a) as shown in Figure 18, suggesting a faster and more effective oxidation process for case (a) (shorter post injection). This phenomenon is confirmed by the contour plots of Figure 22, where the amount of soot mass within the bowl is clearly lower for case (a) respect to case (b). The faster and more effective oxidation of the soot for case (a) can be attributed to a more intense mixing within the bowl, as shown by the higher TKE values in this region in Figure 21, and is further confirmed by the temperature contour plot of Figure 20, which displays a wider area of high temperatures.

- **At θ_4=50 deg aTDCf**–38 deg CA after the end of the 210 μs post injection Finally, when the final stage of the combustion is reached, the soot mass quantity for case (b) (longer post injection) remains higher, because the oxidation process in this case is less effective, as confirmed by the soot mass distributions of Figure 22, where a higher quantity of soot is evident in the upper part of the combustion chamber.

In conclusion, the detailed description of the in-cylinder phenomena, which can be provided by the 3D-CFD numerical simulation, appears to be a powerful tool for a thorough analysis and diagnosis of the combustion process also in case of new combustion chamber designs or in presence of complex injection schedules, which could affect the combustion and emissions formation phenomena significantly.

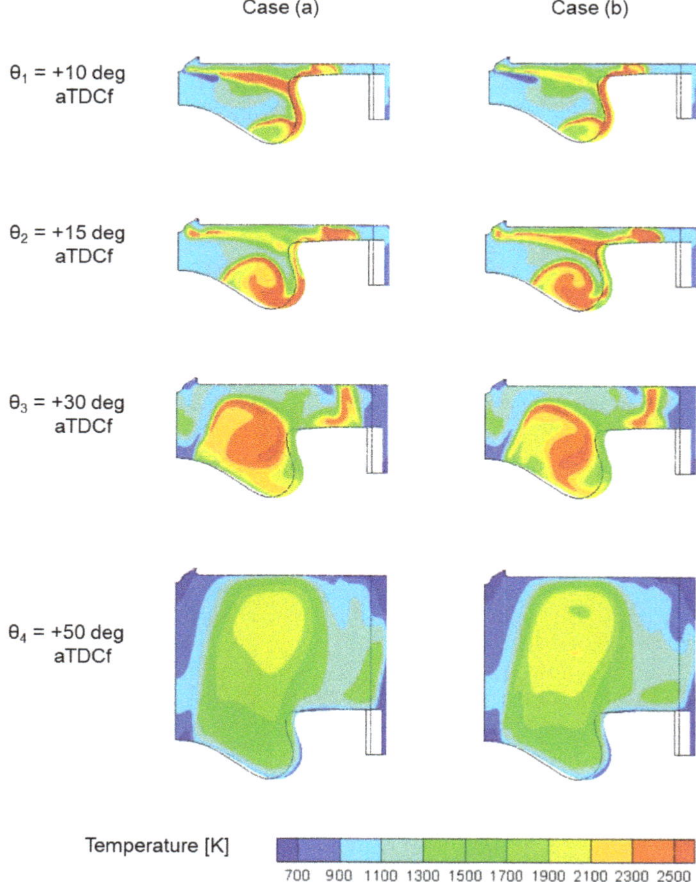

Figure 20. Contour plots of the temperature distribution for the two chosen post injection schedules (case a and case b) at four different crank angle degrees.

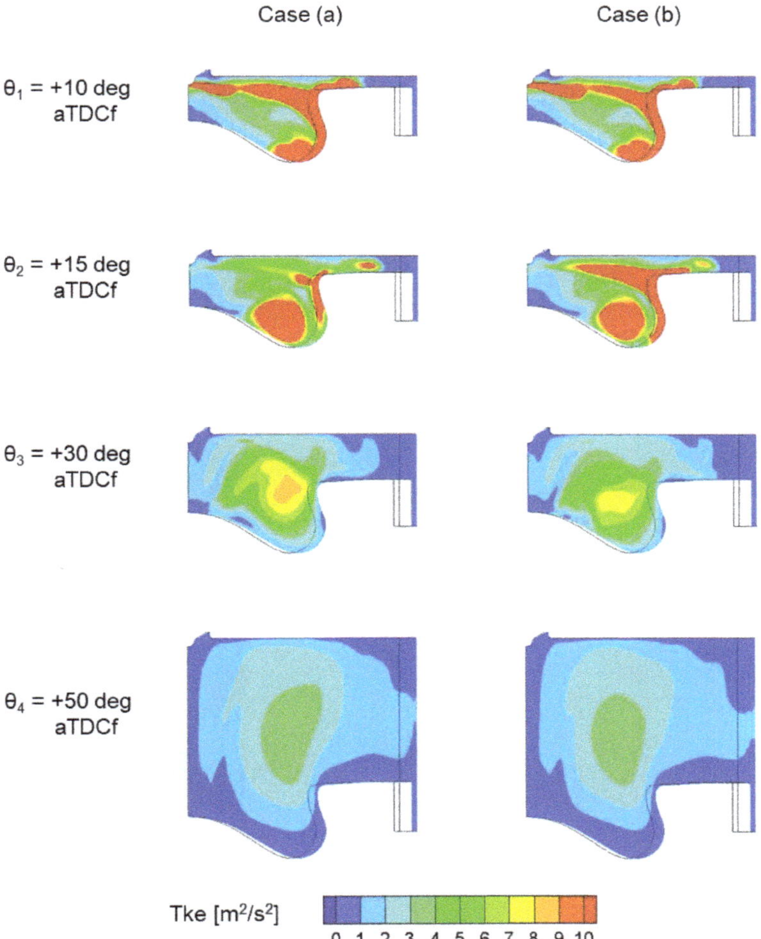

Figure 21. Contour plots of the Turbulent Kinetic Energy (TKE) distribution for the two chosen post injection schedules (case a and case b) at four different crank angle degrees.

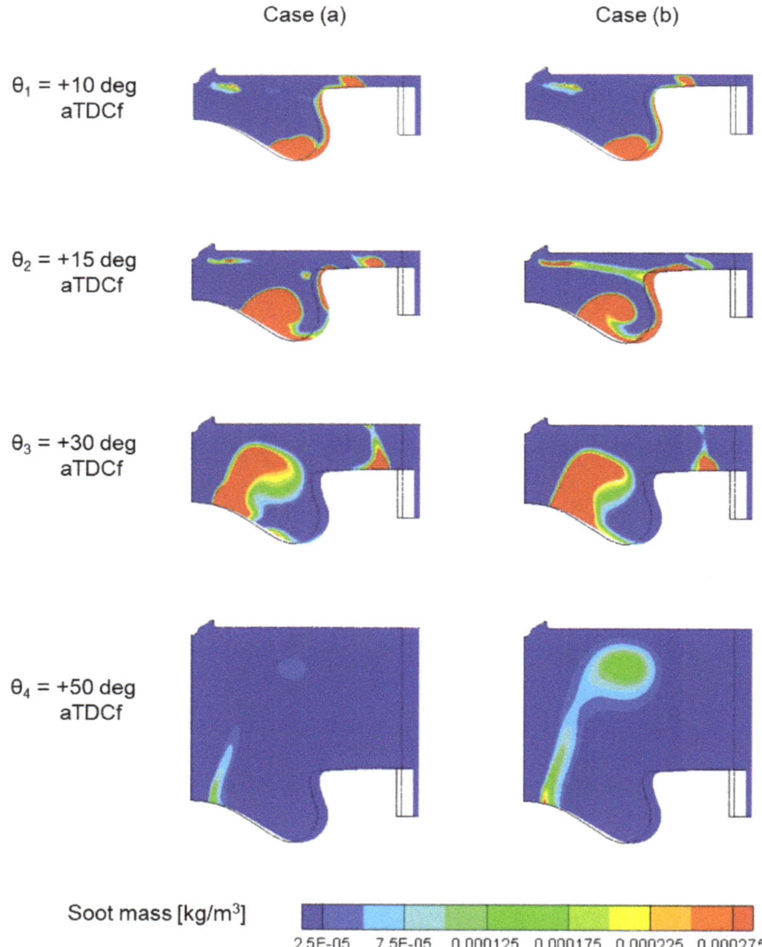

Figure 22. Contour plots of the soot mass distribution for the two chosen post injection schedules (case a and case b) at four different crank angle degrees.

4. Conclusions

In this paper, an integrated methodology for the coupling between 1D- and 3D-CFD simulation codes has been presented. The proposed methodology aims to support the design and calibration of new diesel engines, coupling 1D engine models, usually available in the early stage engine development phases, with 3D detailed and predictive combustion simulations, respectively built in the commercially available software, GT-SUITE and CONVERGE CFD.

An assessment of the developed procedure has been performed by comparing its results with experimental data acquired on an automotive diesel engine, considering different working points, including both part load, and full load conditions. Besides, different multiple injection schedules have been evaluated for part-load operation, including pre- and post-injections and the proposed methodology has been proven to be capable of predicting the combustion and emissions formation processes for both NOx and soot emissions with satisfactory accuracy, also thanks to the integration of a detailed soot model.

The proposed procedure can, therefore, be considered as a suitable methodology to support the design and calibration of new diesel engines, thanks to its capability to provide reliable engine performance and emissions estimations from the early stage of new engine development.

Author Contributions: Conceptualization, F.M.; methodology definition, A.B., A.P.; software, A.B., A.P.; Validation, A.P., B.P.P.; formal analysis, A.P., B.P.P., A.B.; investigation, A.P., B.P.P.; writing—original draft preparation, A.P., B.P.P.; writing—review and editing, F.M., F.C.P., M.R.M.; supervision, F.M., F.C.P.; project administration, F.M., F.C.P.; funding acquisition, F.M. All authors have read and agreed to the published version of the manuscript.

Funding: The activity was financially supported by GM Global Propulsion Systems through the "Predictive Combustion and Fuel Injection System Model" research project (2017).

Acknowledgments: The Authors would like to acknowledge the precious contribution to this work given by the research group of Professor Lucio Postrioti from the University of Perugia and Giacomo Buitoni from STSe – Shot to Shot Engineering for their extensive experimental activity supporting the spray calibration process.

Conflicts of Interest: The authors declare no conflict of interest.

Definitions/Abbreviations

aEVO	After Exhaust Valve Opening
AMR	Adaptive Mesh Refinement
aTDCf	After Top Dead Center Firing
AU	Arbitrary Units
BCs	Boundary Conditions
BMEP	Brake Mean Effective Pressure
BSFC	Brake Specific Fuel Consumption
BSNOx	Brake Specific NOx
CA	Crank Angle
CFD	Computational Fluid Dynamics
CN	Combustion Noise
CPOA	Cylinder Pressure Only Analysis
CPU	Central Processing Unit
deg	Degree
DT	Dwell Time
ECU	Electronic Control Unit
EGR	Exhaust gas Recirculation
ET	Energizing Time
EVO	Exhaust Valve Opening
IR	Injection Rate
IVC	Intake Valve Closing
KH-RT	Kelvin Helmholtz - Rayleigh Taylor
MFB50%	50% Mass Fraction Burned Angle
NTC	No Time Counter
P	Pressure
P1	First Pilot Injection
P2	Second Pilot Injection
PAH	Poly-cyclic Aromatic Hydrocarbon
PM	Particulate Mimic
RANS	Reynolds-Averaged Navier-Stokes
RNG	Re-Normalization Group
SOI	Start of Injection
T	Temperature
TKE	Turbulent Kinetic Energy
VGT	Variable Geometry Turbine
WP	Working Point

References

1. Piano, A.; Millo, F.; Sapio, F.; Pesce, F.C. Multi-Objective Optimization of Fuel Injection Pattern for a Light-Duty Diesel Engine through Numerical Simulation. *SAE Int. J. Engines* **2018**, *11*, 1093–1107. [CrossRef]
2. Millo, F.; Piano, A.; Peiretti Paradisi, B.; Boccardo, G.; Mirzaeian, M.; Arnone, L.; Manelli, S. The Effect of Post Injection Coupled with Extremely High Injection Pressure on Combustion Process and Emission Formation in an Off-Road Diesel Engine: A Numerical and Experimental Investigation. *SAE Tech. Pap.* **2019**. [CrossRef]
3. Sinclair, R.; Strauss, T.; Schindler, P. Code Coupling: A New Approach to Enhance CFD Analysis of Engines. *SAE Tech. Pap.* **2000**. [CrossRef]
4. Riegler, U.G.; Bargende, M. Direct Coupled 1D/3D-CFD-Computation (GT-Power/Star-CD) of the Flow in the Switch-Over Intake System of an 8-Cylinder SI Engine with External Exhaust Gas Recirculation. *SAE Tech. Pap.* **2002**. [CrossRef]
5. Galindo, J.; Tiseira, A.; Fajardo, P.; Navarro, R. Coupling methodology of 1D finite difference and 3D finite volume CFD codes based on the Method of Characteristics. *Math. Comput. Model.* **2010**, *54*, 1738–1746. [CrossRef]
6. Watanabe, N.; Kubo, M.; Yomoda, N. An 1D-3D Integrating Numerical Simulation for Engine Cooling Problem. *SAE Tech. Pap.* **2006**. [CrossRef]
7. Wang, Z.; Scarcelli, R.; Som, S.; McConnell, S.; Salman, N.; Zhu, Y.; Hardman, K.; Freeman, K.; Reese, R.; Senecal, P.K.; et al. Multi-Dimensional Modeling and Validation of Combustion in a High-Efficiency Dual-Fuel Light-Duty Engine. *SAE Tech. Pap.* **2013**. [CrossRef]
8. Abidin, Z.; Florea, R.; Callahan, T. Dual Fuel Combustion Study Using 3D CFD Tool. *SAE Tech. Pap.* **2016**. [CrossRef]
9. Mattarelli, E.; Rinaldini, C.; Savioli, T.; Cantore, G.; Warey, A.; Potter, M.; Gopalakrishnan, V.; Balestrino, S. Scavenge Ports Optimization of a 2-Stroke Opposed Piston Diesel Engine. *SAE Tech. Pap.* **2017**. [CrossRef]
10. Fontanesi, S.; Severi, E.; Siano, D.; Bozza, F.; De Bellis, V. Analysis of Knock Tendency in a Small VVA Turbocharged Engine Based on Integrated 1D-3D Simulations and Auto-Regressive Technique. *SAE Int. J. Engines* **2014**, *7*, 72–86. [CrossRef]
11. Park, S.; Furukawa, T. Validation of Turbulent Combustion and Knocking Simulation in Spark-Ignition Engines Using Reduced Chemical Kinetics. *SAE Tech. Pap.* **2015**. [CrossRef]
12. Priesching, P.; Ramusch, G.; Ruetz, J.; Tatschl, R. 3D-CFD Modeling of Conventional and Alternative Diesel Combustion and Pollutant Formation - A Validation Study. *SAE Tech. Pap.* **2007**. [CrossRef]
13. Som, S.; Ramirez, A.I.; Aggarwal, S.K.; Kastengren, A.L.; El-Hannouny, E.; Longman, D.E.; Powell, C.F.; Senecal, P.K. Development and Validation of a Primary Breakup Model for Diesel Engine Applications. *SAE Tech. Pap.* **2009**. [CrossRef]
14. Magnotti, G.M.; Genzale, C.L. Exploration of Turbulent Atomization Mechanisms for Diesel Spray Simulations. *SAE Tech. Pap.* **2017**. [CrossRef]
15. Piano, A.; Millo, F.; Postrioti, L.; Biscontini, G.; Cavicchi, A.; Pesce, F.C. Numerical and Experimental Assessment of a Solenoid Common-Rail Injector Operation with Advanced Injection Strategies. *SAE Int. J. Engines* **2016**, *9*, 565–575. [CrossRef]
16. Piano, A.; Boccardo, G.; Millo, F.; Cavicchi, A.; Postrioti, L.; Pesce, F.C. Experimental and Numerical Assessment of Multi-Event Injection Strategies in a Solenoid Common-Rail Injector. *SAE Int. J. Engines* **2017**, *10*. [CrossRef]
17. Molina, S.; Desantes, J.M.; Garcia, A.; Pastor, J.M. A Numerical Investigation on Combustion Characteristics with the use of Post Injection in DI Diesel Engines. *SAE Tech. Pap.* **2010**. [CrossRef]
18. Yu, H.; Liang, X.; Shu, G.; Wang, Y.; Sun, X.; Zhang, H. Numerical investigation of the effect of two-stage injection strategy on combustion and emission characteristics of a diesel engine. *Appl. Energy* **2017**, *227*, 634–642. [CrossRef]
19. Leach, F.; Ismail, R.; Davy, M. Engine-out emissions from a modern high speed diesel engine – The importance of Nozzle Tip Protrusion. *Appl. Energy* **2018**, *226*, 340–352. [CrossRef]
20. Leach, F.; Ismail, R.; Davy, M.; Weall, A.; Cooper, B. Comparing the Effect of Fuel/Air Interactions in a Modern High-Speed Light-Duty Diesel Engine. *SAE Tech. Pap.* **2017**. [CrossRef]
21. Leach, F.; Ismail, R.; Davy, M.; Weall, A.; Cooper, B. The effect of a stepped lip piston design on performance and emissions from a high-speed diesel engine. *Appl. Energy* **2018**, *215*, 679–689. [CrossRef]

22. Piano, A.; Millo, F.; Boccardo, G.; Rafigh, M.; Gallone, A.; Rimondi, M. Assessment of the Predictive Capabilities of a Combustion Model for a Modern Common Rail Automotive Diesel Engine. *SAE Tech. Pap.* **2016**. [CrossRef]
23. Piano, A. Analysis of Advanced Air and Fuel Management Systems for Future Automotive Diesel Engine Generations. Ph.D. Thesis, Politecnico di Torino, Turin, Italy, 2018. [CrossRef]
24. Puri lng, T.N.; Soni lng, L.R.; Deshpande, S. Combined Effects of Injection Timing and Fuel Injection Pressure on Performance, Combustion and Emission Characteristics of a Direct Injection Diesel Engine Numerically Using CONVERGE CFD Tool. *SAE Tech. Pap.* **2017**. [CrossRef]
25. Bergin, M.J.; Musu, E.; Kokjohn, S.; Reitz, R.D. Examination of initialization and geometric details on the results of CFD simulations of diesel engines. *J. Eng. Gas. Turbines Power* **2011**, 133. [CrossRef]
26. Richards, K.J.; Senecal, P.K.; Pomraning, E. *CONVERGE 2.3 Manual*; Convergent Science Inc.: Madison, WI, USA, 2016.
27. Orszag, S.A.; Yakhot, V.; Flannery, W.S.; Boysan, F.; Choudhury, D.; Maruzewski, J.; Patel, B. Renormalization Group Modeling and Turbulence Simulations. *Near-Wall Turbul. Flows* **1993**, *13*, 1031–1046.
28. Reitz, R.D.; Bracco, F.V. Mechanisms of Breakup of Round Liquid Jets. *Encycl. Fluid Mech.* **1986**, *3*, 233–249.
29. Schmidt, D.P.; Rutland, C.J. A New Droplet Collision Algorithm. *J. Comput. Phys.* **2000**, *164*, 62–80. [CrossRef]
30. O'Rourke, P.; Amsden, A. The Tab Method for Numerical Calculation of Spray Droplet Breakup. *SAE Tech. Pap.* 1987. [CrossRef]
31. Amsden, A.A.; O'Rourke, P.J.; Butler, T.D. *KIVA-II: A Computer Program for Chemically Reactive Flows with Sprays*; Technical Report Number LA-11560-MS; Department of Energy, Office of Scientific and Technical Information: Washingotn, DC, USA, 1989.
32. Postrioti, L.; Grimaldi, C.N.; Ceccobello, M.; Gioia, R. Di Diesel common rail injection system behavior with different fuels. *SAE Tech. Pap.* **2004**. [CrossRef]
33. Postrioti, L.; Buitoni, G.; Pesce, F.C.; Ciaravino, C. Zeuch method-based injection rate analysis of a common-rail system operated with advanced injection strategies. *Fuel* **2014**, *128*, 188–198. [CrossRef]
34. Zeuch, T.; Moréac, G.; Ahmed, S.S.; Mauss, F. A comprehensive skeletal mechanism for the oxidation of n-heptane generated by chemistry-guided reduction. *Combust. Flame* **2008**, *155*, 651–674. [CrossRef]
35. Kazakov, A.; Frenklach, M. Dynamic modeling of soot particle coagulation and aggregation: Implementation with the method of moments and application to high-pressure laminar premixed flames. *Combust. Flame* **1998**, *114*, 484–501. [CrossRef]
36. Panta Distribuzione, S.p.A. MOL Group Italy. B10 Datasheet. Available online: https://molgroupitaly.it/en/ (accessed on 31 March 2020).
37. Hiroyasu, H.; Kadota, T. Models for Combustion and Formation of Nitric Oxide and Soot in Direct Injection Diesel Engines. *SAE Tech. Pap.* **1976**. [CrossRef]

Article

Investigation on an Injection Strategy Optimization for Diesel Engines Using a One-Dimensional Spray Model

Intarat Naruemon, Long Liu *, Qihao Mei and Xiuzhen Ma

College of Power and Energy Engineering, Harbin Engineering University, Harbin 150001, China; kniisnd@gmail.com (I.N.); mqh1447251725@hrbeu.edu.cn (Q.M.); maxiuzhen@hrbeu.edu.cn (X.M.)
* Correspondence: liulong@hrbeu.edu.cn; Tel.: +86-451-8251-8036

Received: 18 October 2019; Accepted: 30 October 2019; Published: 5 November 2019

Abstract: Common rail systems have been widely used in diesel engines due to the stricter emission regulations. The advances in injector technology and ultrahigh injection pressure greatly promote the development of multiple-injection strategy, leading to the shorter injection duration and more variable injection rate shape, which makes the mixing process more significant for the formation of pollutant emission. In order to study the mixing process of diesel sprays under variable injection rate shapes and find the optimized injection strategy, a one-dimensional spray model was modified in this paper. The model was validated by the measured spray penetrations based on shadowgraphy experiments with the varying injection rate. The simulations were performed with five injection rate shapes, triangle, ramping-up, ramping-down, rectangle and trapezoid. Their spray penetrations, entrainment rates and equivalence ratios along spray axial distance are compared. The potentials of multiple-injection and gas-jet after end-of-injection (EOI) to improve mixing process and emission reduction are discussed finally. The results indicated that ramping-up injection rate obtains the highest entrainment rate after EOI, and it needs 1.5 times of injection duration for the entrainment wave to arrive at the spray tip. For the other four injection rates, the sprays can be treated as a steady-like state, needing twice of injection duration from EOI to the time the entrainment wave reaches the spray tip. The multiple-injection with proper injection rate shape enhanced the entrainment rate, and the gas-jet after EOI affected the mixture distribution and entrainment rate in spray tail under ramping-down injection rate.

Keywords: diesel engines; pollutant emission reduction; mixing process; advanced injection strategy; varying injection rate

---

1. Introduction

With the implementation of more stringent emission regulations, such as US Tier 3 and post-Euro 6, diesel engines are facing unprecedented challenges to reduce their adverse environmental impact in terms of pollutant and greenhouse emissions. To cope with the emission problem and improve the performance of diesel engines, the common rail technology has flourished around the world. The advanced fuel injection technologies in a common rail system, such as the ultrahigh injection pressure [1–4] and the multiple-injection strategy [5–7] as well as the high control accuracy [8], have greatly improved the performance of diesel engines. Under the action of ultrahigh injection pressures, the fuel can be injected into the cylinder in a short time, while the multiple-injection strategy divides the whole injection duration into many parts, creating shorter injection durations. Short injection duration often causes EOI much earlier than ignition under certain conditions, such as premixed charge combustion ignition (PCCI) and low temperature combustion (LTC) cases, resulting in the more complicated and crucial mixing process after EOI. Moreover, it is known that there are the ramping-up part during the initial injection period and the ramping-down part during the late injection period in a typical injection rate profile [9,10]. The shorter the injection duration, the higher the proportion of the ramping part. In the case of a small fuel injection quantity with a short injection duration, the injection rate shape always tends to be triangular [11,12], whose spray mixing characteristics are quite different from that of the commonly used injection rate shapes, like the square, boot or rectangle [9,10,13]. Besides, the advanced innovation of piezoelectric injectors makes a significant contribution to the extremely fast needle response time [14,15], meaning it is easer for diesel engines to adopt highly defined injection rate shapes to improve their performance. In this way, the varying injection rate shape plays a more and more important role in the fuel injection of diesel engine, which deserves in-depth study.

Several achievements over the last decade have shown the great influence that the injection rate shape has on the pollutant emissions and performance of diesel engines. Based on the experimental method, Macian et al. [14] found that the ramp injection rate profiles was able to influence the premixed phase of combustion and reduce the radial position of the ignition. Mohan et al. [16] using the KIVA4 computational fluid dynamics (CFD) code found that injection rate-shaping could not only lower NOx emissions, but also cut down the brake-specific fuel consumption and soot emissions at the same time. Also with the aid of KIVA4-CHEMKIN code, Tay et al. [17] discovered that the start of combustion was subject to the injection rate-shaping, with the combustion phasing and combustion duration controlled simultaneously. Even for the ammonia injection rather than the diesel fuel injection, Lamas et al. [18] found that the varying injection rate shapes, including the rectangular, triangular, or parabolic, took distinct propensities to NOx reduction in a hydrogen-diesel engine by a CFD model. It can be summarized justifiably from these researches that the varying injection rate shape indeed has a potential to optimize the engine combustion and emission characteristics. However, the spray mixing characteristic, which plays a leading role in the combustion process and emission formation, has not been investigated thoroughly. Then another more comprehensive study has been accomplished by Han et al. [19] using CFD simulations. They concluded that benefiting from the longer penetration and smaller SMD, the early injection stage would undergo a more sufficient atomization and mixing process under the left triangle and rectangle injection rate shapes, while more combustible mixture and the longest ignition delay period are formed in the late injection stage of the right triangle injection rate shape. The result reveals that different triangle injection rate shapes have different advantages in the spray mixing characteristic and combustion process. Nevertheless, the research above is aimed at the free-piston engine generator (FPEG), considering the phenomenon of fuel spray impingement. In this way, the complete and detailed spray development process is not available for analysis, but it provides a certain guidance and reference for related investigations on diesel engines. In addition, there are too many complicated factors involving in the CFD simulation, such as the fuel breakup, evaporation, mixing, turbulence dissipation and combustion, which all are sensitive to the computational accuracy. In the meantime, the results of spray mixing characteristic have not been validated by experimental data. Thus, the CFD method is not suitable for the preliminary and robust analysis of spray development. Instead, some simplified spray models, like the quasi-dimensional model of Mrzljak et al. [20], the one-dimensional spray model of Cheng et al. [21], and the two-dimensional spray model of Marčiča et al. [22], have shown superiority in studying the spray characteristics. However it is a pity that those models didn't investigate the influence of various injection rate shapes on the spray characteristics, so a simplified diesel spray model with varying injection rate shape as the core, ignoring some complicated yet uninfluential factors, is still badly needed to study the mixing characteristics of diesel spraying.

Therefore, this paper employed an improved one-dimensional diesel spray model based on that proposed by Musculus et al. [23] to analyze the potential of varying injection rate shapes to lower the pollutant emissions of diesel engine and find the optimized injection strategy. Five injection rate shapes were considered: triangle, ramping-up, ramping-down, rectangle and trapezoid. For model validation, the diesel spray penetration even after EOI was compared with the shadowgraph data of non-combusting diesel sprays in a constant volume chamber. Then the mixing process was studied by calculating and analyzing the spray penetrations, entrainment rates and equivalence ratio distributed along the spray axial distance. After that, the improved mixing characteristics of the multiple-injection strategy that consists of ramping-down and ramping-up injection rates were discussed in depth. Finally, this paper also investigated the effect of the gas-jet after EOI on mixture distribution and emission reduction under the ramping-down injection rate shape.

2. Model Description

2.1. Spray Model

After EOI, the deceleration state of turbulent sprays has been found to travel downstream from the injector nozzle, meanwhile, the entrainment rate of the passing region is greatly enhanced. The propagation of the increased entrainment region is commonly known as the "entrainment wave".

The original one-dimensional discrete spray model, developed by Musculus et al. [23], could not only predict the entrainment wave in diesel spray, but also calculate the spray penetration. The computation results showed that the spray penetration after the entrainment wave arrived at the spray tip was smaller when compared with that of a steady jet. Moreover, under the action of the entrainment wave, the dependence of spray penetration on time transited from the square-root to the fourth-root. This result was validated by several experimental data. It indicates that this model has a significant advantage in mimicking the diesel spray propagation after EOI. However, the original model mainly focused on the constant injection rate case and didn't show any application for the varying injection rate. To overcome this limitation, the original model is improved in this paper to introduce the effect of varying injection rate on the spray propagation by calculating the velocity at the nozzle exit under varying fuel mass flow rate. Because based on the one-dimensional discrete spray model that proposed by Musculus et al. [23], the improved one has the same assumptions as the former. Furthermore, taking the coking and cavitation phenomenon as well as the liquid diesel spray contraction into account, the nozzle diameter is used with a discharge coefficient in computation. Figure 1 shows the schematic diagram of one-dimensional discrete diesel spray model. The fuel mass and axial momentum in each control volume are solved by the transport Equations (1) and (2).

$$\frac{\partial m_f}{\partial t} = \dot{m}_{f,in} - \dot{m}_{f,out} \tag{1}$$

$$\frac{\partial M}{\partial t} = \dot{M}_{in} - \dot{M}_{out} \tag{2}$$

where m_f is fuel mass and M is axial momentum in a control volume. Overdot means integral flux crossing the inlet or outlet of each control volume. Where m_f is fuel mass and M is axial momentum in a control volume. Using an upwind differencing scheme, the discretized equations for fuel mass and momentum are:

$$m_{f,i}(t + \Delta t) = m_{f,i}(t) + \rho_f \left[\beta_{i-1} \cdot \overline{\overline{X}}_{f,i-1}(t) \cdot \overline{\overline{u}}_{i-1}(t) \cdot A_{i-1} - \beta_i \cdot \overline{\overline{X}}_{f,i}(t) \cdot \overline{\overline{u}}_i(t) \cdot A_i \right] \cdot \Delta t \tag{3}$$

$$M_i(t + \Delta t) = M_i(t) + \left\{ \beta_{i-1} \cdot \overline{\overline{\rho}}_{i-1}(t) \cdot \left[\overline{\overline{u}}_{i-1}(t) \right]^2 \cdot A_{i-1} - \beta_i \cdot \overline{\overline{\rho}}_i(t) \cdot \left[\overline{\overline{u}}_i(t) \right]^2 \cdot A_i \right\} \cdot \Delta t \tag{4}$$

where $\overline{\overline{X}}_{f,i}(t)$, $\overline{\overline{u}}_i(t)$ are the fuel volume fraction and fluid velocity averaged over the spray cross section at $i-$ th control volume respectively, ρ_f is the fuel density, β_i is the coefficient which determines the radial velocity profile at $i-$ th control volume, A_i is the cross-sectional area at the downstream surface of $i-$ th control volume, $\overline{\overline{\rho}}_i(t)$ is the cross-section averaged fluid density at $i-$ th control volume. A_i, $\overline{\overline{X}}_{f,i}(t)$, $\overline{\overline{u}}_i(t)$ and $\overline{\overline{\rho}}_i(t)$ are given by the Equations (5)–(8) respectively: align these symbols

$$A_i = \pi \cdot \left[Z_i \cdot \tan(\theta/2) \right]^2 \tag{5}$$

$$\bar{\bar{X}}_{f,i}(t) = \frac{m_{f,i}(t)}{\rho_f \cdot A_i \cdot \Delta Z} \tag{6}$$

$$\bar{\bar{u}}_i(t) = \frac{M_i(t)}{\bar{\bar{\rho}}_i(t) \cdot A_i \cdot \Delta Z} \tag{7}$$

$$\bar{\bar{\rho}}_i(t) = \rho_f \cdot \bar{\bar{X}}_{f,i}(t) + \rho_a(t) \cdot [1 - \bar{\bar{X}}_{f,i}(t)] \tag{8}$$

where ΔZ is the length of a control volume, θ is the spreading angle of diesel spray, $\rho_a(t)$ is the density of surrounding gas. A_i is evaluated at the axial midpoint of each control volume in the spray model of Musculus et al. [23], however, A_i defined at the downstream surface area of a control volume gives almost the same results when a sufficiently small ΔZ is used.

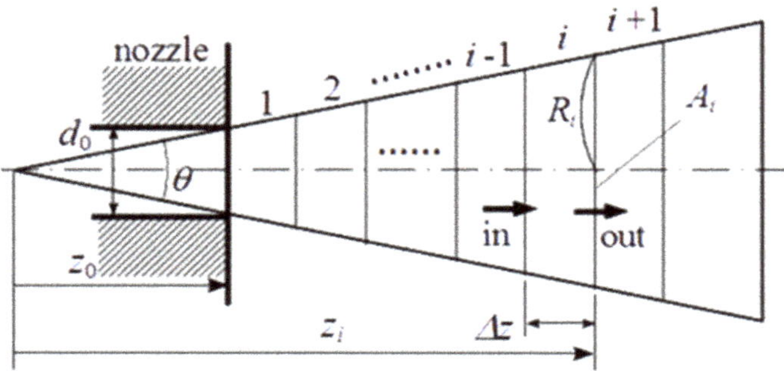

Figure 1. Schematic diagram of one-dimensional discrete diesel spray model.

2.2. Varying Injection Rate Calculation

Boundary conditions for solving these equations above are according to Equations (9) and (10):

$$\bar{\bar{u}}_0(t) = u_f(t) \tag{9}$$

$$\bar{\bar{X}}_{f,0}(t) = 1 \tag{10}$$

where the subscript "0" means the inlet surface of the first control volume, i.e., the nozzle exit. $u_f(t)$ is a cross-section averaged fuel velocity at the nozzle exit and is calculated by Equation (11):

$$u_f(t) = \frac{\dot{m}_{f,0}(t)}{\rho_f \cdot \mu \cdot \pi \cdot d_n^2/4} \tag{11}$$

where $\dot{m}_{f,0}(t)$ is the injection mass rate, μ is the flow coefficient of the nozzle orifice, d_n is the geometric orifice diameter.

$\dot{m}_{f,0}(t)$ is assumed to have a trapezoidal pattern, as shown in Figure 2, and it can be calculated according to Equation (12).

$$\dot{m}_{f,0}(t) = \begin{cases} \dot{m}_{f,max} \cdot t / \Delta t_r & (0 \le t < \Delta t_r) \\ \dot{m}_{f,max} & (\Delta t_r \le t < \Delta t_{js}) \\ \dot{m}_{f,max} \cdot [1 - (t - \Delta t_{js}) / \Delta t_d] & (\Delta t_{js} \le t < \Delta t_{js} + \Delta t_d) \\ 0 & (\Delta t_{js} + \Delta t_d \le t) \end{cases} \tag{12}$$

where the injection mass rate peak $\dot{m}_{f,max}$ is obtained by Equation (13):

$$\dot{m}_{f,max} = m_{f,total} / [\Delta t_{js} + (\Delta t_d - \Delta t_r)/2] \tag{13}$$

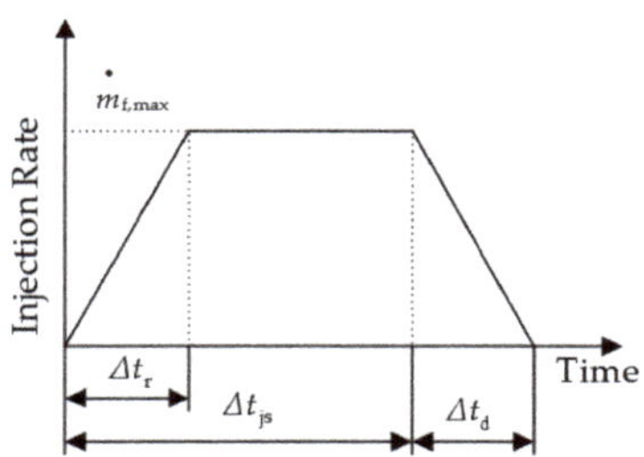

Figure 2. Injection rate pattern.

3. Results and Discussion

3.1. Validation of the Spray Model with Short Injection Duration

For model validation, the diesel spray tip penetration even after EOI with short injection duration was calculated and compared with the shadowgraph data of non-combusting diesel sprays in a constant volume chamber. The experimental setup and measuring equipment are the same as that in our previous study [11]. The constant volume chamber is in cylindrical shape with a volume of 150 cm^3. Its diameter and depth are 80 mm and 30 mm, respectively. In the shadowgraph imaging system, a Xeon lamp is used as the light source. The focal lengths of the two same concave mirrors used are 1910 mm. The high-speed camera has a shooting speed of 50,000 fps and the exposure time was 5 ms. With the aid of a common-rail injection system, an advanced injector with a discharge coefficient of 0.8 is employed to inject diesel fuel into the chamber. The operating conditions are listed in Table 1, and the calculation was also performed based on these conditions.

Table 1. Calculation conditions for validation.

Items	Parameters
Ambient pressure	4.0 MPa
Ambient temperature	1015 K
Fuel	JIS 2#
Cetane number	54.2
Fuel density at 15 °C	832.7 kg/m^3
Ignition temperature	72.0 °C
Dynamic viscosity at 30 °C	3.431 mm^2/s
Calorific value	45.82 MJ/kg
Injection pressure	180 MPa
Injection quantity	2.5 mg, 5.0 mg

Table 1. *Cont.*

Items	Parameters
Injection duration	0.175 ms, 0.297 ms
Nozzle diameter	0.18 mm
Nozzle holes number	6
Ambient gas composition by volume before injection	O_2: 0.5%, N_2: 87.4%, CO_2: 4.8%, H_2O: 7.3%

Figure 3 shows the comparison of spray tip penetrations between the simulation result and the experiment data from shadowgraph with the injection quantity of 0.417 mg and 0.833 mg for single nozzle hole under 180 MPa injection pressure, respectively. The spray tip penetrations and time are expressed on a logarithmic scale. As the injection rates in Figure 3 show, the actual injection rates have a shape like a triangle, therefore the calculations were based on the triangle injection rate and the injection durations were kept the same with that of the measurements. The results reveal that the simulation results are slightly smaller than the experiment data, which can be ascribed to the slightly lower measurement accuracy of injection rate and spray tip penetration under a very small amount of fuel injected, especially in the initial injection period. The simplification of the triangle injection rate for simulation further increases such a difference. However, the simulation results of spray tip penetration show similar tendencies to the measured data. Furthermore, the spray penetrations are able to capture the relation of square-root dependence on time around the moment of twice the injection duration. Then the spray penetrations change to the relation of fourth-root dependence on time. This phenomenon is coherent with that affected by the entrainment wave [24]. Thus, it is reasonable to apply this spray model to analyze the diesel spray mixing process.

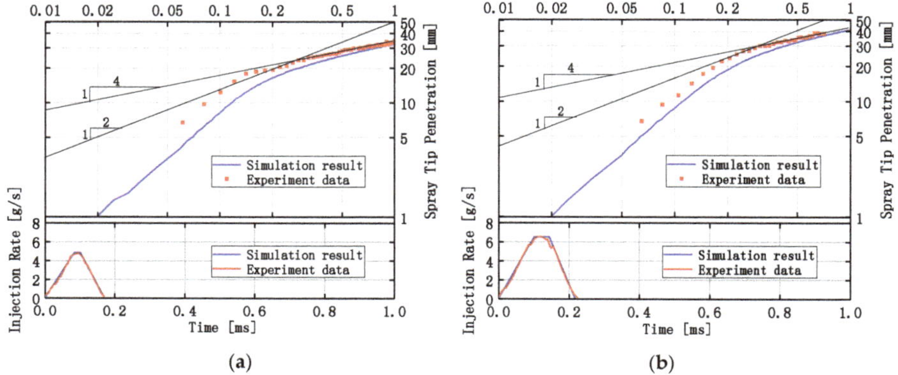

Figure 3. Comparison of spray tip penetrations between the simulation result and experiment data under (**a**) the injection quantity of 0.417 mg and (**b**) injection quantity of 0.833 mg.

3.2. Calculation of Spray Tip Penetrations Under Varying Injection Rate

As has been found in a previous study [19], the ramping-up and ramping-down injection rate shapes both have a significant influence on FPEG. Due to the similarity between FPEG and the conventional diesel engine, these two injection rate shapes must also play an important role in the latter. However, there is little relative research about this. Inversely, as the common used and typical injection rate shapes in diesel engines, scholars have devoted a great deal of effort to the triangle and rectangle as well as trapezoid shapes [9–13]. In order to compare the potential of varying injection rate shapes to lower the pollutant emissions of diesel engines more objectively and comprehensively and find the optimized injection strategy, five injection rate shapes (including triangle, ramping-up, ramping-down, rectangle and trapezoid) were selected in this paper, as shown in Figure 4. To simulate the diesel spray propagation under the conditions extant in engines' cylinders, the ambient gas conditions are set as the actual in-cylinder pressure and temperature, so the calculation conditions are the same as given in Table 1, and the injection rates have been shown in a previous study [11].

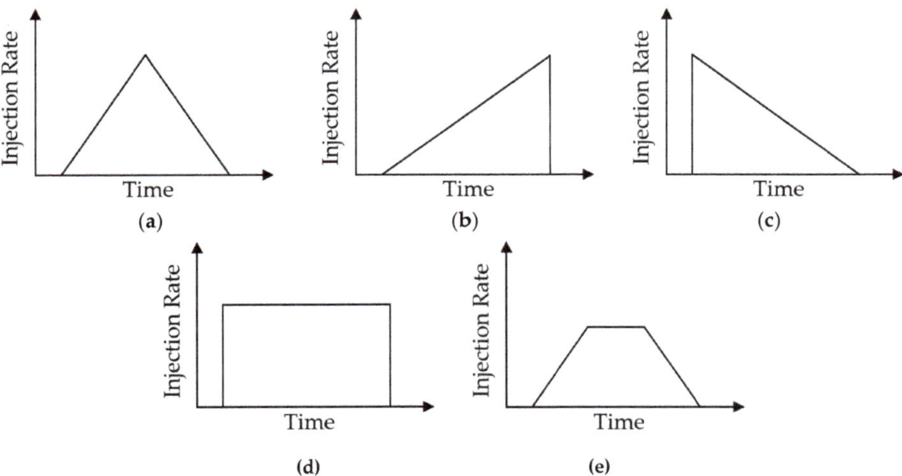

Figure 4. Schematic diagram of five injection rate shapes: (**a**) Triangle; (**b**) Ramping-up; (**c**) Ramping-down; (**d**) Rectangle; (**e**) Trapezoid.

Figure 5 shows the spray tip penetrations with the injection quantity of 0.417 mg and 0.833 mg for a single nozzle hole, respectively. Both Figure 5a,b reveal the same tendency that the initial spray tip penetration depends on the injection rate. A higher injection rate leads to a larger spray tip penetration during the initial injection.

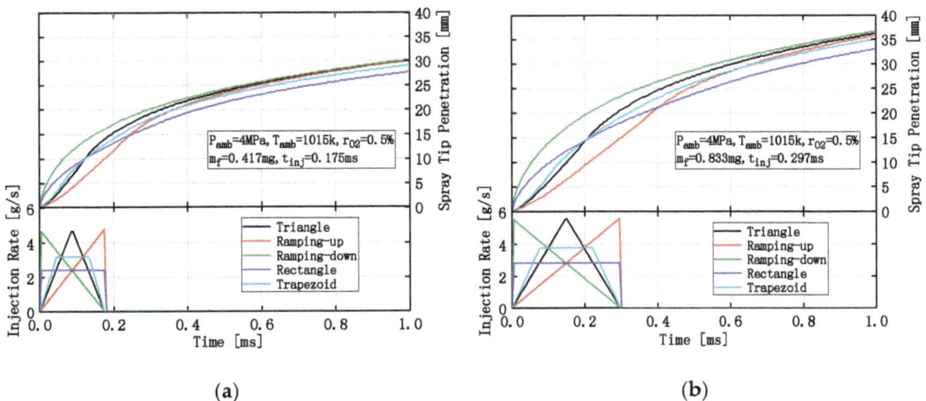

Figure 5. Spray tip penetration under (**a**) the injection quantity of 0.417 mg and (**b**) injection quantity of 0.833 mg NOTE. P_{amb}: ambient pressure; T_{amb}: ambient temperature; r_{O2}: volume fraction of O_2 in the ambient gas; m_f: injection quantity of fuel for single nozzle hole; t_{inj}: injection duration.

A higher maximum injection rate finally leads to a larger spray tip penetration, which is different from the previous study [19] and can be seen from two aspects. On the one hand, the triangle, ramping-up and ramping-down injection rate shapes have the same maximum injection rate, although their spray tip penetrations evolve with different growth rates in the early injection duration, they all gradually tend to the same penetration value and slope over time. On the other hand, compared with the rectangle and trapezoid injection rate shapes, the triangle injection rate reaches the larger spray tip penetration finally due to its higher maximum injection rate though its injection rate is smaller than that of the other two during the initial injection. However, in spite of the fact that the triangle and ramping-up as well as ramping-down injection rate shapes reach the final same spray penetration, the ramping-down injection rate has been keeping the greater spray tip penetration until the differences between them can be ignored, meaning that its overall spray gets more time to mix with air. Therefore, the ramping-down injection rate shape could obtain the better fuel-air mixing if the ignition delay is long enough.

3.3. Analysis of Entrainment Rates Under Varying Injection Rate

One of the indicators that allow for the insightful analysis on mixing process of diesel spray is the entrainment rate. Figure 6 shows the entrainment rates with varying injection rate shapes. Comparing the entrainment rates for the same injection rate shape under different fuel injection quantity, the larger fuel injection quantity case (Figure 6b) visibly gets much higher entrainment rate, with the variation tendency extremely similar to the smaller fuel injection quantity case (Figure 6a). For the same fuel injection quantity, it can be found that the higher injection rate causes the higher entrainment rate during the initial injection. Each injection rate shape has the special entrainment rate pattern and the entrainment rate comes to the peak soon after EOI. The entrainment rate peaks (from large to small) are in the order: ramping-up, triangle, trapezoid, ramping-down and rectangle. It is the entrainment wave after EOI that should be responsible for such difference.

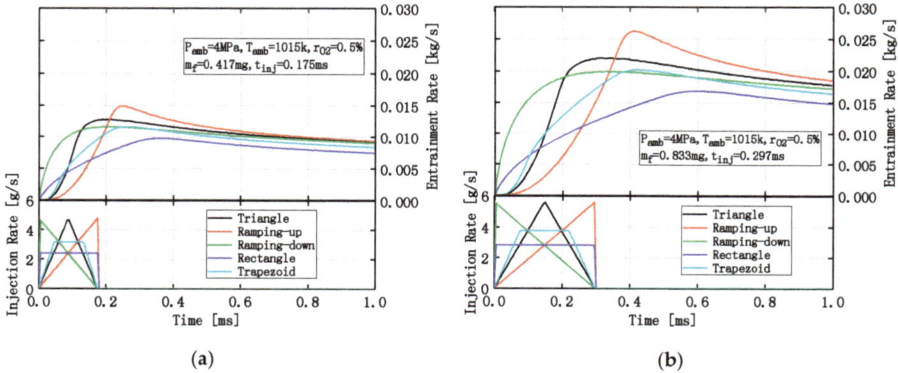

Figure 6. Entrainment rate under (**a**) the injection quantity of 0.417 mg and (**b**) injection quantity of 0.833 mg.

As discussed in the previous study [2], the velocity at the nozzle exit is high enough to affect the spray when the injection rate just begins to decrease for the ramping-down, triangle and trapezoid injection rates. Once the velocity at the nozzle exit is too low to affect the spray, the real time of the EOI (t_{EOI}) can be defined. t_{EOI} is the start of entrainment wave and the velocity at the nozzle exit at t_{EOI} (U_{EOI}) determines the spray penetration level. In fact, the peak of entrainment rate occurs when the entrainment wave arrives at the spray tip. Thus, the entrainment rate peak is decided by t_{EOI} and U_{EOI}. To calculate t_{EOI} and U_{EOI} under varying injection rate shapes, the Equation (14) proposed by Liu et al. [11] is used in this paper:

$$U_{EOI} = (1 - R_{IV} \cdot t_{decay} / t_{inj}) \cdot u_{max} \tag{14}$$

where R_{IV} is a dimensionless ratio and is selected to be 0.65 in this paper, t_{decay} is the period that injection velocity decays, t_{inj} is the injection duration, u_{max} is the maximum injection velocity.

The calculation results of t_{EOI} and U_{EOI} for the injection quantity of 0.417 mg and 0.833 mg are shown in Tables 2 and 3 respectively. t_{peak} denotes the time at the entrainment rate peak.

Table 2. Calculation results of t_{EOI} and U_{EOI} for injection quantity of 0.417 mg.

Injection Rate Shape	t_{EOI} (ms)	U_{EOI} (m/s)	t_{peak}(ms)/2 · t_{EOI} (ms)
Triangle	0.116	188.12	0.21/0.232
Ramping-up	0.175	278.70	0.25/0.35
Ramping-down	0.1138	97.55	0.215/0.2276
Rectangle	0.175	139.30	0.355/0.35
Trapezoid	0.1401	156.74	0.265/0.2802

Table 3. Calculation results of t_{EOI} and U_{EOI} for injection quantity of 0.833 mg.

Injection Rate Shape	t_{EOI} (ms)	U_{EOI} (m/s)	t_{peak}(ms)/2 · t_{EOI} (ms)
Triangle	0.195	219.92	0.355/0.39
Ramping-up	0.297	325.80	0.42/0.594
Ramping-down	0.195	114.03	0.385/0.39
Rectangle	0.297	165.00	0.595/0.594
Trapezoid	0.23	183.34	0.42/0.46

A same law can be discovered from Tables 2 and 3, that is, t_{peak} in the other four injection rate shapes, except the ramping-up, is almost twice that of t_{EOI}, which is because the entrainment wave travels downstream at twice the initial spray penetration rate. As reference [24] demonstrated, the typical phenomenon is revealed in the case of rectangle injection rate, the diesel jet with rectangle injection rate can be treated as a steady jet, and the entrainment wave front arrives at the jet tip at twice of injection duration. Figure 7 shows the mean velocity over the cross-sectional area against the axial position in the diesel spray under five injection rate shapes with time varying. As shown in Figure 7d (rectangle injection rate), the same axial position has the same velocity during injection due to the steady state. There is an envelope velocity curve for the steady spray, and the velocity at the position that belong to the spray steady part after EOI is on the envelope velocity curve. After the entrainment wave front arriving at the spray tip, the total spray transfers to the decelerating state, which causes the velocities at any axial position are lower than that of steady state.

For the trapezoid injection rate, the constant injection rate takes main injection process, so that the spray can be treated as the steady state. According to the previous study [11], the twice of t_{EOI} is close to the t_{peak}. As shown in Figure 7e, the envelope velocity curve starts from the steady part and it extends to the t_{peak}. After t_{peak}, the spray gets into the decelerating state. In Figure 7c, although the spray is not in the steady state due to the ramping-down injection rate, the spray tip velocities at different time before t_{peak} are on the same envelope velocity curve. After t_{peak}, the spray tip velocities depart from the envelope velocity curve, and the values are lower than that on the envelope curve. It might be reasonable to use the twice of U_{EOI} to represent the average entrainment wave speed, and the U_{EOI} can be treated as the average effective fuel injection velocity that considers the fuel penetration as the spray penetration. Then the t_{peak} is similar with twice of t_{EOI} even for the unsteady spray with ramping-down injection rate.

However, because the momentum flux at each axial position of spray needs a certain response time to the change of the injected velocity, the velocities at some positions still increase even the injection velocity turns to decrease or stop, which can be found in Figure 7a,b. In Figure 7a, the envelope velocity curve can cover the spray tip velocity up to t_{peak} from 0.16 ms. That means the injection velocity increase effect has been reached spray tip, the decrease of injection velocity effect mainly control the spray. Meanwhile, the t_{peak} is almost equal to the twice of t_{EOI}, which is indicated the effectiveness of the average entrainment wave and spray penetration speed concept.

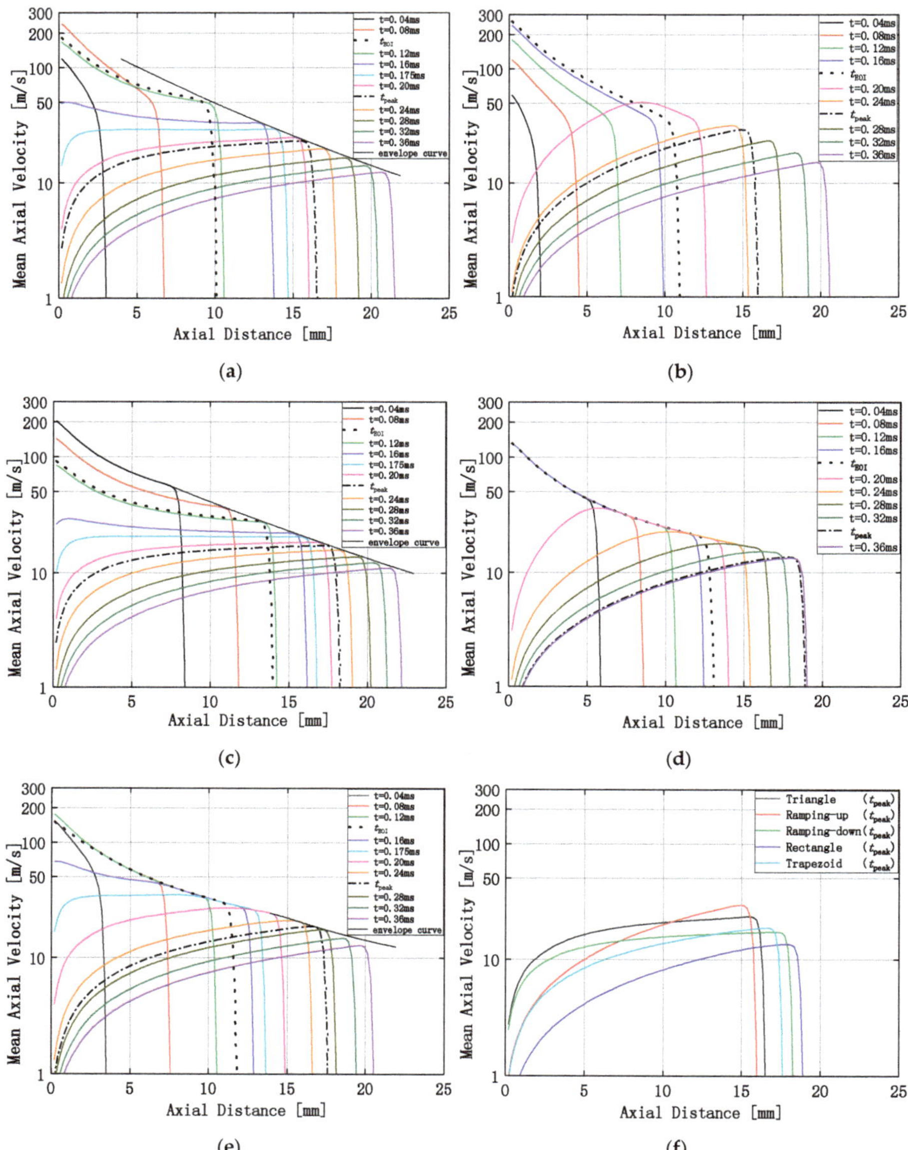

Figure 7. Mean axial velocity under five injection rate shapes: (**a**) Triangle; (**b**) Ramping-up; (**c**) Ramping-down; (**d**) Rectangle; (**e**) Trapezoid and (**f**) Comparison at t_{peak}.

Regarding the ramping-up injection rate, there is no steady-like state as shown in Figure 7b, because the injection rate is reduced to zero sharply. At the t_{EOI}, the U_{EOI} is the maximum injection velocity, the average entrainment wave speed can be calculated by twice of U_{EOI}. However, the increase of the injection velocity effect has not traveled to spray tip, and the U_{EOI} cannot represent the average spray penetration speed. Similar to the assumption of Liu et al. [11], a same parameter $R_{\text{IV}}(0.65)$ is selected in this paper as a factor of U_{EOI} to calculate the average fuel penetration speed, which represents the average spray penetration speed and one third of average entrainment wave speed. As shown in Equation (15), t_1 is the duration from the t_{EOI} to the time of entrainment wave reaching the spray tip:

$$2 \cdot U_{EOI} \cdot t_1 = 0.65 \cdot U_{EOI} \cdot (t_{EOI} + t_1) \tag{15}$$

Equation (16) can be simplified as:

$$3 \cdot U_{EOI} \cdot t_1 = U_{EOI} \cdot (t_{EOI} + t_1) \tag{16}$$

then the following expression is obtained:

$$t_1 = 0.5 \cdot t_{EOI} \tag{17}$$

Thus, the entrainment wave arrives at the spray tip at $t = 1.5\ t_{EOI}$, which is consistent with the calculation result of t_{peak} and t_{EOI} under the ramping-up injection rate, t_{peak} (0.25 ms) is about 1.5 times (0.2625 ms) of t_{EOI} for the injection quantity of 0.417 mg, while t_{peak} (0.42 ms) is also about 1.5 times (0.4455 ms) of t_{EOI} for the injection quantity of 0.833 mg. In addition, as shown in Figure 7b, the entrainment wave takes 0.075 ms to arrive at the spray tip with the spray penetration of 15.8 mm at 0.25 ms. The average entrainment wave traveling speed is 210.67 mm/ms, and the average spray penetration rate is about 63.2 mm/ms. The average entrainment wave traveling speed is 3.33 times of the average spray penetration rate, which is similar with the result calculated using the U_{EOI}.

Moreover, due to the largest U_{EOI}, the spray of ramping-up injection rate penetrates fast after EOI, compared with the triangle and ramping-down injection rates (Figure 5), so the terminal spray penetration is a combined effect result of injection rate before EOI and U_{EOI}. The triangle, ramping-up and ramping-down injection rate shapes tend to the same spray penetration finally just because an opportune tradeoff is reached.

As shown in Figure 7, there are some similarities between five injection rate shapes: the mean axial velocity decreases along the axial distance before EOI and increases along the axial distance after EOI, but the velocity gradients are different. When fuel injection supply stops after EOI, the mass conservation and continuity theorem require a fast gas entrainment to compensate for the decreasing fuel mass flux. Therefore, the fuel mass flux, or speaking directly, the velocity gradient determines the intensity of entrainment wave. As mentioned above, the entrainment rate peak occurs when the entrainment wave arrives at the spray tip and t_{peak} is decided by t_{EOI} as well as the speed of entrainment wave. In order to explain the difference of entrainment rate peak for five injection rate shapes, the velocity gradient at t_{peak} should be taken into account. Shown in Figure 7f, it is obvious that five injection rate shapes have different velocity gradients and velocity magnitude at t_{peak}. Because the entire spray has been enhanced by the entrainment wave at this moment, associated with the cone structure of spray, the entrainment effect of downstream spray accounts for a larger proportion of the entire spray due to its larger volume. Thus, the ramping-up injection rate has the highest velocity gradient and the fast mean axial velocity in downstream, which leads to the biggest entrainment rate peak in Figure 6, followed by the triangle injection rate. The trapezoid injection rate gets the slightly larger entrainment rate peak than the ramping-down on account of its higher velocity gradient, too. While for the rectangle injection rate, its velocity gradient and velocity magnitude are both the lowest, resulting in a much weaker entrainment effect than the other four injection rate shapes, so its entrainment rate peak is the lowest.

3.4. Analysis of Mixing Process Based on Equivalence Ratios

Spray tip penetration and entrainment rate can estimate the diesel spray mixing process in the macroscopic view. However, the microscopic mixing process is also a very significant factor for diesel spray combustion. To analyze the microscopic mixing process, the equivalence ratio along the spray axial position with varying time was calculated using the spray model. Figure 8 shows the equivalence ratio distributions under five injection rate shapes. Figure 8a,c,e show the similar equivalence ratio level after EOI in the cases of triangle, ramping-down as well as trapezoid injection rates, the equivalence ratio is almost evenly distributed along the spray axial distance after EOI.

Especially at the time of 0.36 ms, the equivalence ratio is around 1.0. The fuel could be burned completely if the ignition occurred at that time, as well as there is no enough O_2 remained for NO_x formation. Therefore, the fuel consumption and NO_x emission can decrease simultaneously. But compared with the triangle and trapezoid injection rates, the ramping-down is able to reach much longer spray penetration before EOI, while the downstream spray has lower equivalence ratio, indicating a stronger mixing process. So the ramping-down injection rate has an advantage over the other two in creating homogeneous mixture. However, due to the low velocity near the nozzle exit, the fuel-rich mixture formation near the nozzle exit, it usually leads to the high soot formation.

Figure 8b,d show the similar results of equivalence ratio along the spray axial distance with varying time in the cases of ramping-up and rectangle injection rates, the mixture near the nozzle exit quickly gets leaner after EOI, and the equivalence ratio increases with the axial distance increasing. It means that the both two injection rates with longer ignition delay can decrease the soot formation near the nozzle exit on certain conditions. It can be observed that the ramping-up injection rate obtains much leaner mixture than that of rectangle injection rate after EOI. This phenomenon indicates that the ramping-up injection rate is a better choice to realize LTC, because the ignition is occurred after EOI (with the overlong ignition delay not considered). In summary, the spray characteristics among the penetration and entrainment rate as well as equivalence ratio under varying injection rate shapes can be synoptically shown in Table 4, where I to V denote the maximum to the minimum of each parameter. From the practical point of view, these spray characteristics can provide some important guidance and reference. For example, the long penetration is always expected in the low-speed diesel engine with big bore and long stroke, so the ramping-down injection rate is an optimal choice. Instead, the spray-wall impingement process is tried to be avoided sometimes, hence a relatively short penetration is needed and the rectangle injection rate should be taken into account first. In some advanced combustion modes, such as PCCI and LTC, the ramping-up injection rate can be used to greatly promote the mixing of fuel and the cold ambient air or exhaust gas, then quickly decrease the temperature of mixture as well as the formation of NO_x. As for homogeneous charge compression ignition (HCCI), the uniform charge in cylinder is indispensable, so that all points in diesel spray are ignited at the same time and the heat is evenly distributed throughout the cylinder. In this way, the triangle, ramping-down as well as trapezoid injection rates are the theoretically feasible alternatives because the equivalence ratio is almost evenly distributed along the spray axial distance after EOI in these cases.

Table 4. Comparison of spray characteristics under varying injection rate shapes.

Injection Rate Shape	Penetration After EOI	Entrainment Rate Peak	Equivalence Ratio Near Nozzle
Triangle	II	II	II
Ramping-up	III	I	V
Ramping-down	I	IV	I
Rectangle	V	V	IV
Trapezoid	IV	III	III

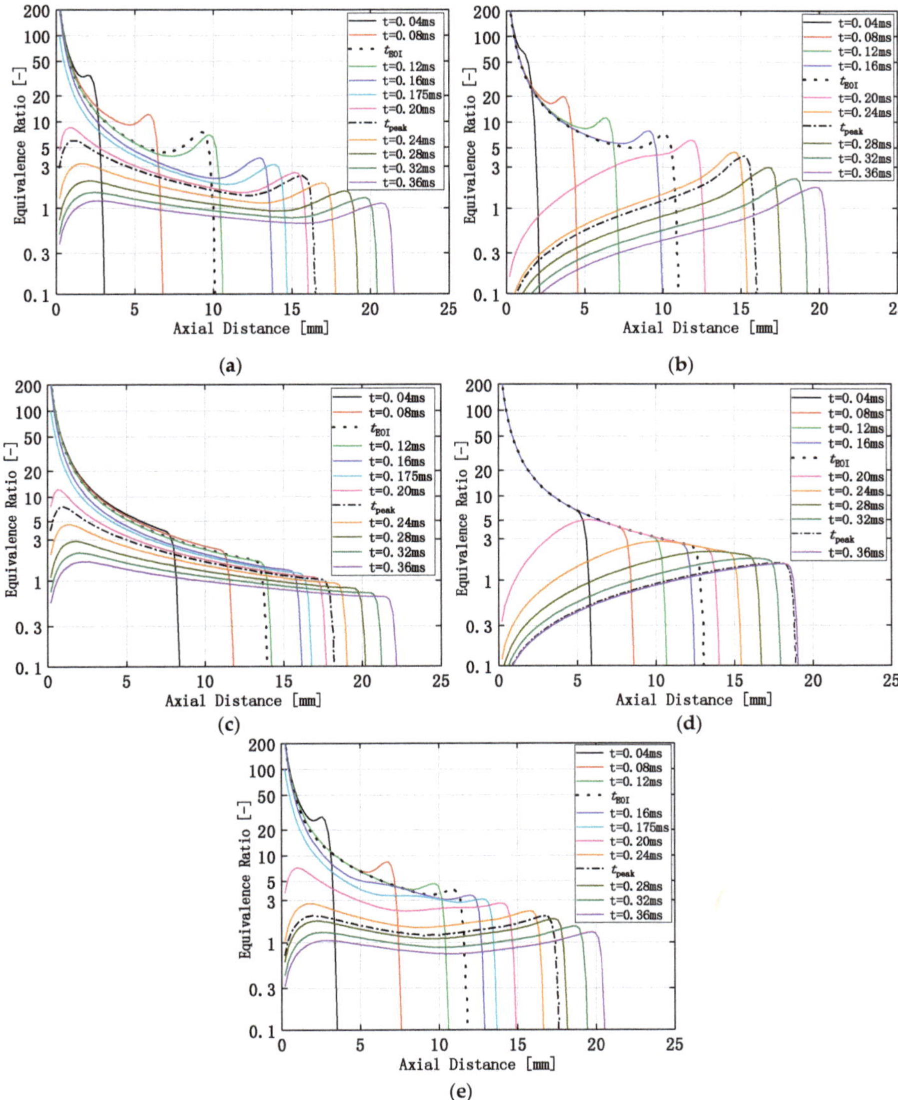

Figure 8. Equivalence ratio after EOI under five injection rate shapes: (**a**) Triangle; (**b**) Ramping-up; (**c**) Ramping-down; (**d**) Rectangle; (**e**) Trapezoid.

3.5. Effect of the Multiple-Injection

According to the information in Table 4, the ramping-down injection rate could reach the longest penetration but also the highest equivalence ratio near the nozzle, while the ramping-up injection rate obtains the largest entrainment rate peak and the lowest equivalence ratio near the nozzle. So it is a good idea to develop a new multiple-injection strategy that both contains the ramping-down and ramping-up injection rates. Figure 9 shows the comparisons of spray tip penetration and entrainment rate between the ramping-down single injection and the multiple-injection. In Figure 9a, the spray tip penetration of the multiple-injection is equal to that of the ramping-down single injection before

$t = 0.1$ ms. After that, the spray tip penetration of the multiple-injection becomes smaller than that of the ramping-down single injection due to the sharply decrease of injection rate. As shown in Figure 9b, although the entrainment rate of the multiple-injection is smaller than that of the ramping-down single injection from $t = 0.05$ ms to $t = 0.3$ ms, it increases rapidly over the latter after $t = 0.3$ ms because the ramping-up sub-injection begins to play a leading role in the entrainment process.

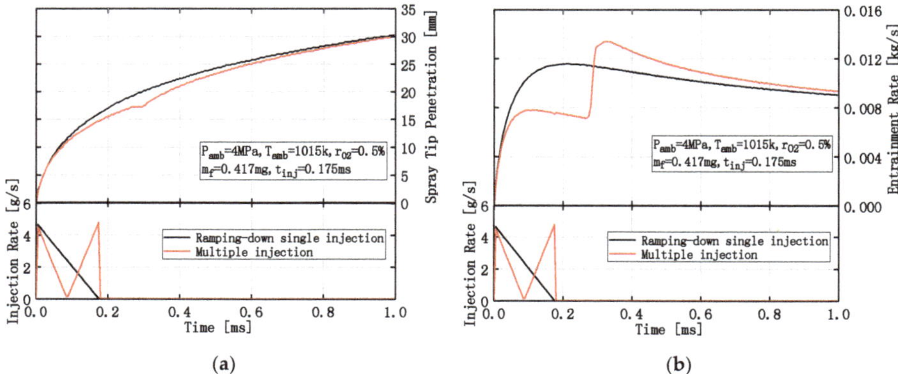

(a) (b)

Figure 9. Comparisons of spray tip penetration (**a**) and entrainment rate (**b**) between the ramping-down single injection and the multiple-injection.

Figure 10 shows the comparison of entrainment rate after EOI between the multiple-injection and the ramping-down single injection. The spray head of the multiple-injection is a little shorter than that of the ramping-down single injection, which is in agreement with Figure 9a. When it is at $t = 0.18$ ms and $t = 0.24$ ms, since the sub-spray generated by the ramping-up sub-injection has not overtaken the sub-spray generated by the ramping-down sub-injection, and much fuel accumulates in the former spray head, so the high-concentration fuel causes little gas to be entrained. Due to the slower injection velocity, the entrainment rate of the ramping-down sub-injection in the multiple-injection is smaller than that of the ramping-down single injection even though they have the similar injection rate shape. However, the upstream sub-spray generated by the ramping-up sub-injection of the multiple-injection has much higher entrainment rate than that of the ramping-down single injection. After $t = 0.24$ ms, the upstream sub-spray overtakes the downstream sub-spray. Because the entrainment wave of the ramping-up sub-injection plays a dominant role in the whole spray, the entrainment rate of the multiple-injection is much higher than that of the ramping-down single injection, indicating that the multiple-injection can get better mixing effect.

Figure 11 shows the comparisons of equivalence ratio after EOI between the multiple-injection and the ramping-down single injection. In the multiple-injection case, the equivalence ratio after EOI remains the characteristic of ramping-up injection rate shape, resulting from the major influence of the ramping-up sub-injection. The variation of equivalence ratio of the multiple-injection is seen to be quite different from the ramping-down single injection. The equivalence ratio of the multiple-injection is much lower near the nozzle exit and increases gradually along the axial distance. In addition, it is obvious that the spray tail of the multiple-injection moves downstream quickly, while the ramping-down single injection consistently keeps the high equivalence ratio near the nozzle exit. Therefore, the multiple-injection is beneficial to improve the working environment of the nozzle and decrease the soot emission.

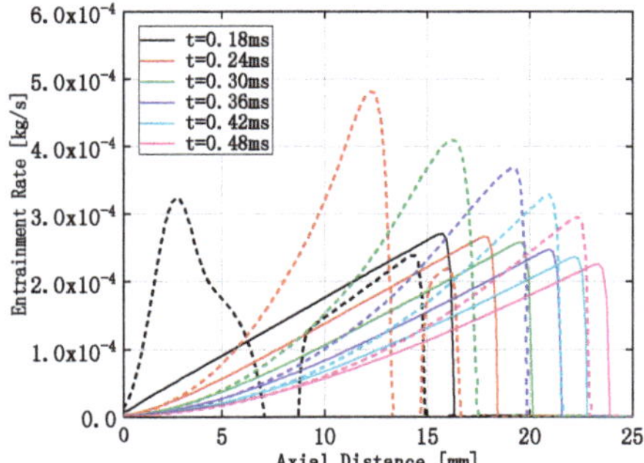

Figure 10. Comparisons of entrainment rate after EOI between the ramping-down single injection and the multiple-injection (solid lines: ramping-down single injection; dotted lines: multiple-injection).

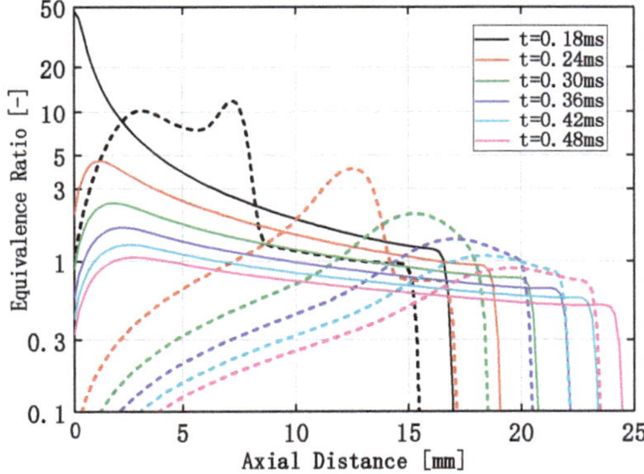

Figure 11. Comparisons of equivalence ratio between the ramping-down single injection and the multiple-injection (solid lines: ramping-down single injection; dotted lines: multiple-injection).

3.6. Effect of the Gas-Jet After EOI

However, incomplete combustion will occur in the fuel-lean zone if the equivalence ratio is too low, so the originality of combining the ramping-down and ramping-up injection rates to optimize the working conditions of the nozzle and emission performance still has some limitations. To overcome the problem, another improvement approach is presented in this paper.

The main idea of the new method is to use the gas-jet after EOI to blow the mixture near the nozzle exit downstream under the ramping-down injection rate shape, as shown in Figure 12. In this way, the stagnant fuel near the nozzle exit due to the influence of entrainment wave could be carried by the gas-jet to join in the combustion of downstream spray, which can not only improve the working conditions of the nozzle, but also decrease the soot emission.

This paper studied the influence of gas jet with two different velocities on the spray evolution, the selected gas-jet velocities (100 m/s, 200 m/s) are both lower than the local speed of sound (648.89 m/s), so the influence of shock wave is inexistent. The calculation conditions for gas-jet are listed in Table 5. The comparisons of spray tip penetration and entrainment rate between the ramping-down injection without gas-jet and the ramping-down injection with gas-jet are shown in Figure 13. It seems that whether there is gas-jet after EOI or not has no effect on the spray tip penetration and entrainment rate, which is related to the limited effect range of gas-jet.

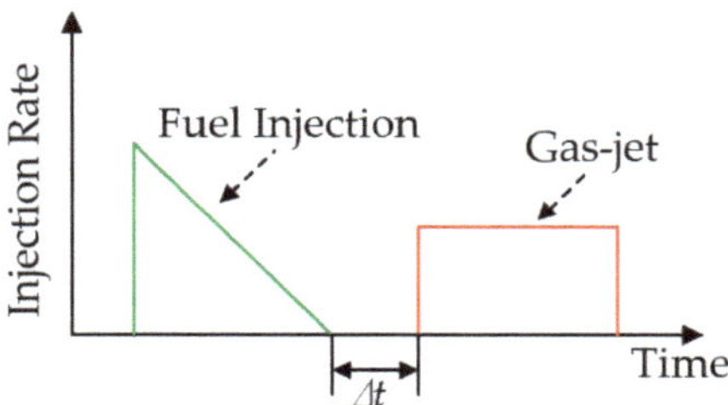

Figure 12. Schematic diagram of the ramping-down injection with gas-jet.

Table 5. Calculation conditions for gas-jet.

Items	Parameters
Gas-jet composition by volume	O2: 0.5%, N2: 87.4%, CO2: 4.8%, H2O: 7.3%
Injection pressure	4.0 MPa
Gas temperature	1015 K
Gas density	13.30 kg/m^3
Injection interval	0.05 ms
Injection duration	0.2 ms
Injection velocity	100 m/s, 200 m/s

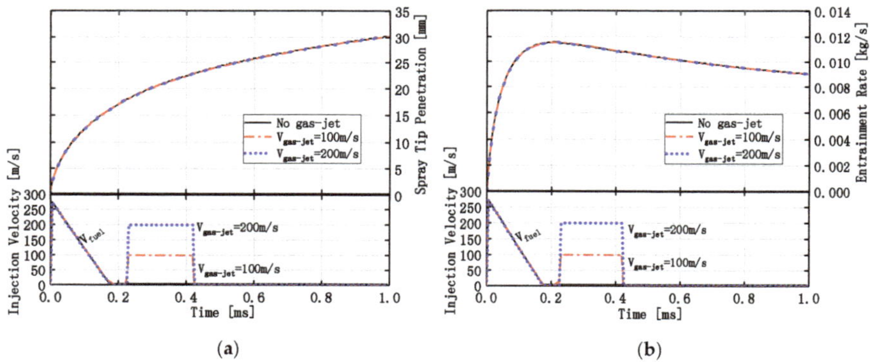

Figure 13. Spray tip penetrations (**a**) and entrainment rates (**b**) of the ramping-down injection with gas-jet and the ramping-down injection without gas-jet (black solid lines: ramping-down injection without gas-jet; red chain lines: ramping-down injection with gas-jet of 100 m/s; blue dotted lines: ramping-down injection with gas-jet of 200 m/s).

Figure 14 shows the variations of equivalence ratio of the ramping-down injection without gas-jet and the ramping-down injection with gas-jet. It is obvious that the gas-jet after EOI has a great impact on the equivalence ratio of spray tail mixture. On the one hand, the gas-jet makes the spray tail move downstream, and the higher the velocity of gas-jet, the farther the spray tail leaves from the nozzle exit and the larger the zone affected by the gas-jet. For instance, when it is at t = 0.48 ms, the spray tail is around 4 mm away from the nozzle exit and the zone from the nozzle exit to 7.5 mm downstream is affected by the gas-jet with velocity of 100 m/s, while the spray tail is near 6 mm away from the nozzle exit and the zone from the nozzle exit to 10 mm downstream is affected in the case of gas-jet with velocity of 200 m/s. Under the action of the gas-jet, the stagnant fuel near the nozzle exit due to the influence of entrainment wave can flow downstream quickly, so the formation of fuel-lean zone near the nozzle exit is reduced and the following combustion environment is improved, decreasing the generation of soot and HC. Nevertheless, because of the rapid momentum exchange with low velocity mixture in spray tail, the gas-jet is not strong enough to affect the downstream zone any more. Thus, the ramping-down injection with gas-jet keeps the same equivalence ratio in the spray downstream as the ramping-down injection without gas-jet. In addition, owing to the limited effect range of gas-jet, there is almost no difference in spray tip penetration and overall entrainment rates between the ramping-down injection without gas-jet and the ramping-down injection with gas-jet as Figure 13 shows.

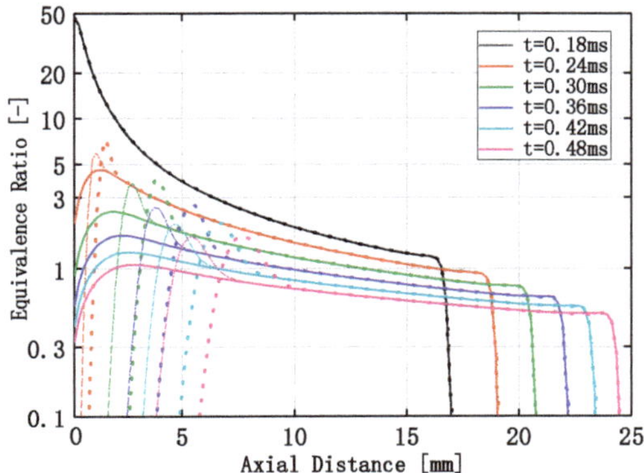

Figure 14. Equivalence ratio of the ramping-down injection with gas-jet and the ramping-down injection without gas-jet (solid lines: ramping-down injection without gas-jet; chain lines: ramping-down injection with gas-jet of 100 m/s; dotted lines: ramping-down injection with gas-jet of 200 m/s).

On the other hand, since the fuel in the spray tail flows downstream quickly due to the gas-jet, so more fuel accumulates in the new spray tail, which makes the equivalence ratio peak increase slightly. But the velocity of gas-jet has little impact on the equivalence ratio peak. Therefore, compared with the ramping-down injection without gas-jet, the one with gas-jet has more potential to achieve complete combustion and reduce the emission of soot and HC by organizing the ideal fuel-air distribution.

4. Conclusions

To find the optimized injection strategy for lower pollutant emissions, the mixing process of diesel spray under varying injection rate shapes has been studied in this paper by using a modified one-dimensional discrete spray model. The calculations were performed with five injection rate shapes: triangle, ramping-up, ramping-down, rectangle and trapezoid. Then the spray penetrations, entrainment rates and equivalence ratio distributed along the spray axial distance were analyzed. And this paper also discussed the ameliorative mixing characteristic of the multiple-injection strategy that was combined by the ramping-down and ramping-up injection rates, with the impact of the gas-jet after EOI on mixture distribution and pollutant emission reduction under the ramping-down injection rate shape investigated too. The conclusions are summarized as follows:

(1) The higher injection rate leads to the larger spray tip penetration during the initial injection and the higher maximum injection rate leads to the larger spray tip penetration finally.

(2) The ramping-up injection rate needs 1.5 times the injection duration for the entrainment wave to arrive at the spray tip. For the other four injection rates, the sprays can be treated as steady-like state, needing twice the injection duration from EOI to the time the entrainment wave reaches the spray tip. The concept using U_{EOI} to represent the average entrainment wave and spray penetration speed is easy and effective to estimate the time when the entrainment wave arrives at the spray tip.

(3) The ramping-down injection rate enhances the spray penetration and mixing rate during the initial period, and obtains an almost even distribution of the equivalence ratio along the spray axis after EOI. The ramping-up injection rate leads to the leaner mixture after EOI and even much leaner near the nozzle exit, which are useful to control the combustion process with longer ignition delay and decrease the soot emissions.

(4) The spray characteristic summaries about the penetration and entrainment rate as well as equivalence ratio can provide a certain guidance and reference for practical application.

(5) The multiple-injection consisting of the ramping-down and ramping-up injection rates can obtain much higher entrainment rates than the ramping-down single injection, and the equivalence ratio of the former is much lower near the nozzle exit than that of the latter.

(6) The gas-jet after EOI has no obvious effect on the spray tip penetration and overall entrainment rate because the momentum of gas-jet is limited and the affected zone is small. However, the gas-jet after EOI has a great impact on the equivalence ratio of spray tail mixture. The spray tail will move downstream quickly under the action of the gas-jet. The higher the velocity of gas-jet, the farther the spray tail leaves from the nozzle exit and the larger the zone affected by the gas-jet. With the gas-jet after EOI, the equivalence ratio in spray tail will increase slightly due to the more fuel accumulated.

Author Contributions: Conceptualization, L.L.; methodology, I.N.; software, Q.M.; validation, I.N. and Q.M.; formal analysis, L.L. and X.M.; data curation, I.N.; writing—original draft preparation, Q.M.; writing—review and editing, L.L.; project administration, X.M.

Funding: This research was supported by National Natural Science Foundation of China (Grant No. 51509051), Natural Science Foundation of Heilongjiang Province of China (Grant No. LC2015017), and Marine Power Research and Development Program (grant number DE0302).

Conflicts of Interest: The authors declare no conflicts of interest.

References

1. Miyaki, M.; Takeuchi, K.; Kojima, A.; Uchiyama, K.; Nakagawam, M.; Herrmann, O.E.; Maassen, F.; Laumen, H.J. Fulfilling Euro 6 Emission Regulations for Heavy Duty Engines without SCR-System-A Challenge to the FIE System. *Fortschritt-Berichte VDI Reihe 12. Verkehrstechnik-Fahrzeugtechnik* **2011**, *2*, 196–211. Available online: http://d.oldg.wanfangdata.com.cn/NSTLQK_NSTL_QKJJ0222007139.aspx (accessed on 17 October 2019).

2. Matsumoto, S.; Klose, C.; Schneider, J.; Nakane, N.; Ueda, D.; Kondo, S. 4th Generation Diesel Common Rail System: Realizing Ideal Structure Function for Diesel Engine. *SAE Tech. Pap.* **2013**, *2*. [CrossRef]

3. Wang, L.; Lowrie, J.; Ngaile, G.; Fang, T. High Injection Pressure Diesel Sprays from a Piezoelectric Fuel Injector. *Appl. Therm. Eng.* **2019**, *152*, 807–824. [CrossRef]

4. Boccardo, G.; Millo, F.; Piano, A.; Arnone, L.; Manelli, S.; Fagg, S.; Gatti, P.; Herrmann, O.E.; Queck, D.; Weber, J. Experimental investigation on a 3000 bar fuel injection system for a SCR-free non-road diesel engine. *Fuel* **2019**, *243*, 342–351. [CrossRef]

5. Peng, Z.; Liu, B.; Wang, W.; Lu, L. CFD Investigation into Diesel PCCI Combustion with Optimized Fuel Injection. *Energies* **2011**, *4*, 517–531. [CrossRef]

6. Kim, G.; Moon, S.; Lee, S.; Min, K. Numerical Analysis of the Combustion and Emission Characteristics of Diesel Engines with Multiple Injection Strategies Using a Modified 2-D Flamelet Model. *Energies* **2017**, *10*, 1292. [CrossRef]

7. Lee, C.; Chung, J.; Lee, K. Emission Characteristics for a Homogeneous Charged Compression Ignition Diesel Engine with Exhaust Gas Recirculation Using Split Injection Methodology. *Energies* **2017**, *10*, 2146. [CrossRef]

8. Liu, E.; Su, W. Study on Effects of Common Rail Injector Drive Circuitry with Different Freewheeling Circuits on Control Performance and Cycle-by-Cycle Variations. *Energies* **2019**, *12*, 564. [CrossRef]

9. Yang, S.; Moon, H.; Lee, C. A Study of Spill Control Characteristics of JP-8 and Conventional Diesel Fuel with a Common Rail Direct Injection System. *Energies* **2017**, *10*, 2104. [CrossRef]

10. Yang, S.; Lee, C. Experimental Research on the Injection Rate of DME and Diesel Fuel in Common Rail Injection System by Using Bosch and Zeuch Methods. *Energies* **2018**, *11*, 273. [CrossRef]

11. Liu, L.; Peng, Y.; Ma, X.; Horibe, N.; Ishiyama, T. Phenomenological modeling of diesel spray with varying injection profile. *J. Proc. Inst. Mech. Eng. Part D* **2018**, *233*, 2780–2790. [CrossRef]
12. Bao, Z.; Horibe, N.; Ishiyama, T. A Study on Diesel Spray Characteristics for Small Quantity Injection. *SAE Tech. Pap.* **2018**. [CrossRef]
13. Jafarmadar, S.; Shafee, S.; Barzegar, R. Numerical investigation of the effect of fuel injection mode on spray/wall interaction and emission formation in a direct injection diesel engine at full load state. *Therm. Sci.* **2010**, *14*, 1039–1049. [CrossRef]
14. Macian, V.; Payri, R.; Ruiz, S.; Bardi, M.; Plazas, A.H. Experimental Study of the Relationship between Injection Rate Shape and Diesel Ignition Using a Novel Piezo-actuated Direct-acting Injector. *Appl. Energy* **2014**, *118*, 100–113. [CrossRef]
15. Plamondon, E.; Seers, P. Development of a Simplified Dynamic Model for a Piezoelectric Injector Using Multiple Injection Strategies with Biodiesel/Diesel Fuel Blends. *Appl. Energy* **2014**, *131*, 411–424. [CrossRef]
16. Mohan, B.; Yang, W.; Yu, W.; Tay, K.L.; Chou, S.K. Numerical investigation on the effects of injection rate shaping on combustion and emission characteristics of biodiesel fueled CI engine. *Appl. Energy* **2015**, *160*, 737–745. [CrossRef]
17. Tay, K.L.; Yang, W.; Zhao, F.; Yu, W.; Mohan, B. Effects of triangular and ramp injection rate-shapes on the performance and emissions of a kerosene-diesel fueled direct injection compression ignition engine: A numerical study. *Appl. Therm. Eng.* **2017**, *110*, 1401–1410. [CrossRef]
18. Lamas, M.I.; Rodriguez, C.G. NOx Reduction in Diesel-Hydrogen Engines Using Different Strategies of Ammonia Injection. *Energies* **2019**, *12*, 1255. [CrossRef]
19. Han, C.; Yuan, C.; He, Y.; Liu, Y. Effect of Fuel Injection Rate Shapes on Mixture Formation and Combustion Characteristic in a Free-piston Diesel Engine Generator. *Adv. Mech. Eng.* **2018**, *10*. [CrossRef]
20. Mrzljak, V.; Medica, V.; Bukovac, O. Volume agglomeration process in quasi-dimensional direct injection diesel engine numerical model. *Energy* **2016**, *115*, 658–667. [CrossRef]
21. Cheng, Q.; Tuomo, H.; Kaarioa, O.T.; Marttia, L. Spray dynamics of HVO and EN590 diesel fuels. *Fuel* **2019**, *245*, 198–211. [CrossRef]
22. Marčič, S.; Marčič, M.; Wensing, M.; Vogel, T.; Praunseis, Z. A simplified model for a diesel spray. *Fuel* **2018**, *222*, 485–495. [CrossRef]
23. Musculus, M.P.B.; Kattke, K. Entrainment Waves in Diesel Jets. *SAE Int. J. Engines* **2009**, *2*, 1170–1193. [CrossRef]
24. Musculus, M.P.B. Entrainment Waves in Decelerating Transient Turbulent Jets. *J. Fluid Mech.* **2009**, *638*, 117–140. [CrossRef]

Article

A New Physics-Based Modeling Approach for a 0D Turbulence Model to Reflect the Intake Port and Chamber Geometries and the Corresponding Flow Structures in High-Tumble Spark-Ignition Engines

Yirop Kim, Myoungsoo Kim, Sechul Oh, Woojae Shin, Seokwon Cho and Han Ho Song *

Department of Mechanical and Aerospace Engineering, Seoul National University, Seoul 08826, Korea; yirop@snu.ac.kr (Y.K.); rlaaudtn1214@snu.ac.kr (M.K.); sc150@snu.ac.kr (S.O.); woojae9308@snu.ac.kr (W.S.); seokwon@snu.ac.kr (S.C.)
* Correspondence: hhsong@snu.ac.kr; Tel.: +82-2-880-1651

Received: 30 March 2019; Accepted: 13 May 2019; Published: 18 May 2019

Abstract: Turbulence is one of the most important aspects in spark-ignition engines as it can significantly affect burn rates, heat transfer rates, and combustion stability, and thus the performance. Turbulence originates from a large-scale mean motion that occurs during the induction process, which mainly consists of tumble motion in modern spark-ignition engines with a pentroof cylinder head. Despite its significance, most 0D turbulence models rely on calibration factors when calculating the evolution of tumble motion and its conversion into turbulence. In this study, the 0D tumble model has been improved based on the physical phenomena, as an attempt to develop a comprehensive model that predicts flow dynamics inside the cylinder. The generation and decay rates of tumble motion are expressed with regards of the flow structure in a realistic combustion chamber geometry, while the effects of port geometry on both charging efficiency and tumble generation rate are reflected by supplementary steady CFD. The developed tumble model was integrated with the standard k-ε model, and the new turbulence model has been validated with engine experimental data for various changes in operating conditions including engine speed, load, valve timing, and engine geometry. The calculated results showed a reasonable correlation with the measured combustion duration, verifying this physics-based model can properly predict turbulence characteristics without any additional calibration process. This model can suggest greater insights on engine operation and is expected to assist the optimization process of engine design and operating strategies.

Keywords: 0D model; predictive model; tumble; turbulent intensity; spark-ignition engine; engine geometry

---

1. Introduction

As modern engines increasingly become diverse and complex in their configuration, component design, and manufacturing process for multiple purposes of high efficiency, low emission, and/or high power, it is accordingly being complicated to optimize their design as well as driving strategy. Thus, the predictive engine model will be an essential technology in the near future as it can alleviate such challenges the manufacturers confront during the engine development process. Since such model should cover a wide operating range, a 0D/1D model is more suitable for its fast computation speed compared to the 3D model. The speed of low-dimensional models is a benefit obtained from many assumption and simplification, which in return, compromises the accuracy. Excessive simplifications tend to increase the model's reliance on modeling constants, which can overlook some essential physics. Therefore, it is imperative to develop a predictive 0D model capable of capturing sufficient physical phenomena associated with modern engine technologies.

Among the various attributes, turbulence plays a critical role in determining the engine performance, especially in spark-ignition engines where the combustion takes place based on flame propagation. This is mainly because a higher turbulence level can increase the turbulent flame speed, which may result in improved thermal efficiency (by bringing it close to the constant-volume combustion) [1] as well as mitigation of knocking phenomena (in case the flame propagates ahead of the auto-ignition of the end gas) [2,3]. Moreover, it has been demonstrated in many researches that turbulence enhancement can also increase the EGR tolerance [3,4], which is favorable for engines to meet the emission regulations. For all these reasons, turbulence is a primary aspect that the 0D model should correctly predict.

Turbulence originates from the large-scale flow motion generated by the intake flow, and among different types of large-scale motion, tumble possesses several advantageous structural characteristics for turbulence enhancement: (1) Tumble is spontaneously generated with pentroof cylinder head, which is most common for recent four-valve engines, (2) as a relatively stable mean motion, tumble can store a certain amount of the intake kinetic energy, and (3) having rotation axis perpendicular to the cylinder axis, even well-ordered tumble cannot avoid its breakdown into small-scale motions (or turbulence) near the end of compression, at which high turbulence is desired [4–8]. In summary, the intake-generated tumble can preserve the turbulence potential and release it in a timely manner. Hence, tumble motion is increasingly being utilized in modern engines, encouraging the 0D model to take into account the tumble-related physics.

Modulation of tumble strength can be implemented by altering the head design (e.g., straight intake port [7,9], intake port angle [10], valve masking [9,11], etc.) or by using some auxiliary device (shrouded valve [12], tumble flap [4], etc.). In spite of very complicated flow dynamics there must be involved, most 0D tumble models available have used a single coefficient obtained from steady rig test to represent all the influence on valve flow and resultant tumble [6,13–15]. Additionally, in case of tumble decay, fundamental analysis on the effect of chamber geometry and flow structure within it has been hardly implemented, but instead, decay functions for each particular engine are often exploited [14–16]. In order for the model to be predictive, it should rely more on the physical observations and understanding rather than modeling constants or engine-specific analysis.

Therefore, the objective of this study is to develop a fully-predictive 0D tumble/turbulence model via further investigation of physical phenomena and corresponding model improvements. The resultant model is expected to serve as a sub-model of a virtual engine, responding to various operating conditions and design modifications. The details in the modeling method, as well as its validation process, are discussed in the following sections.

2. Tumble Model Development

2.1. Generation of Tumble Motion

2.1.1. Mechanism of Tumble Generation

When a flow is introduced into a constrained volume, namely the engine cylinder, it soon encounters the boundary wall and creates circular motions [17]. Depending on its direction of rotation, the circular charge motion can be classified as swirl, tumble or cross-tumble (see Figure 1). It is important for an engine how the charge motion is built inside the cylinder because even the same amount of intake mass can have different influences on the engine combustion.

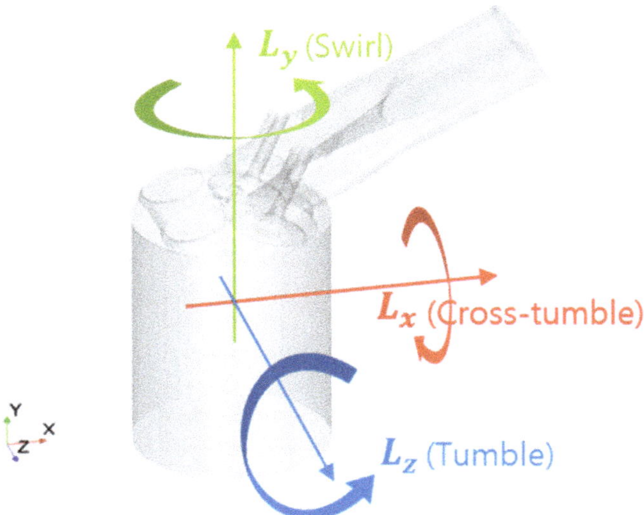

Figure 1. Classification of charge motions.

Figure 2 shows an example of the mass flux field on a valve curtain area, and an apparent variation can be seen over the surface. This distribution is what eventually determines the structure of charge motion inside the cylinder. Stronger tumble, for instance, can be created when more intake flow is concentrated toward the exhaust side (or upper-side of the valve) rather than the wall-side (or lower-side) [16,18,19]. The major cause of such distribution is understood to be the flow separation at the intake port, which is sensitively affected by port geometry and valve lift. Therefore, in order for a 0D model to correctly predict the characteristics of valve flow, it is necessary to investigate the effects of port geometry and valve lift on flow distribution and apply them in the model. Furthermore, these detailed flow characteristics should be properly taken into account in the tumble calculations.

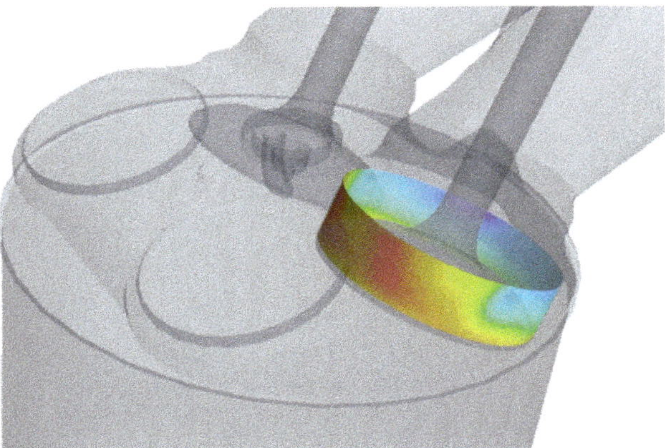

Figure 2. Distribution of mass flux on valve curtain area.

2.1.2. Modeling Concept

In 0D tumble models found in previous researches [6,13–15], angular momentum is employed to represent tumble, with the typical expression of its generation rate similar to:

$$\dot{L}_{in} = C_T \dot{m}_{in} v_{in} r \tag{1}$$

where $\dot{L}_{in}$, $\dot{m}$, v_{in}, and r denotes the tumble generation rate, mass flow rate, flow inlet velocity, and the tumble moment arm, respectively. Here, intake mass flow is assumed to be introduced at a single point in a single direction, and all inaccuracies from this assumption are practically compensated by a single coefficient C_T.

In order to simulate the valve flow more closely to the actual physical phenomenon, our previous work [16] suggested dividing the valve curtain area into multiple divisions (each division representing flow in different directions) and treating them separately in the calculations. As it can consider the direction of the valve flow, this model can be considered as quasi-dimensional (QD) tumble model, and the somewhat ambiguous tumble coefficient can be eliminated from the expression of tumble generation rate:

$$\dot{L}_{in} = \sum_i \dot{m}_i v_{tum,i} r_i \tag{2}$$

where i is the division number, and $\dot{m}_i$, $v_{tum,i}$, and r_i are the mass flow rate, the tumble-effective velocity and the moment arm of each division, respectively. Figure 3 shows the schematics of the two calculation approaches of tumble generation rate. As illustrated in Figure 3 (right), the unique flow inlet points are defined to be the center point of each division surface, in the new QD approach, and the moment arms are the perpendicular distance between tumble-effective velocity and the tumble center C, assumed to be at the midpoint of cylinder total height.

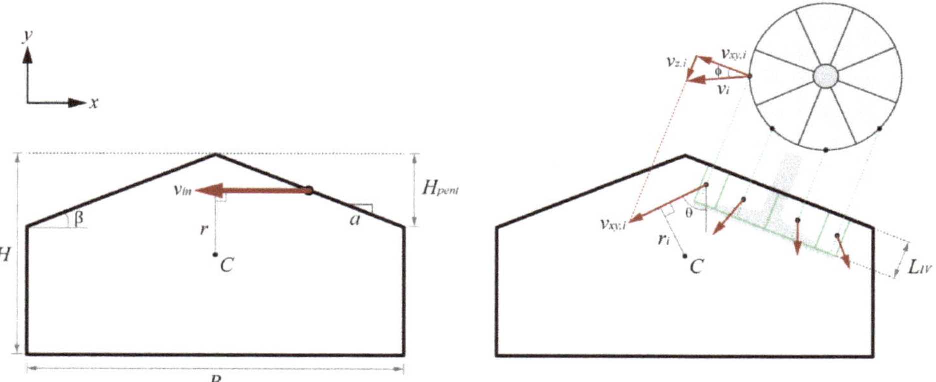

Figure 3. The sectional view of a pentroof cylinder showing the approaches of the tumble generation rate estimation: typical 0D tumble model (**left**) and the suggested QD tumble model (**right**).

This method requires detailed information on valve flow across the curtain area, and a simple steady CFD calculation is utilized to obtain the resultant flow characteristics at certain port geometries and valve lifts. For each case of steady CFD, the intake flow is simulated until it reaches a steady-state and the detailed flow characteristics of each curtain division are evaluated. Figure 4 shows examples of qualitative mass flux field at five different valve lifts on the unrolled curtain area, to demonstrate the change in mass flow distribution with the valve lift increase. The divisions are consecutively numbered from 1 to 8 that the uppermost divisions being divisions 4 and 5. It can be observed that the mass flux is fairly uniform at lower lifts, but as the valve lift increases, the upper-side remains highly concentrated (red) while the lower-side reveals some region of deficiency (blue). The result of steady CFD provides an intuition of detailed valve curtain flow being introduced into the cylinder, however, is yet too complex to be utilized in a QD model. Therefore, they need to be further processed into a form friendly to QD. The detailed explanations on each term in Equation (2) are given in the following paragraphs.

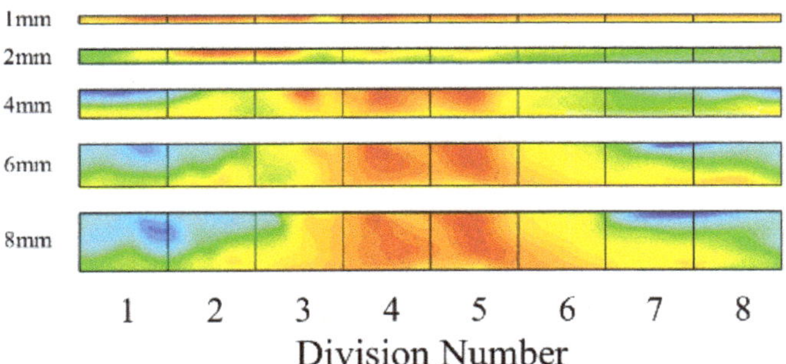

Figure 4. Qualitative mass flow distribution across the valve curtain area at different valve lifts.

2.1.3. Head Characterization

From the steady CFD results, mass flow rates of each division can be computed simply by taking the surface-integral of mass flux from the steady CFD results, and the fractional distribution can be obtained by dividing them by their sum. This allows the mass flux field in Figure 4 to be reproduced into simple intuitive plot shown in Figure 5, in which each dot represents for individual divisions at given valve lift. One noteworthy founding was that this distribution tends to be retained despite the alternation in pressure conditions, leaving the valve lift as the only variable affecting it [16]. Also, the fraction shows a nearly monotonic trend of increase or decrease within each division, which justifies the fraction value being interpolated for other valve lifts. These indicate that a single set of CFD simulation can cover any operating conditions with a certain cylinder head.

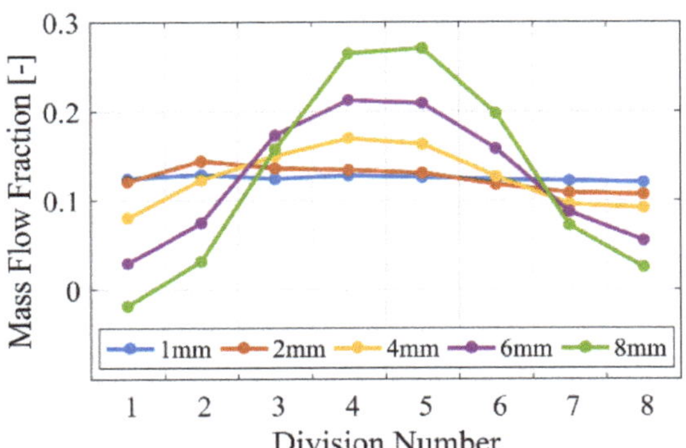

Figure 5. Fractional distribution of intake mass flow rate through each curtain area division.

Since the fractional distribution can serve its role only when the correct total mass flow rate is provided, the discharge coefficient is also a major factor of concern. As mentioned previously, a stronger tumble is attained when the lower-side flow is restricted [11], implying the trade-off relationship between the tumble strength and flow coefficient [7]. Thus, when port geometry is altered to manipulate tumble, its impact on charging efficiency must be considered concurrently. Since this mass flow rates calculated from steady CFD results (before normalization) already reflect the effect of port geometry, the appropriate discharge coefficient curve can be obtained by dividing with the theoretical value from isentropic compressible flow model.

With the discharge coefficient and fractional distribution characterized, it is possible to predict the total and divisional mass flow rates within the engine cycle. The characterization results are integrated into the GT-Power, a commercial 1D software for engine simulation, and preliminary simulation is performed with a reference engine. Figure 6 depicts the results of mass flow rate calculated in GT-Power model, compared with the results of transient CFD performed with identical conditions. Except for some pulsation near the opening of intake valve, the total mass flow rate ($\dot{m}_{in}$) in Figure 6 (left) showed very high agreement with the transient CFD, verifying the reliability of the discharge coefficient obtained by the head characterization process. By combining the fractional distribution in Figure 5 with the total mass flow rate, the divisional mass flow rates ($\dot{m}_i$) can be obtained and they are illustrated in Figure 6 (right). One deficiency of GT-Power results is that the flow inversion occurs at the identical instant for all divisions, but other than that, it showed an excellent correspondence with the CFD. The similar level of predictabilities was observed in additional simulations implemented with varying engine speed, load, and port design, which confirms this QD model can respond to any change in head geometry and predict the detailed valve flow comparable to transient CFD, once characterization results are supplemented. Plus, it could also be concluded that a few runs of steady CFD (five, in this case) are sufficient to characterize a given cylinder head.

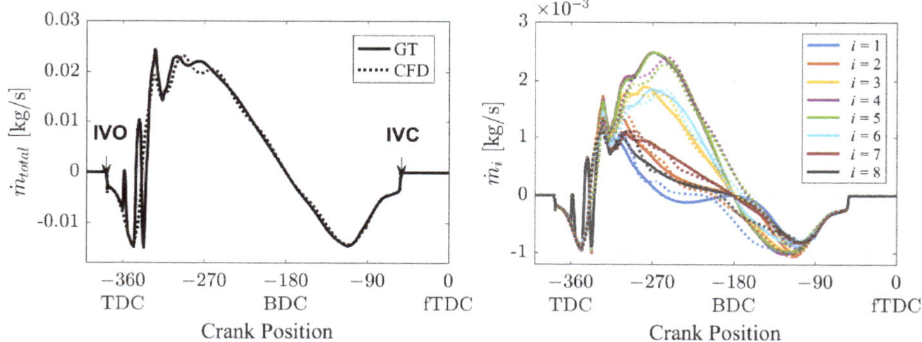

Figure 6. Total (**left**) and divisional (**right**) mass flow rate over the gas exchange process (solid lines: GT-Power, dotted lines: transient CFD).

In addition to the fractional distribution and discharge coefficient, the average velocity is another parameter that can be extracted from steady CFD simulation. Knowing x-, y- and z-components of the average velocity, the representative flow angles θ and ϕ can be drawn for each flow division (the angle configuration shown in Figure 3). The angle ϕ is used to decompose the intake velocity into xy- and z-components, each of which constitutes tumble motion and non-tumble mean motion, respectively, while the angle θ is used to calculate tumble moment arms.

Prior to the decomposition, the intake velocity should be found. The velocity across each valve curtain division can simply be estimated for given mass flow rate as:

$$v_i = \frac{\dot{m}_i}{\rho A_i} \tag{3}$$

Since the curtain area is evenly divided, the area of each division (A_i) is simply $\pi D_{IV} L_{IV} / n_d$, where D_{IV}, L_{IV}, and n_d are intake valve diameter, intake valve lift, and the number of divisions, respectively. Compared to the mean velocity from the steady CFD simulation, it was shown that the representative velocity calculated by Equation (3) falls within a reasonable range despite the assumption of constant density (see Figure 7). The biggest error is observed in the lower side at high lift condition, and the major reason is the simultaneity of incoming and outgoing flow motion in these divisions. Since the net mass flow rate is used in the calculation, the representative velocities tend to be underestimated for divisions with such bidirectional flow, compared to the CFD results. However, since these velocities are multiplied by the relatively small mass flow rates in further calculations, this error will only have a minor impact on the final results.

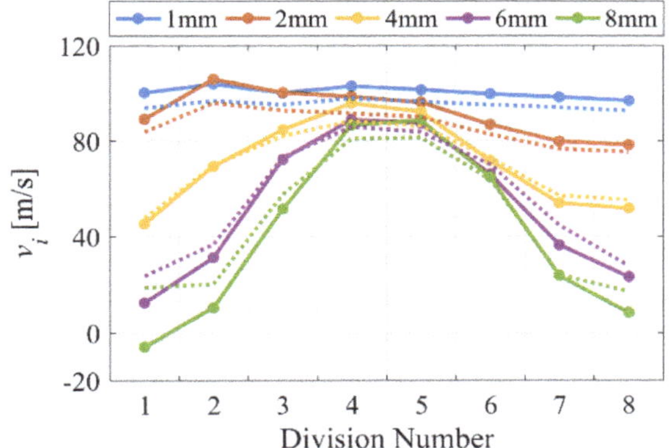

Figure 7. Divisional flow velocities passing valve curtain area (solid: QD, dotted: steady CFD).

One thing to note is that these velocities at the valve opening do not wholly contribute to the tumble generation. The flow experiences an abrupt expansion of the flow path immediately after passing through the valve opening, thereby the velocity decreases. A coefficient (denoted here as C_{KE}) is typically applied to the kinetic energy flux term ($\dot{E}_{in}$) to take account of such velocity loss:

$$\dot{E}_{in} = \frac{1}{2}C_{KE}\dot{m}v^2 \tag{4}$$

Since this loss is due to the decrease of velocity, the effective velocity can be interpreted to be $(C_{KE})^{1/2}v$. Therefore, the xy-component of effective velocity, which contributes to tumble generation, can be decomposed written as:

$$v_{tum,i} = (C_{KE})^{1/2}v_i \cos\phi \tag{5}$$

Similarly, the z-component can be obtained by multiplying $\sin\phi$. Controlling of C_{KE} will be discussed in later sections along with other modeling coefficients.

2.1.4. Application of Pentroof Geometry

As mentioned earlier, the model is aimed to be also considerate of the effects of combustion chamber geometry because it can influence both the generation and decay of the tumble motion. In most 0D models, the combustion chamber is assumed to be a pancake shape, which could be misleading in some respects. For example, the postulated locations of representative inlet point for each division can be inadequate when pancake shape is assumed, resulting in the errors on the calculation of tumble generation. In addition, pancake shape with uniform cross-sectional area underestimates the combustion chamber height, especially near top dead center (TDC), and this significantly affects the flow field and corresponding tumble decay rate (further discussed in the later section). Therefore, a correct reflection of the pentroof geometry is strongly encouraged within the calculation process.

As the example shown in Figure 8 (left), the geometry of actual pentroof head is rather complicated for various reasons such as manufacturing process, cooling channels, slots for valves, spark plug, and/or injector, etc. In this model, a smooth, symmetrical pentroof was assumed for the sake of simplicity (illustrated in right schematic in Figure 8), and then it could be defined only by the pentroof angle (β), as shown in Figure 3.

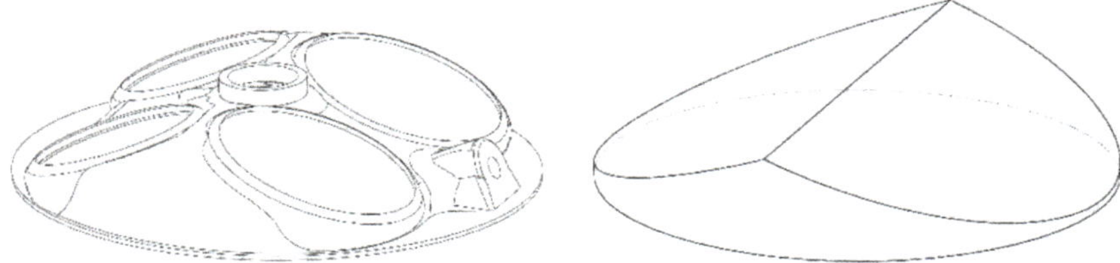

Figure 8. Outline of actual (**left**) and simplified (**right**) pentroof geometry.

To assess this simplified pentroof geometry, the moment of inertia, which is the quality that reflects the rotational characteristic in the three-dimensional space, has been chosen for comparison. For typical pancake-shaped combustion chamber, the moment of inertia (I) about the z-axis with the origin at tumble center C is written as:

$$I_z = \iiint_{vol} \rho\left(x^2 + y^2\right)dV$$
$$= m\left(\frac{B^2}{16} + \frac{H^2}{12}\right) \tag{6}$$

where B and H are cylinder bore and height, respectively. When the simplified pentroof geometry is applied, the volume integral becomes:

$$I_{z,pent} = 8 \int_0^R \int_0^{\frac{H}{2}} \int_0^{\sqrt{R^2-x^2}} \rho\left(x^2 + y^2\right)dzdydx - 4 \int_0^R \int_{\frac{H}{2}+ax}^{\frac{H}{2}} \int_0^{\sqrt{R^2-x^2}} \rho\left(x^2 + y^2\right)dzdydx$$
$$= \rho\left[\frac{\left(64a^3+192a\right)R^5+\left(45\pi a^2\right)HR^4+120aH^2R^3+30\pi H^3R^2}{360}\right] \tag{7}$$

where a is the pentroof slope defined as $-H_{pent}/R$. Also, note that the volume is no more equal to $\pi R^2 H$. The moment of inertia calculated for each geometric assumption are plotted in Figure 9 along with that calculated by the 3D CFD.

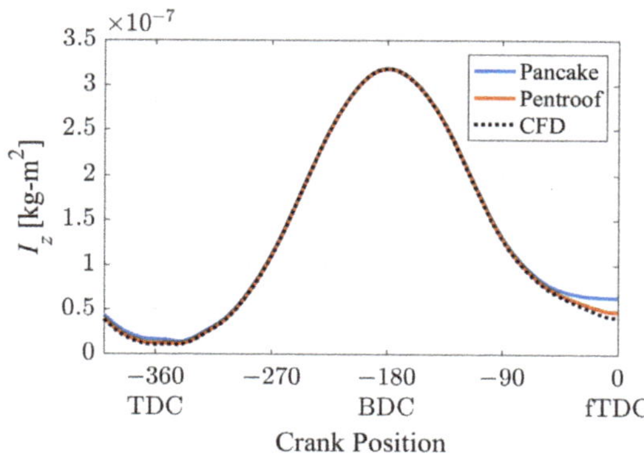

Figure 9. Comparison of the moment of inertia calculated by 0D and 3D model.

For a sufficiently large cylinder volume, the assumption of pancake geometry provides a highly accurate moment of inertia, but a distinct error increase was observed as compression proceeds toward TDC (~55% of error at TDC). Such overestimation in the moment of inertia can be a critical issue because a significant error in the calculation of tumble decay rate would follow. But the Equation (7) yields a result with the error considerably relieved (~70% reduction), from which the simplified pentroof head can be justified.

2.2. Decay of Tumble Motion

2.2.1. Modeling Concept

It is quite common to directly relate tumble ratio and combustion speed, but in fact, it is not the tumble itself that boosts flame propagation. Large-scale tumble motion should be converted into small-scale turbulent eddies to wrinkle the flame sheet and enhance flame propagation speed. Therefore, the decay of tumble motion into turbulent kinetic energy (TKE) is as important as its generation. Several different approaches to model tumble decay rate could be found in literature, but most of them require a calibration process corresponding to any change in engine geometry [14,15]. This indicates that they are not fully-predictive with changes in engine designs, which is the major intent of this research, so a different, somewhat comprehensive approach is necessitated.

According to the energy cascade concept, larger scales of motion transform into smaller scales. By applying this concept to in-cylinder charge motion, the TKE production can be directly related to the decay of tumble rotational energy. Then, the decay rate of tumble angular momentum can be written as:

$$\dot{L}_\Psi = -\frac{I}{L}\frac{dE_{rot}}{dt} = -\frac{I}{L}\iiint_{vol} \rho P_\Psi dV \tag{8}$$

where P_Ψ is the TKE production rate by internal shear, which is essentially determined by the flow structure. Typically, 0D models cannot take into account the flow structure, but if velocity field is defined within the combustion chamber, the local TKE production rate can be found by utilizing the expression suggested by Boussinesq's eddy viscosity hypothesis:

$$P_\Psi = \nu_t\left(\frac{\partial \overline{U}_i}{\partial x_j} + \frac{\partial \overline{U}_j}{\partial x_i}\right)\frac{\partial \overline{U}_i}{\partial x_j} \tag{9}$$

2.2.2. Flow Field Definition

In several previous researches on tumble modeling found in literature, a 2D linear velocity profile is assumed to represent the flow field inside the cylinder [5,20,21]. These studies postulated that boundary velocities at the top, bottom, and wall side are all equal regardless of the piston position, which the authors suspect to be inconsistent with the real situation. Seeing from the perspective of the continuity about rotation, the velocity of the shorter side should be greater than the longer side to allow the same mass flow rate. Hence, the velocity profile is modified so that the boundary velocities at top/bottom sides and wall side are related using the ratio between cylinder height and width [22]. The prescript of modified velocity is as below, and this is also illustrated in Figure 10:

$$\begin{cases} U_x = -U\left(1 - \frac{z^2}{R^2}\right)\frac{y}{h} \\ U_y = \frac{h}{w}U\left(1 - \frac{z^2}{R^2}\right)\frac{x}{w} = \frac{Uh}{R^2}x \\ U_z = 0 \end{cases} \tag{10}$$

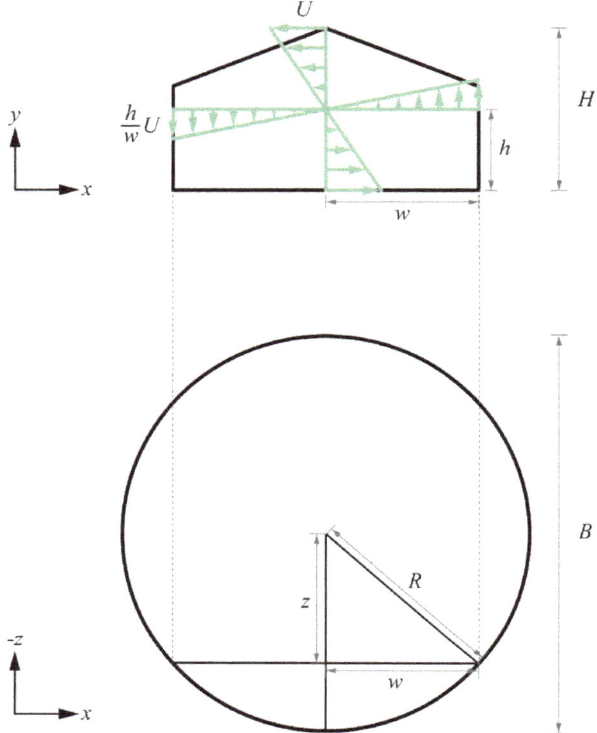

Figure 10. Modified velocity profile.

2.2.3. Governing Equations

Having the velocity field defined, the overall angular momentum of in-cylinder charge can be calculated by taking volume integral as:

$$L = \iiint\limits_{vol} \rho\left(-U_x y + U_y x\right)dV$$

$$= \tfrac{1}{4}\rho U \pi R^2 H^2 + \frac{\rho U R^3\left(256a^3 R^2 + 175\pi a^2 HR + 672aH^2\right)}{840H}$$

(11)

By solving Equation (11) for U, the characteristic velocity can be expressed as a function of density, angular momentum and geometric dimensions only:

$$U = \frac{840H}{\rho R^2(256a^3 R^3 + 175\pi a^2 HR^2 + 672aH^2 R + 210\pi H^3)}L$$

(12)

Similarly, overall TKE production within the cylinder is:

$$\dot{m}k_\Psi = \iiint\limits_{vol} \rho P_\Psi dV = \rho v_t \iiint\limits_{vol}\left[\left(\frac{\partial U_x}{\partial y}\right)^2 + \left(\frac{\partial U_y}{\partial x}\right)^2 + \left(\frac{\partial U_x}{\partial z}\right)^2 + \left(\frac{\partial U_y}{\partial z}\right)^2 + 2\left(\frac{\partial U_x}{\partial y}\right)\left(\frac{\partial U_y}{\partial x}\right)\right]dV$$

$$= \rho v_t U^2\left[\pi R^2 H\left(\frac{5}{2H^2} - \frac{7}{6R^2} + \frac{H^2}{4R^4}\right) + \left(\frac{(128a^2 + 1152)aR^3}{315H^2} + \frac{\pi a^2 R^2}{3H} - \frac{16aR}{15} + \frac{aH^2}{3R}\right)\right]$$

(13)

Finally, by combining Equations (8) and (13), the corresponding tumble decay rate can be expressed in terms of angular momentum, characteristic velocity, turbulent viscosity, and geometric parameters:

$$\dot{L}_\Psi = -\frac{I}{L}m\dot{k}_\Psi \tag{14}$$

3. New Turbulence Model

3.1. Integration into Turbulence Model

The developed tumble model is integrated into the *k-ε* model to complete the new turbulence model and Figure 11 describes the overall concept of energy flow. In this study, large-scale motions, generally classified as mean kinetic energy (MKE) in 0D turbulence model, are divided into the tumble and the non-tumble components. A certain amount of non-tumble component participates in instantaneous TKE production, and the rest creates minor mean motions. Loss of each quality occurs along with the outgoing flow, and the TKE is dissipated into internal energy at the rate of *ε*.

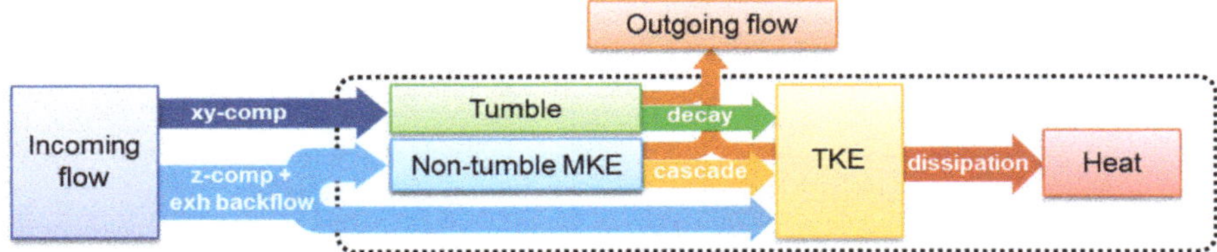

Figure 11. Flow chart of the new turbulence model.

The new turbulence model solves four differential equations of *L*, *K*, *k*, and *ε*, each representing the tumble, non-tumble mean kinetic energy, turbulent kinetic energy, and dissipation rate, respectively [16]:

$$\frac{dL}{dt} = \dot{L}_{in} - C_{out}L\frac{\dot{m}_{out}}{m} - \dot{L}_\Psi \tag{15}$$

$$\frac{dK}{dt} = (1 - C_\alpha)\frac{\dot{E}_{non}}{m} - K\frac{\dot{m}_{out}}{m} - P_k \tag{16}$$

$$\frac{dk}{dt} = C_\alpha\frac{\dot{E}_{non}}{m} - k\frac{\dot{m}_{out}}{m} + P_k - \varepsilon + \dot{k}_\Psi \tag{17}$$

$$\frac{d\varepsilon}{dt} = \frac{\varepsilon}{k}\frac{\dot{E}_{non}}{m} - \varepsilon\frac{\dot{m}_{out}}{m} + P_\varepsilon - 1.92\frac{\varepsilon^2}{k} + \frac{\varepsilon}{k}C_3\dot{k}_\Psi \tag{18}$$

The non-tumble mean kinetic energy (MKE) is treated as typical MKE in *K-k* model [15,23], except the fact that it is limited to the kinetic energy by the non-tumble velocity component ($\dot{E}_{non} = \sum \frac{1}{2}\dot{m}_i v_{eff,z,i}^2$). The TKE production rate from non-tumble MKE and the corresponding term in *ε* equation are:

$$P_k = C_\beta \nu_t \frac{2mK}{L_g^2} - \frac{2}{3}\nu_t\left(-\frac{\dot{\rho}}{\rho}\right)^2 - \frac{2}{3}k\left(-\frac{\dot{\rho}}{\rho}\right) \tag{19}$$

$$P_\varepsilon = \frac{\varepsilon}{k}\left[2.88C_\beta\nu_t\frac{2mK}{L_g^2} - 0.88\nu_t\left(-\frac{\dot{\rho}}{\rho}\right)^2 - 2k\left(-\frac{\dot{\rho}}{\rho}\right)\right] \tag{20}$$

Note that the coefficients in Equations (18) and (20) are from the standard value for unidirectional compression with reference to [24].

3.2. Modeling Constants

3.2.1. Definition

As can be seen in Equations (15)–(20), there exist some modeling constants in addition to C_{KE}. The presence of any modeling constant involves calibration, any of which is not favored for a predictive engine model. Hence, it is sought to eliminate calibration process by either assigning physical meaning to the coefficients or using a comprehensive value. As the basis for such a predictive model, the coefficients are designed as follows:

- C_{KE} is the coefficient to take into account the flow velocity decrease as it passes the valve opening. Since the cause is interpreted to be the change of flow area, it may be correlated to the ratio of the valve curtain area to the cylinder bore area, as in [23]. This coefficient must be less than 1 because the velocity cannot be increased after expansion.
- C_α is the split factor of non-tumble intake kinetic energy between non-tumble MKE and the instantaneous TKE. The instantaneous TKE is interpreted as a consequence of significant shearing of inflow at valve opening [14,15], so it seems plausible to express it as a function of valve diameter and/or lift. As a split factor, it should lie between 0 and 1, as well.
- C_β is the coefficient for the cascade rate of non-tumble MKE into TKE. It basically controls the residence time of the minor mean motion and the corresponding TKE production period. As it only comprises of minor mean motions, the cascade is presumed to be rather quick. Plus, since the extent of non-tumble MKE is dependent on C_α, the influence of C_β diminishes with the increase of C_α. Therefore, C_β can be considered as a subsidiary coefficient.
- C_3 is the coefficient of the dissipation rate corresponding to tumble-generated TKE, which is newly added in the epsilon equation in our proposed model. This is adopted for structural consistency with the other terms in the standard k-ε model, so it is expected to also be universal, once specified. The coefficient for the non-tumble MKE term is 2.88 (Equation (20)), and the reasonable range for C_3 is supposed to be a similar order of magnitude.
- C_{out} is the multiplier of loss term caused by outgoing flow (e.g., back-flow into the intake port). This is applied particularly to the angular momentum because of the flow structure inside the cylinder. In case of a high tumble engine, it is unquestionable that flow in the outer side has greater velocity, and the role of C_{out} is to compensate for the fact that outgoing mass has relatively higher velocity compared to the mass-averaged value. It would be possibly related to the boundary velocity and/or chamber dimension, which may affect the velocity gradient. A rough range of 1 to 5 seemed to be reasonable for this coefficient.

3.2.2. Influence on Tumble and Turbulence

Figure 12 shows the results of tumble angular momentum (L) and turbulent intensity (u') from individual sweeps of each modeling constant. Note that the turbulent intensity is an indicator of the TKE because $u' = \sqrt{2k/3}$. A reasonable range has been set for each constant except for the C_β, which is tested with extreme range to demonstrate its insensitivity.

First, the effective flow velocity increases with greater C_{KE} as inferred in Equation (5), which enhances the initial build-up rate of both angular momentum and turbulent intensity. The split factor C_α also affects the initial build-up of turbulent intensity, although it does not have a direct impact on the angular momentum. However, the enhanced TKE leads to an increase of turbulent viscosity ($= 0.09k^2/\varepsilon$), which in turn affects the tumble decay rate and reduces overall angular momentum level. In addition, an insufficient TKE build-up was observed with C_α below 0.7. This sets the lower boundary of C_α, and the instantaneous TKE is within similar order of magnitude with that reported in other researches [14,15] in terms of the fraction of total intake kinetic energy (including tumble component).

Next, smaller C_β value causes an increase of TKE level after the initial peak, but the effectiveness is relatively minor for the inputted value. It is also observed that C_β exerts negligible impact above a certain level, and this is because all of non-tumble MKE is cascaded immediately after being introduced. Both C_3 and C_{out} are related to TKE production from tumble decay, but C_3 adjusts the sensitivity of tumble decay, so it influences the TKE over the entire range where angular momentum exists. Meanwhile, C_{out} controls the amount of tumble angular momentum itself, and the height of second TKE peak is decided proportionally to the remaining tumble at intake valve closing (IVC).

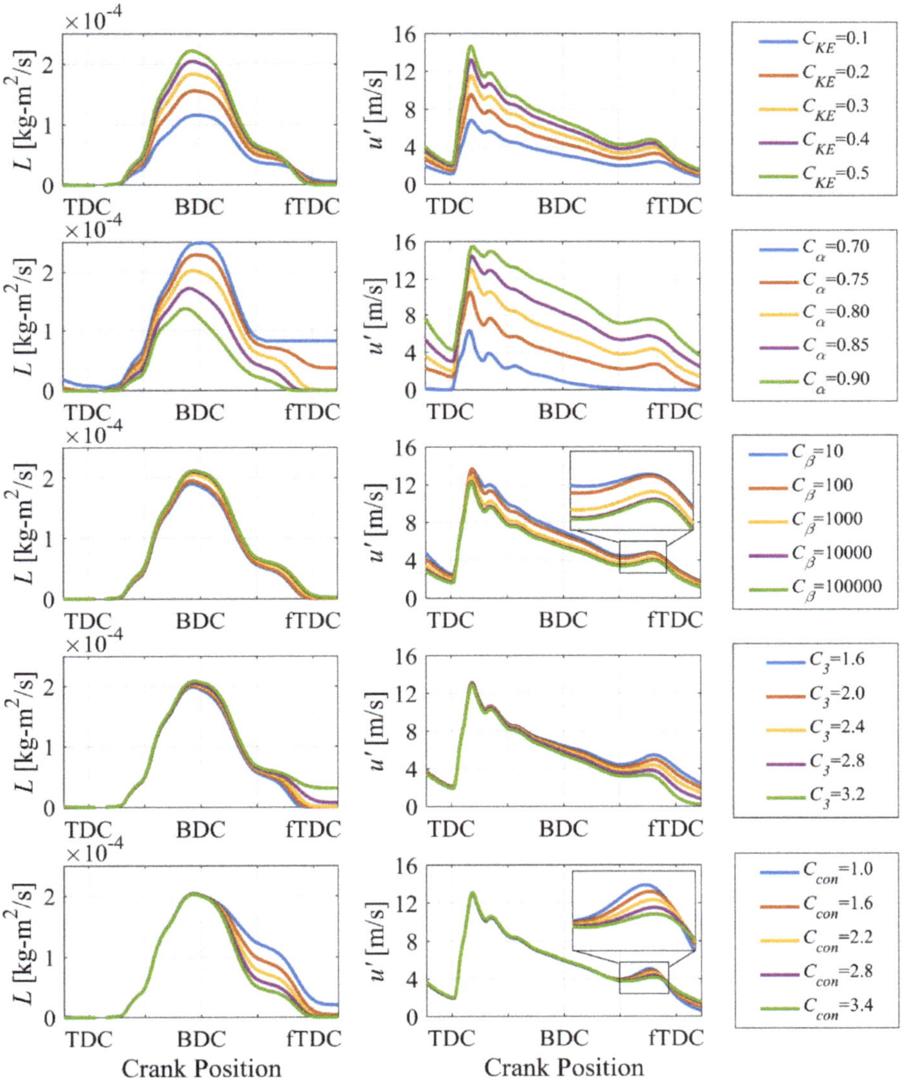

Figure 12. Results of angular momentum and turbulent intensity under modeling constants sweep.

3.2.3. Calibration of Modeling Constants

The five modeling constants were to be determined so that the developed model would correctly predict the tumble and turbulent intensity. The optimization of constants was carried out with the optimization target to minimize the error in angular momentum and turbulent intensity between calculation results of 0D turbulence model and 3D CFD. The optimizer provided in GT-Power is utilized, and the turbulent intensity near the end of the compression process was more weighted as it is the most important model results related to the combustion behavior. The CFD results reported in authors' previous modeling study [16] cover variations in intake port geometry, engine speed and load, which make them good reference data for model constant optimization.

From the optimization, the best set of modeling constants are found to be as listed in Table 1. Figure 13 shows the temporal history of angular momentum, turbulent intensity of a sample case at 1600 rpm with an intake pressure of 85 kPa. In general, the predictions of the developed model were highly satisfactory, especially for the turbulent intensity near the end of compression. Given the physical meanings explained in Section 3.2.1, each constant is expected to be independent of changes in engine speed, load, and cam timing shift, and this is supported by the additional comparative results as shown in Figure 14. Therefore, the optimized values in Table 1 are deemed as universal and used throughout the rest of this study.

Table 1. Modeling constants calibrated for reference CFD data.

C_{KE}	0.5
C_α	0.8273
C_β	500
C_3	1.8125
C_{out}	3.5

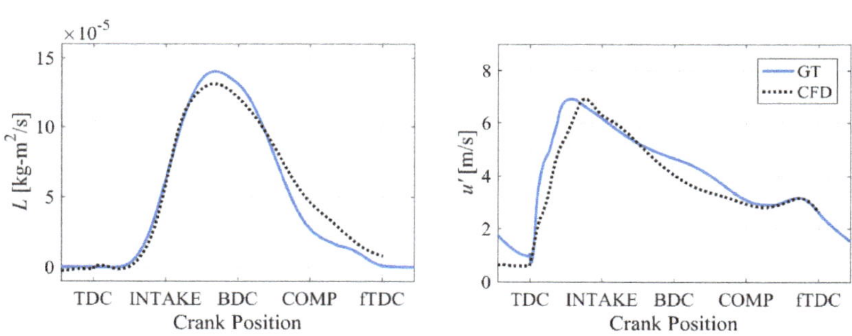

Figure 13. Comparison of angular momentum (**left**) and turbulent intensity (**right**) after calibration. CFD data is reproduced from [16], SAE International: 2018.

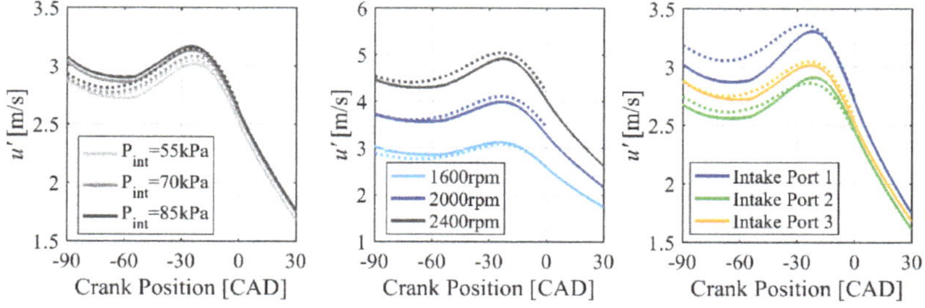

Figure 14. Comparison of near-TDC turbulent intensity under variations in intake pressure, engine speed, and intake port curvature ((solid lines: GT-Power, dotted lines: transient CFD).

4. Model Validation

4.1. Experimental Data

A comparative analysis with the experimental data has been planned to validate the predictability of the developed model. The experimental data of Oh et al. was chosen for the validation because their experiments cover not only a variety of operating parameters including engine speed, load, and intake/exhaust valve timing, but also different port and chamber geometries [25]. The three engines used in this study have different stroke-to-bore ratios (SBR) while displacement volumes and compression ratio kept equivalent. In addition, the same cylinder head is shared for Engines B and C, while Engine A has different head (i.e., port) geometry than the others. More details on engine specifications and operating conditions are listed in Tables 2 and 3.

Table 2. Engine specifications.

Parameter	Engine A	Engine B	Engine C
Displacement (cc)	-	500	
Bore (mm)	86	81	75.6
Stroke (mm)	86	97	111
Conrod Length (mm)	211.65	207.65	199.15
Stroke-to-Bore Ratio (-)	1.0	1.2	1.47
Compression Ratio (-)		12 ± 0.1	
Intake Valve Diameter (mm)	33	29	29
Exhaust Valve Diameter (mm)	27	27	27
Pentroof Angle (degree)		15	
Number of data points	98	40	56

Table 3. Experimental conditions.

Engine speed (rpm)	1500, 2000
IMEP (bar)	4.5–10.5
Valve open duration (In/Ex) (CAD)	280/240
InCam shift range (CAD)	−40–0(default[1])
ExCam shift range (CAD)	0(default[1])–30
Valve Overlap (CAD)	35–105

[1] Default valve timing: Intake valve open at 10 CAD before TDC, exhaust valve close at 1 CAD after TDC.

4.2. Validation Method

Using the developed model, the turbulent intensity can be predicted for any given engine geometry and operating condition. The characterization results and modeling constants in Table 1 are input into the GT-Power model, and simulation is carried out to reproduce the experimental results. Figure 15 shows the comparison of 50% burn duration calculated by the developed model with that measured from the experiment.

Although the model prediction appears to follow the right trend, the accuracy is not quite satisfactory, with an R-squared value of 0.6655. However, such poor correlation does not necessarily imply the model prediction was inaccurate. This study specifically aims to predict the flow dynamics during engine cycle rather than the consequent combustion behavior, thus, any further development and/or precise calibration of combustion model, other than the minimal calibration of the built-in combustion model of GT-Power, is deemed to be beyond the research scope. The inaccuracy of the combustion model possibly is the major cause of low agreement, and another validation approach was considered to validate the calculated turbulent intensity with no direct involvement of the combustion model.

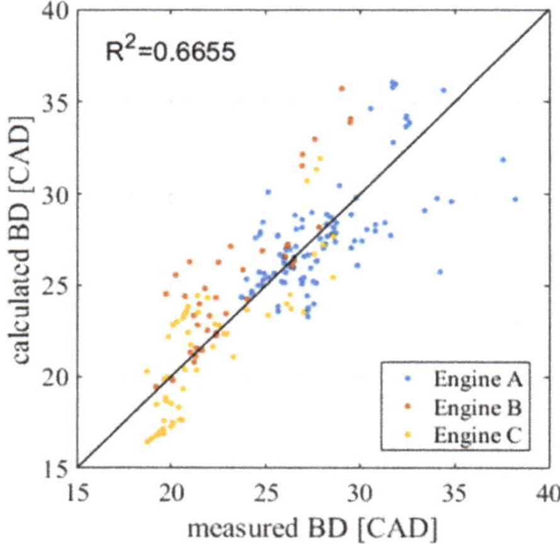

Figure 15. Measured and calculated duration between ignition and 50% burn angle (BD0050).

Greater turbulent intensity is interpreted to enhance wrinkling of the flame sheet, thus the flame propagation speed. Since the flame propagation speed is the flame travel distance divided by the burn duration, the measured burn duration and calculated turbulent intensity would possibly be correlated. The relationship between turbulent flame speed (S_T) and turbulent intensity can be described with a general formula of:

$$\frac{S_T}{S_L} = 1 + C\left(\frac{u'}{S_L}\right)^n \tag{21}$$

In order to approximate the average turbulent flame speed for a span of interest, the corresponding flame travel distance and burn duration are necessary, but the flame travel distance (R_f) is not easily measurable in practice. Therefore, a rough estimation was to be made to quantify the flame travel distance, and it is implemented by utilizing the flame radius calculated from GT-Power. Figure 16 (left) depicts the temporal history of the relative flame radius (R_f/R) of each of all tested cases.

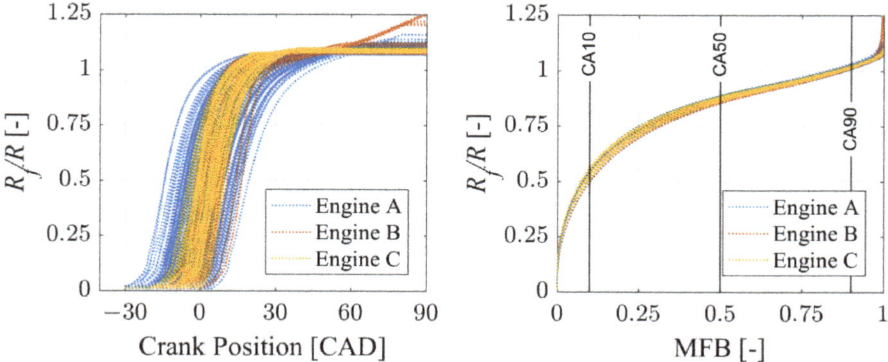

Figure 16. Development of flame radius.

As the operating conditions, including ignition timing, differ among the cases, the graph shows quite diverse flame radius profiles. But if the same relative radii are plotted over the fraction of burned mass (MFB), all profiles nearly collapse regardless of operating conditions as seen in Figure 16 (right). Taking the average suggests the flame radii at 10, 50, and 90 percent burn angles (CA10, CA50, and CA90) to be 0.538R, 0.876R, and 1.019R, respectively.

Next, a reasonable range for validation was needed to be determined. It is well-known that flame kernels initially show laminar-like development, and gradually reach to fully-developed turbulent flame over a certain time scale. This infers that the turbulence does not wholly participate during this early stage of the combustion process, and makes it inappropriate to be used for validation of turbulence prediction. In the later stage of combustion, the combustion can be disturbed by the expansion of cylinder volume, especially under high engine speed or slow-burn conditions. Therefore, the range for validation is somewhat arbitrarily chosen to be between 10 and 50 percent burn angles (CA10-50), and Equation (21) then becomes:

$$\frac{S_{T,1050}}{S_{L,1050}} = \frac{\left(R_{f,50} - R_{f,10}\right)/BD1050}{S_{L,1050}} = 1 + C\left(\frac{u'_{1050}}{S_{L,1050}}\right)^{n} \tag{22}$$

where the subscript 1050 indicates the average value of simulation result of over the same CA10-50 range. Then, the indirect validation according to Equation (22) was implemented to verify the correlation between calculated turbulent intensity and measured burn duration.

4.3. Correlation between Model Prediction and Experimental Data

The scatter plot of dimensionless turbulent flame speed (S_T/S_L) and dimensionless turbulent intensity (u'/S_L) in Figure 17 summarizes the validation results of Engines A, B, and C. All 194 data points could adequately be described by a single curve with the C and n of 8.983 and 0.6405, respectively, with a much stronger correlation (R-squared value of 0.8328) than the direct comparison of burn duration in Figure 15. This verifies that the developed model with fixed modeling constants is fairly adaptive to changes in the chamber and head design as well as various types of engine operating conditions including engine speed, load, and valve actuation.

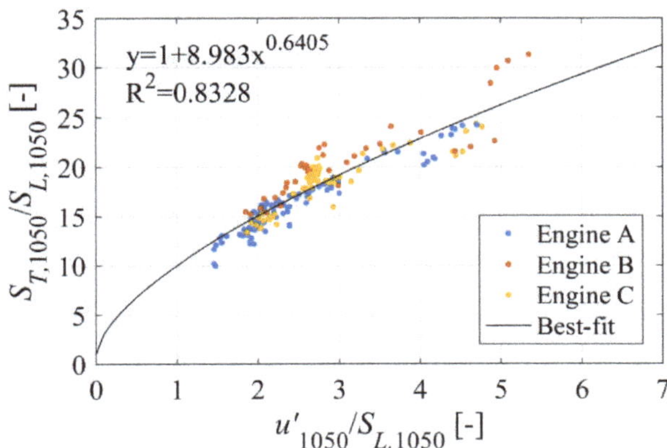

Figure 17. Correlation between simulation and experiment results (Engines A, B, and C).

The improved correlation, on the other hand, evidences that the combustion model is erroneous. It should be noted that, despite the effort of the indirect comparison method, the overall validation results still are influenced by the combustion model because the combustion process is determinative to pressure, temperature, and consequently the laminar flame speed. Among others, the high RMF operation demonstrated the lowest accuracy in prediction. With no external exhaust gas recirculation (EGR) employed, the amount of burned gas in the cylinder at IVC is determined majorly by the valve overlap. In the simulation results, a combination of long overlap duration and low load condition caused excessive RMFs near 30%, which would have caused inaccurate laminar flame speed. The gray 'x' markers in Figure 18 indicates data points with RMF above 25%, and just by excluding these

13 points, the R-squared value could be improved from 0.8328 to 0.8605. This infers that the variance of the scatter plot could be further alleviated by improving the combustion model.

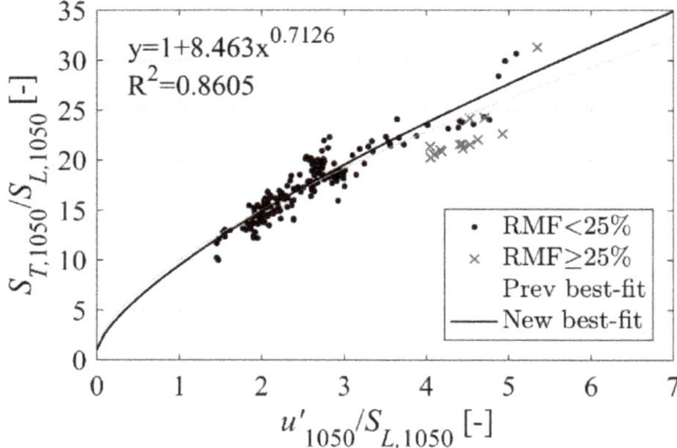

Figure 18. Correlation between simulation and experiment results (Engines A, B, and C, high RMF cases excluded).

4.4. Detailed Analysis on Model Prediction

4.4.1. Effect of Intake Valve Timing Sweep

Figure 19 shows the results of intake valve timing sweep at 2000 rpm and 4.5 bar indicated mean effective pressure (IMEP). As seen in this figure, the advancement of intake valve timing increased the valve overlap duration and caused greater backflow of the burned gas from the exhaust manifold to the cylinder and intake manifold. This reverse flow caused a rapid increase of turbulent intensity during the valve overlap, however, the angular momentum is not much influenced due to the very small cylinder mass at this period. On the other hand, the advancement of intake valve timing also causes an earlier IVC. Such earlier IVC traps more the fresh mixture in the cylinder, hence the energy loss due to outgoing mass flow is reduced. Due to the breakdown of greater remaining tumble, the turbulent intensity rise near the firing TDC is greater in the case with advanced intake valve timing.

As illustrated in Figure 20, the increased burned gas backflow caused a decrease in the average laminar flame speed, and the greater trapped tumble caused the increase in average turbulent intensity. Figure 21 shows these particular intake valve timing sweep cases from the scatter plot in Figure 17, and it can be inferred that the valve strategies and the consequent gas exchange dynamics are included in the model prediction, which shows a high agreement with the measured burn durations.

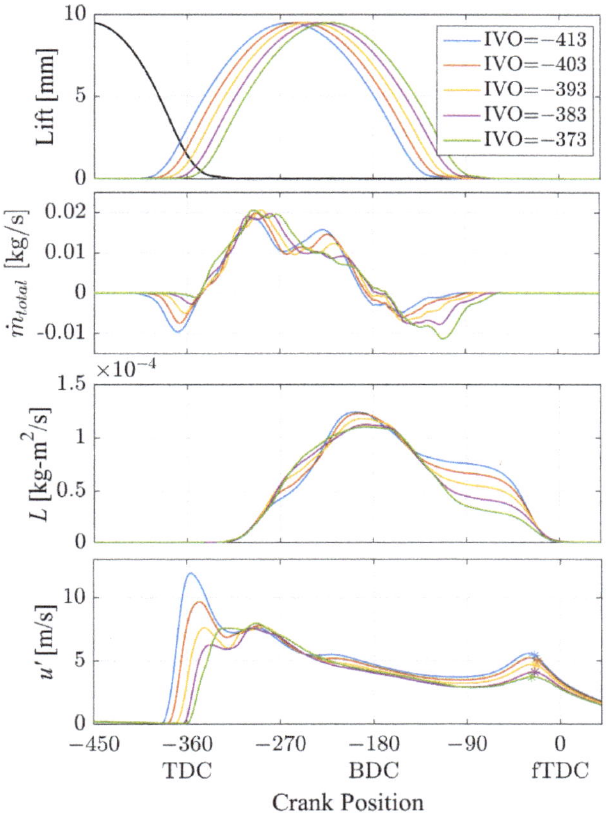

Figure 19. Model prediction with intake valve timing sweep (Engine A, 2000 rpm, 4.5 bar IMEP).

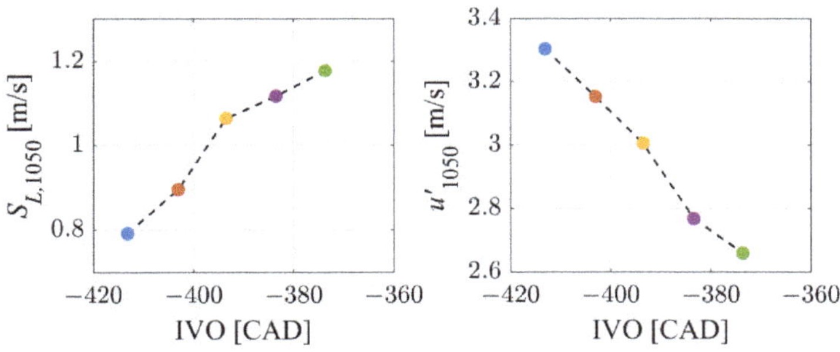

Figure 20. Averaged laminar flame speed and average turbulent intensity with intake valve timing sweep.

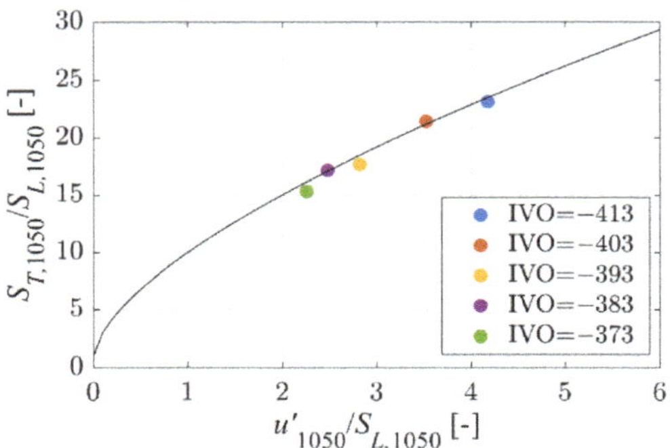

Figure 21. Correlation between simulation and experiment results (intake valve timing sweep).

4.4.2. Effect of Engine Geometry

Figure 22 are the model prediction for Engines A, B, and C, at fixed engine speed, load, and valve timings. Here, it can be seen that the tumble build-up reveals some differences while the mass flow rate is nearly equivalent except for some pulsation effect. Higher SBR means the cylinder is narrower and longer, and this causes the tumble moment arm to be longer for upper side flow, and shorter for lower side flow. Consequently, the tumble generation rate is greatest for Engine C.

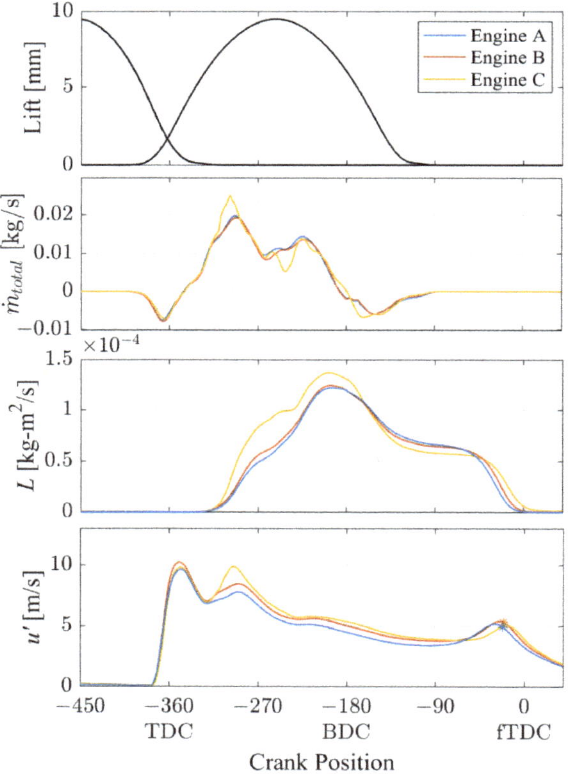

Figure 22. Model prediction with intake valve timing sweep (2000 rpm, 4.5 bar IMEP).

Another effect of SBR is tumble breakdown behavior. Due to its geometry, the engine with a higher SBR undergoes a less severe distortion as the piston compresses the cylinder volume. This yields a milder breakdown of the tumble for Engine C, and the angular momentum drop and turbulent intensity rise are observed to occur at later timing. Therefore, for moderate or late spark timings, the higher SBR engines show greater average turbulent intensity (as shown in Figure 23).

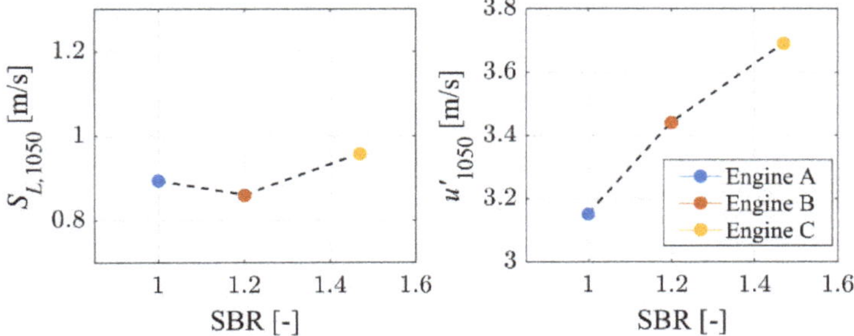

Figure 23. Averaged laminar flame speed and average turbulent intensity with intake valve timing sweep.

Although not as significant as in the intake valve timing sweep, some difference in the laminar flame speed is also witnessed. This is mainly because different intake manifold pressures were necessary for each engine design to match IMEP of 4.5 bar. Again, the model prediction with regards to engine geometry and the subsequent flow behavior showed a good agreement with the experimental data (Figure 24).

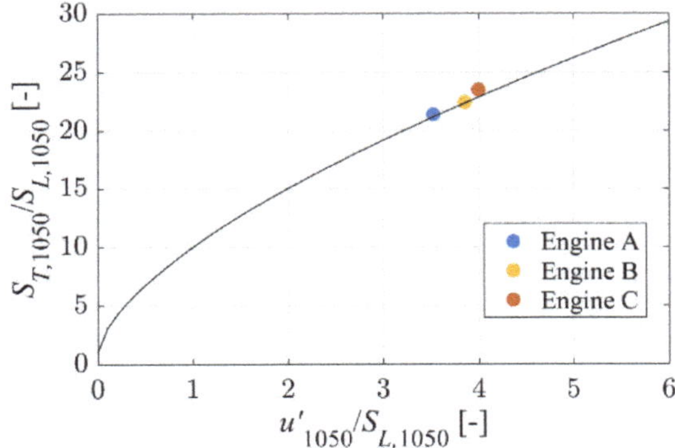

Figure 24. Correlation between simulation and experiment results (intake valve timing sweep).

5. Conclusions

As an effort to develop a calibration-free predictive 0D turbulence model, the previous 0D tumble model has been advanced to reflect more realistic geometry and physical phenomena supposedly occurring inside it. In summary, this study:

- describes the port characterization methodology in detail, which is verified via comparing the model prediction with the transient CFD results. This port characterization facilitates 0D model to predict the detailed flow characteristics with accuracy comparable to 3D CFD regardless of operating condition, which is quite noteworthy for a 0D model.

- suggests a physics-based approach to calculate tumble generation and decay rates for given engine geometry, utilizing the pre-obtained detailed flow characteristics. This QD tumble model is integrated into the k-ε model to complete the new predictive turbulence model.
- demonstrates the validation of the developed model with engine experiment results to verify whether the model correctly considers for port and cylinder geometry and whether it is applicable for various operating conditions. An adequate correlation between the model and experiment results has been observed with fixed modeling constants, implying that the developed model could sufficiently capture the core physics related to engine geometry and operating condition.

The final validation results support the predictability of the new turbulence model, thus its potential as a part of a virtual engine. Moreover, it could be glimpsed that the model reliability can possibly be improved with the further advancement of other elements such as combustion model.

Author Contributions: Conceptualization, Y.K. and M.K.; methodology, Y.K. and M.K.; software, Y.K. and W.S.; resources, S.C.; validation, Y.K.; Investigation, Y.K., M.K., S.O., W.S. and S.C.; data curation, S.O.; writing—original draft preparation, Y.K.; writing—review and editing, Y.K. and H.H.S.; supervision, H.H.S.

Funding: This research was funded by the Ministry of Trade, Industry & Energy (MOTIE, Korea) (grant number 20002762, Development of RDE DB and Application Source Technology for Improvement of Real Road CO2 and Particulate Matter). The APC was funded by the Department of Mechanical and Aerospace Engineering of Seoul National University.

Acknowledgments: The supports from the Institute of Advanced Machinery and Design (IAMD) and Advanced Automotive Research Center (AARC) of Seoul National University are also acknowledged.

Conflicts of Interest: The authors declare no conflict of interest.

Nomenclature

A	area
a	pentroof slope
B	cylinder bore
C	tumble center
CA10, CA50, CA90	10, 50, 90% burn angle
C_T	tumble coefficient
C_{KE}, C_α, C_β, C_3, C_{out}	model constants
$\dot{E}_{non}$	non-tumble intake kinetic energy flux
H	cylinder height
h, w	vertical/horizontal distance from tumble center
I	moment of inertia
K	non-tumble mean kinetic energy
k	turbulent kinetic energy
L	angular momentum
L_g	geometric length scale
L_{IV}	intake valve lift
$\dot{m}$	mass flow rate
n_d	number of divisions
P_k, P_ε	TKE and dissipation production from non-tumble MKE
R	cylinder radius
R_f	flame radius
r	moment arm length
S_T, S_L	turbulent and laminar flame speed
U_x, U_y, U_z	axial velocities
U	characteristic velocity
u'	turbulent intensity
V	volume
v	velocity

Greek Letters

β	pentroof angle
ε	dissipation rate
$\phi,\ \theta$	flow angles
ν_T	turbulent viscosity
ρ	gas density
Ψ	term related to tumble decay

Subscripts

1050	averaged value over CA10-50
eff	effective
i	division number
pent	pentroof
tum	tumble-effective

Abbreviations

0D	zero-dimensional
1D	one-dimensional
3D	three-dimensional
BD	burn duration
BDC	bottom dead center
CAD	crank angle degree
CFD	computational fluid dynamics
EGR	exhaust gas recirculation
fTDC	firing top dead center
IMEP	gross indicated mean effective pressure
IVC	intake valve closing
IVO	intake valve opening
MFB	mass fraction burned
MKE	mean kinetic energy
RMF	residual mass fraction
TDC	top dead center
TKE	turbulent kinetic energy
QD	quasi-dimensional

References

1. Ikeya, K.; Takazawa, M.; Yamada, T.; Park, S.; Tagishi, R. Thermal Efficiency Enhancement of a Gasoline Engine. *SAE Int. J. Engines* **2015**, *8*, 1579–1586. [CrossRef]
2. Hirooka, H.; Mori, S.; Shimizu, R. Effects of High Turbulence Flow on Knock Characteristics. *SAE Trans.* **2004**, *113*, 651–659.
3. Berntsson, A.W.; Josefsson, G.; Ekdahl, R.; Ogink, R.; Grandin, B. The Effect of Tumble Flow on Efficiency for a Direct Injected Turbocharged Downsized Gasoline Engine. *SAE Int. J. Engines* **2011**, *4*, 2298–2311. [CrossRef]
4. Ogink, R.; Babajimopoulos, A. Investigating the Limits of Charge Motion and Combustion Duration in a High-Tumble Spark-Ignited Direct-Injection Engine. *SAE Int. J. Engines* **2016**, *9*, 2129–2141. [CrossRef]
5. Achuth, M.; Mehta, P.S. Predictions of Tumble and Turbulence in Four-Valve Pentroof Spark Ignition Engines. *Int. J. Engine Res.* **2001**, *2*, 209–227. [CrossRef]
6. Bozza, F.; Teodosio, L.; De Bellis, V.; Fontanesi, S.; Iorio, A. Refinement of a 0D Turbulence Model to Predict Tumble and Turbulent Intensity in SI Engines. Part II: Model Concept, Validation and Discussion. SAE Technical Paper Series. **2018**. [CrossRef]
7. Yoshihara, Y.; Nakata, K.; Takahashi, D.; Omura, T.; Ota, A. Development of High Tumble Intake-Port for High Thermal Efficiency Engines. SAE Technical Paper. **2016**. [CrossRef]
8. Heywood, J.B. *Internal Combustion Engine Fundamentals*; McGraw-Hill Education: Columbus, OH, USA, 1988.
9. Li, Y.-f.; Liu, S.-l.; Shi, S.; Feng, M.; Sui, X. An investigation of in-cylinder tumbling motion in a four-valve spark ignition engine. *J. Proc. Inst. Mech. Eng. Part D J. Automob. Eng.* **2001**, *215*, 273–284. [CrossRef]

10. Falfari, S.; Forte, C.; Brusiani, F.; Bianchi, G.M.; Cazzoli, G.; Catellani, C. Development of a 0D Model Starting from Different RANS CFD Tumble Flow Fields in Order to Predict the Turbulence Evolution at Ignition Timing. SAE Technical Paper Series. **2014**. [CrossRef]

11. Kent, J.; Mikulec, A.; Rimal, L.; Adamczyk, A.; Mueller, S.; Stein, R.; Warren, C. Observations on the effects of intake-generated swirl and tumble on combustion duration. *SAE Trans.* **1989**, *98*, 2042–2053.

12. Gosman, A.; Tsui, Y.; Vafidis, C. Flow in a Model Engine with a Shrouded Valve-A Combined Experimental and Computational Study. SAE Technical Paper. **1985**. [CrossRef]

13. Dai, W.; Newman, C.E.; Davis, G.C. Predictions of in-cylinder tumble flow and combustion in SI engines with a quasi-dimensional model. *SAE Trans.* **1996**, *105*, 2014–2025.

14. Grasreiner, S.; Neumann, J.; Luttermann, C.; Wensing, M.; Hasse, C. A quasi-dimensional model of turbulence and global charge motion for spark ignition engines with fully variable valvetrains. *Int. J. Engine Res.* **2014**, *15*, 805–816. [CrossRef]

15. Fogla, N.; Bybee, M.; Mirzaeian, M.; Millo, F.; Wahiduzzaman, S. Development of a K-k-ε Phenomenological Model to Predict In-Cylinder Turbulence. *SAE Int. J. Engines* **2017**, *10*, 562–575. [CrossRef]

16. Kim, Y.; Kim, M.; Kim, J.; Song, H.H.; Park, Y.; Han, D. Predicting the Influences of Intake Port Geometry on the Tumble Generation and Turbulence Characteristics by Zero-Dimensional Spark Ignition Engine Model. SAE Technical Paper. **2018**. [CrossRef]

17. Kent, J.; Haghgooie, M.; Mikulec, A.; Davis, G.; Tabaczynski, R. Effects of Intake Port Design and Valve Lift on In-Cylinder Flow and Burnrate. SAE Technical Paper. **1987**. [CrossRef]

18. Baratta, M.; Misul, D.; Spessa, E.; Viglione, L.; Carpegna, G.; Perna, F. Experimental and numerical approaches for the quantification of tumble intensity in high-performance SI engines. *Energy Convers. Manag.* **2017**, *138*, 435–451. [CrossRef]

19. Kuwahara, K.; Watanabe, T.; Takemura, J.; Omori, S.; Kume, T.; Ando, H. Optimization of in-cylinder flow and mixing for a center-spark four-valve engine employing the concept of barrel-stratification. *SAE Trans.* **1994**, *103*, 1502–1513.

20. Benjamin, S. Prediction of barrel swirl and turbulence in reciprocating engines using a phenomenological model. In Proceedings of the Institute of Mechanical Engineers Conference on Experimental and Predictive Methods in Engine Research and Development, Birmingham, UK, 17 November 1993.

21. Ramajo, D.; Zanotti, A.; Nigro, N. Assessment of a zero-dimensional model of tumble in four-valve high performance engine. *Int. J. Numer. Methods Heat Fluid Flow* **2007**, *17*, 770–787. [CrossRef]

22. Kim, M.; Kim, Y.; Song, H.H. Development of Zero-Dimensional Spark Ignition Engine Model Considering Turbulence Formation in Various Intake Pressure Conditions and Intake Manifold Designs. In Proceedings of the 56th KOSCO Symposium, Jeonju, Korea, 10–12 May 2018.

23. Kim, N.; Ko, I.; Min, K. Development of a zero-dimensional turbulence model for a spark ignition engine. *Int. J. Engine Res.* **2018**, *20*, 441–451. [CrossRef]

24. Morel, T.; Mansour, N. Modeling of Turbulence in Internal Combustion Engines. SAE Technical Paper. **1982**. [CrossRef]

25. Oh, S.; Cho, S.; Seol, E.; Song, C.; Shin, W.; Min, K.; Song, H.H.; Lee, B.; Jinwook, S.; Woo, S.H. An Experimental Study on the Effect of Stroke-to-Bore Ratio of Atkinson DISI Engines with Variable Valve Timing. *J. SAE Int. J. Engines* **2018**, *11*, 1183–1193. [CrossRef]

Article

Comparison of Physics-Based, Semi-Empirical and Neural Network-Based Models for Model-Based Combustion Control in a 3.0 L Diesel Engine

Song Hu [1,2], Stefano d'Ambrosio [1], Roberto Finesso [1,*], Andrea Manelli [1], Mario Rocco Marzano [1], Antonio Mittica [1], Loris Ventura [1], Hechun Wang [2] and Yinyan Wang [2]

[1]   Department of Energy, Politecnico di Torino, Corso Duca degli Abruzzi 24, 10129 Torino, Italy
[2]   College of Power and Energy Engineering, Harbin Engineering University, Harbin 150001, China
*   Correspondence: roberto.finesso@polito.it; Tel.: +39-011-090-4493

Received: 1 August 2019; Accepted: 2 September 2019; Published: 5 September 2019

Abstract: A comparison of four different control-oriented models has been carried out in this paper for the simulation of the main combustion metrics in diesel engines, i.e., combustion phasing, peak firing pressure, and brake mean effective pressure. The aim of the investigation has been to understand the potential of each approach in view of their implementation in the engine control unit (ECU) for onboard combustion control applications. The four developed control-oriented models, namely the baseline physics-based model, the artificial neural network (ANN) physics-based model, the semi-empirical model, and direct ANN model, have been assessed and compared under steady-state conditions and over the Worldwide Harmonized Heavy-duty Transient Cycle (WHTC) for a Euro VI FPT F1C 3.0 L diesel engine. Moreover, a new procedure has been introduced for the selection of the input parameters. The direct ANN model has shown the best accuracy in the estimation of the combustion metrics under both steady-state/transient operating conditions, since the root mean square errors are of the order of 0.25/1.1 deg, 0.85/9.6 bar, and 0.071/0.7 bar for combustion phasing, peak firing pressure, and brake mean effective pressure, respectively. Moreover, it requires the least computational time, that is, less than 50 μs when the model is run on a rapid prototyping device. Therefore, it can be considered the best candidate for model-based combustion control applications.

Keywords: model-based; control; diesel engine; ANN; physics-based model; semi-empirical model

1. Introduction

Nowadays, emission and fuel consumption reductions are the two main challenges for internal combustion engines, and in particular for diesel technology [1–3]. Many techniques have been proposed to achieve this aim, such as variable geometry turbocharger (VGT), high-pressure common rail systems [4–8], exhaust gas recirculation (EGR) [9], innovative combustion concepts such as HCCI and PCCI [10], and innovative combustion controls [11–14]. As far as these solutions are concerned, model-based combustion control is expected to make a significant contribution in the near future, thanks to the possibility of its integration with the emerging Vehicle-to-Everything (V2X) architectures. This conclusion is also supported by recent international projects, such as IMPERIUM H2020 [14], in which a significant reduction in CO_2 emissions will be achieved (−20%) for heavy-duty trucks, thanks to the use of model-based controllers coupled with V2X systems. The model-based control technology will also be boosted by the development of new sensors and by the increasing computational performance of the new multi-core processors that are now available for mobility applications.

Several advantages may be obtained from adopting a model-based controller instead of a conventional map-based one, such as the possibility of realizing a real-time optimization of the engine parameters, and the need for less calibration effort. It should, in fact, be considered that modern internal combustion engines (ICEs), especially of the diesel type, require a heavy tuning procedure at the engine test bench [12,15] when a conventional map-based controller is adopted. Therefore, interest in model-based combustion control for diesel ICEs has been increasing more and more over the last few years. It is generally believed that the diesel technology will remain the best solution for light-duty and heavy-duty vehicles over the next decades [14], and research efforts to reduce fuel consumption and pollutants are therefore still required.

As a result of the increasing interest in the development of model-based combustion controllers, there is a need for accurate and low computational time demanding models. However, several types of modeling approaches, which can roughly be divided into physics-based (white box), and mathematical (gray-box or black-box) methods, can be used to this purpose. A detailed comparison of the different approaches is therefore required in order to identify the advantages and drawbacks of each modeling approach in view of their use for control-oriented applications.

As far as physics-based models are concerned, previous studies carried out by the authors [11,12,16] have shown that zero-dimensional approaches are the best solution for the development of model-based controllers. A zero-dimensional combustion model, which is capable of simulating the heat release rate, on the basis of the accumulated fuel mass approach and the in-cylinder pressure, on the basis of a single zone thermodynamic model, was presented in [17]. The accumulated fuel mass approach [18–21] is more physically consistent than the classical Wiebe function methodology [22] for heat release prediction. Highly accurate simulation results have been obtained using this model in many types of single- and multi-injection diesel engines [11–13,17,22].

Apart from physics-based models, mathematical models are also widely used for control-oriented applications. The aim of purely mathematical models is to identify the correlations between the input and output variables, without having knowledge of the physics of the system.

A first category of mathematical models includes empirical or semi-empirical approaches. These methods are easy to calibrate and can provide accurate results, if robust input variables are selected. Moreover, they require a short computational time, which makes them suitable for control applications. An example of a semi-empirical model for the estimation of combustion metrics is reported in [23].

Artificial intelligence systems, which include methods such as support vector machine (SVM) and artificial neural networks (ANNs) [13,24–28], can also be adopted for control-oriented applications. The use of ANNs has increased more and more in the last few years, due to their capacity to accurately predict the behavior of complex systems with short computational times. Many applications based on ANNs can be found in the literature on engine control and diagnostics [13,26–34]. From an analysis of the literature, it may be observed that the predictive performance of ANNs is influenced to a great extent by the selection of their main parameters, such as the number of layers, the training algorithm and the number of neurons [31–37]. The study reported in [37] pointed out that networks that are too small result in underfitting. However, larger networks may result in overfitting. Therefore, a sensitivity analysis is always needed in order to identify what the optimal number of neurons is for a specific application.

Although the previously mentioned modeling approaches have been reported in the literature in detail, to the best of the authors' knowledge there is a lack of studies concerning the comparison of the performances of physics-based, semi-empirical and ANN-based models, for the specific type of application investigated in this paper. The present study therefore addresses this research need and compares the performance of four different modeling approaches for the prediction of the main combustion metrics which are generally considered for combustion control applications, i.e., crank angle at which 50% of fuel mass has burnt (MFB50), peak firing pressure (PFP) and brake mean effective pressure (BMEP):

- Baseline physics-based model: this model was previously presented by the authors in [12] for control-oriented applications. However, in the present study, the model calibration procedure has been refined and the performance of the model has been improved with respect to the previously reported version. In this approach, the tuning parameters are modeled by means of power-law functions, in which the input parameters are the main engine operating variables.

- ANN physics-based model: this is a new modeling approach, which is based on the use of ANNs to predict the tuning parameters of the aforementioned physics-based model.
- Direct semi-empirical model: in this approach, semi-empirical correlations, which are constituted by power-law functions, are used to directly estimate MFB50, PFP, and BMEP.
- Direct ANN model: this approach exploits feed-forward artificial neural networks to directly estimate MFB50, PFP, and BMEP. Details of the methodology used for the training and optimal selection of the number of neurons are also provided in this study.

The four different models have been assessed considering the same experimental dataset, for a 3.0 L FPT diesel engine for light-duty applications, and their performances have been compared under steady-state conditions and in transient operation over a WHTC.

In addition to the comparison of the performance of the previous models, another innovative contribution introduced in this paper concerns the methodology used for the optimal selection of the input parameters of the models. The proposed method is based on the sequential use of the Pearson correlation and partial correlation coefficients, which are used to identify the least and most robust correlation variables that should be excluded or included as model inputs. The use of the Pearson correlation and partial correlation analysis allows the computational effort required for the identification of the input parameters to be reduced significantly.

The paper is organized as follows. Section 2 summarizes the experimental setup and the engine specifications. Section 3 reports the description of the four investigated models. Section 4 is focused on the presentation of the methodology used for the selection of the model input variables. Finally, Section 5 reports the main results, which include:

- A comparison of the performance of the physics-based model tuned using the methodology introduced in this paper (based on Pearson correlation and partial correlation analysis) with that of the previous version reported in [12].
- A comparison of the accuracy of the four investigated models under steady-state conditions and in transient operation over a WHTC.
- A comparison of the performance of the four investigated models, in terms of required computational time, when they are run on an ETAS ES910 (ETAS, Stuttgart, Germany) rapid prototyping device.

This research is a first step towards the development of a model-based combustion controller, which will be developed and tested on the real engine, through rapid prototyping, in the near future.

2. Experimental Setup and Engine Conditions

The experimental tests were run on a Euro VI FPT F1C 3.0 L diesel engine (FPT Motorenforschung AG, Arbon, Switzerland), within a research project conducted in collaboration with FPT Industrial [11]. The activity was carried out in the dynamic test bench facility at the Politecnico di Torino. The main technical specifications of the engine are reported in Table 1.

Table 1. Main technical specifications of the FPT F1C 3.0 L diesel engine.

Engine Type	Number of Cylinders	Displace Ment	Bore × Stroke	Rod Length	Compres Sion Ratio	Valves per Cylinder	Turbo- Charger	Fuel Injection System
FPT F1C Euro VI diesel engine	4	2998 cm^3	95.8 mm × 104 mm	160 mm	17.5	4	VGT type	High pressure Common Rail

The fuel is characterized by a density of 835 kg/m^3 at 14 °C, a viscosity of 2 mm^2/s at 40 °C and a cetane number equal to 43. The considered injection pattern features two pilot shots and a main shot.

The engine is shown in Figure 1. It is equipped with a VGT turbocharger, a short-route cooled EGR system and a flap valve located on the exhaust side downstream from the turbine. Details about the layout of the engine, the test bench and the used sensors (including the main accuracy) have already been described in detail in [38] and are not repeated in this paper for the sake of brevity.

Figure 1. Picture of the FPT F1C 3.0 L Euro VI diesel engine installed at the Politecnico di Torino. The rapid prototyping device is located on the right side.

An ETAS ES910 rapid prototyping device, which is equipped with a main Freescale PowerQUICCTM III MPC8548 processor with an 800 MHz clock, was used to verify the computational time required for the models investigated in this paper.

The experimental tests were conducted under steady-state and transient conditions. The steady-state tests have been divided into three categories (Figure 2):

- a complete engine map (123 tests, indicated with the blue circles in Figure 2).
- EGR-sweep tests (162 tests, carried out on the points indicated with the red diamonds in Figure 2).
- sweep tests of the main injection timing (SOI_{main}) and injection pressure (p_f) (125 tests, carried out on the points indicated with the black circles in Figure 2).

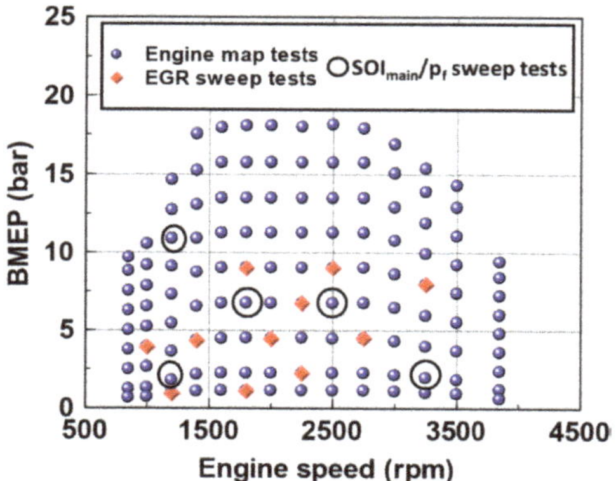

Figure 2. Experimental tests acquired at the test bench.

Validation of the methods was carried out over a WHTC. Details on these validations are given in Section 5.

3. Description of the Models

3.1. Physics-Based Model

Although the physics-based combustion model has already been presented in other studies, a summary of the approach and of the main equations is reported hereafter.

A scheme of the model is shown in Figure 3.

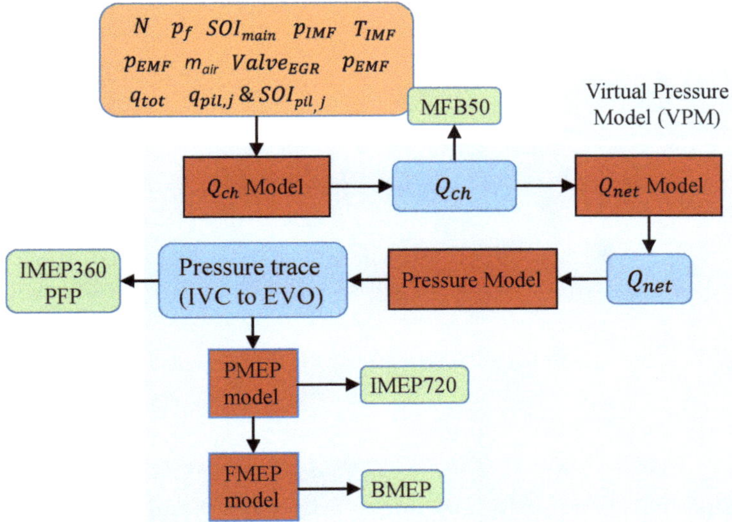

Figure 3. Scheme of the physics-based combustion model [12].

N in Figure 3 indicates the engine speed, while p_{IMF} and T_{IMF} the intake manifold pressure and temperature, respectively, p_{EMF} the exhaust manifold pressure, SOI_{main} the start of injection of the main pulse, $SOI_{pil,j}$ the start of injection of the pilot pulse j, $q_{pil,j}$ the injected fuel quantity of pilot injection j, q_{tot} the total injected fuel quantity, $Valve_{EGR}$ the opening position of the high pressure EGR valve and m_{air} the fresh air trapped mass per cycle/cylinder.

The first step consists in the evaluation of the chemical heat release, according to the following equations (AFM approach):

$$\frac{dQ_{ch,pil,j}}{dt}(t) = K_{pil,j}[Q_{fuel,pil,j}(t - \tau_{pil,j}) - Q_{ch,pil,j}(t)] \tag{1}$$

$$\frac{dQ_{ch,main}}{dt}(t) = K_{1,main}[Q_{fuel,main}(t - \tau_{main}) - Q_{ch,main}(t)] + K_{2,main}\frac{dQ_{fuel,main}(t - \tau_{main})}{dt} \tag{2}$$

$$Q_{fuel,j}(t) = \int_{t_{SOI,j}}^{t} \dot{m}_{f,inj}(t)H_L dt \qquad t \leq t_{EOI,j} \tag{3}$$

$$Q_{fuel,j}(t) = \int_{t_{SOI,j}}^{t_{EOI,j}} \dot{m}_{f,inj}(t)H_L dt \qquad t > t_{EOI,j} \tag{4}$$

$$Q_{ch} = \sum_{j=1}^{n} Q_{ch,j} \tag{5}$$

where j indicates the generic injection pulse, H_L indicates the lower heating value of the fuel, and K and τ are the combustion rate coefficient and ignition delay coefficients, respectively.

Subsequently, the net energy of the charge (Q_{net}) is calculated as follows:

$$Q_{net,ht} \cong Q_{ch} \frac{m_{f,inj} H_L - Q_{ht,glob}}{m_{f,inj} H_L} \tag{6}$$

$$Q_{net} \cong Q_{net,ht} - Q_{f,evap} \tag{7}$$

where $Q_{f,evap}$ and $Q_{ht,glob}$ indicate the fuel evaporation heat from SOI to SOC and the heat exchanged by the charge with the walls over the combustion cycle; $m_{f,inj}$ is the total injected fuel mass per cyc/cyl. Q_{net} is then used to derive the in-cylinder pressure, which is based on the inversion of a single-zone thermodynamic model [39]:

$$p^i = \frac{\Delta Q_{net} - \frac{p^{i-1}}{2}(V^i - V^{i-1}) + \frac{1}{\gamma-1} p^{i-1} V^{i-1}}{\frac{V^i - V^{i-1}}{2} + \frac{V^i}{\gamma-1}} \quad (\gamma = 1.4, \; valid \; from \; SOC \; to \; EOC) \tag{8}$$

The compression and expansion phases are modeled with polytropic processes:

$$pV^{\underline{n}} = const \; (valid \; from \; IVC \; to \; SOC) \tag{9}$$

$$pV^{\underline{n'}} = const \; (valid \; from \; EOC \; to \; EVO) \tag{10}$$

while the pressure at IVC (i.e., the starting condition) is correlated with the pressure in the intake manifold:

$$p_{IVC} = p_{IMF} + \Delta p_{IMF} \tag{11}$$

The main combustion metrics, that is, MFB50, IMEP360 (gross Indicated Mean Effective Pressure) and PFP (Peak Firing Pressure), can now be evaluated. Finally, friction (FMEP) and pumping (PMEP) models are used, and the net IMEP (IMEP720) and BMEP (Brake Mean Effective Pressure) are estimated:

$$IMEP360 = \frac{\int_{0}^{360} pdV}{V_0} \tag{12}$$

$$IMEP720 = IMEP360 - PMEP \tag{13}$$

$$BMEP = IMEP720 - FMEP \tag{14}$$

The tuning parameters of the physics-based model are underlined in Equations (1)–(14).

The Chen-Flynn approach [40] was adopted to estimate FMEP:

$$FMEP = x_1 + x_2 \cdot N + x_3 \cdot N^2 + x_4 \cdot PFP \tag{15}$$

where x_{1-4} are fitting parameters.

The intake oxygen concentration (O_2) is an operating variable that is closely correlated with the combustion process, and it is therefore used as an input variable in several sub-models. It is measured, on the experimental test bench, by means of a paramagnetic sensor, which is included in the test cell gas analyzer. However, if the developed combustion model is intended to be used for model-based control, this sensor is not available onboard, and O_2 therefore needs to be estimated by means of a specific sub-model.

The following function has been used to estimate O_2 [11,41]:

$$O_2 = x_1 \frac{X_{r,EGR}}{RAF} + x_2 \tag{16}$$

where x_{1-2} are fitting parameters, RAF is the relative air-to-fuel ratio and $X_{r,EGR}$ is the EGR rate.

Therefore, it is necessary to estimate RAF and $X_{r,EGR}$ in order to evaluate the intake O_2 concentration. The relative air-to-fuel ratio is defined as follows:

$$RAF = \frac{m_{air}}{m_{f,inj}AF_{st}} \tag{17}$$

where m_{air} is the trapped air mass, $m_{f,inj}$ is the injected fuel mass and AF_{st} is the stoichiometric air-to-fuel ratio (taken as 14.4 in this work).

The accuracy of the estimation of RAF depends to a great extent on the accuracy of the estimation of m_{air}. If m_{air} is obtained from the engine air flow sensor, its accuracy is generally not very high (the error is typically of the order of 5%).

An alternative way of estimating RAF (the method that was adopted in this study) is to develop a semi-empirical correlation, such as:

$$RAF = x_1 \cdot p_{IMF}^{x_2} \cdot T_{IMF}^{x_3} \cdot m_{f,inj}^{x_4} \tag{18}$$

where x_{1-4} are fitted parameters. The correlation is fitted using a calibration dataset, in which the experimental values of RAF are obtained from steady-state measurements carried out using an accurate sensor (Bosch Lambda sensor LSU 4.9 (Bosch, Gerlingen, Germany)).

With reference to the EGR rate $X_{r,EGR}$, it is defined as follows:

$$X_{r,EGR} = \frac{m_{EGR}}{m_{EGR} + m_{air}} \tag{19}$$

where m_{EGR} is the trapped EGR mass.

The EGR rate can in general be derived experimentally on the basis of the measured emissions and of the measured intake CO_2 concentration. However, if the combustion model is intended to be applied for control-oriented applications, a real-time estimation of the EGR rate is needed, through the use of a dedicated submodel. To this aim, several approaches can be used (e.g., see [41]).

The simplest method to estimate the EGR rate (which was adopted in this paper) is to resort to look-up tables that depend on the engine speed and pedal position. These look-up tables were derived on the basis of the analysis of steady-state measurements.

As far as the calibration procedure is concerned, as reported in the previous sections, the physics-based model is characterized by several tuning parameters, which are mentioned in Table 2.

Table 2. Tuning parameters of the physics-based model.

Submodel	Calibration Parameter
Heat release model	$K_{pil,j}$; $K_{1,main}$; $K_{2,main}$; $\tau_{pil,j}$; τ_{main}
Net energy release model	$Q_{f,evap}$; $Q_{ht,glob}$
Pressure model	Δp_{IMF}; n; n'
BMEP model	FMEP, PMEP

The calibration of these parameters is carried out in two steps.

In the first step, the optimal values of all the parameters are identified, test by test, on the basis of the available experimental data. The process used for the identification of these values is reported in detail in [16], and is based on the best matching between the experimental and predicted heat release and in-cylinder pressure curves. At the end of this first step, a dataset of the optimal values of the model calibration parameters is available.

In the second step, on the basis of the dataset identified in step 1, the model calibration parameters are correlated with input variables, which are generally related to the engine operating conditions (e.g., intake manifold pressure/temperature, intake oxygen concentration, engine speed and load …), using such mathematical methods as look-up tables, empirical or semi-empirical functions or machine learning tools. Semi-empirical correlations, based on power-law functions, are used in the baseline version of the physics-based model (see [12]), since they are capable of capturing the non-linear behavior of the correlations. However, in previous studies, the input parameters for these correlations were identified on the basis of a trial and error approach.

In the next sections, two aspects are investigated in order to improve the physics-based model with respect to previously reported versions:

1. A new procedure for the identification of the optimal set of input parameters is developed. This procedure is based on the joint use of the Pearson correlation analysis, partial correlation analysis, and sensitivity analysis, and it allows the performance of the baseline physics-based model to be improved with respect to the previously developed versions.

2. An alternative mathematical method (i.e., ANNs) is used to estimate the model calibration parameters:

$$K_{pil}, K_{1,main}, \ldots = \text{ANN}(x_0, x_1, \ldots, x_n) \tag{20}$$

   where $x_0, x_1, \ldots, x_n$ are correlation variables. This approach has never been investigated before, and is denoted as "ANN physics-based model".

3.2. Direct Semi-Empirical Models

In addition to the baseline and ANN physics-based models, semi-empirical correlations were also developed in order to directly predict the MFB50, PFP, and BMEP metrics.

These correlations are based on power-law functions of the type:

$$\text{MFB50}, \text{PFP}, \text{BMEP} = A \cdot x_0^{a_0} \cdot x_1^{a_1} \cdot \ldots \cdot x_n^{a_n} \tag{21}$$

The identification of the $x_0, x_1 \ldots x_n$ correlation variables and the fitting of the model are analyzed in the next sections. The semi-empirical models are trained on the basis of the same experimental data used for the physics-based model calibration.

3.3. Direct Artificial Neural Networks

The final method investigated to predict MFB50, PFP, and BMEP is based on feed-forward artificial neural networks with one inner layer:

$$\text{MFB50}, \text{PFP}, \text{BMEP} = \text{ANN}(x_0, x_1, \ldots, x_n) \tag{22}$$

where $x_0, x_1, \ldots, x_n$ are the input variables of the neural networks.

The networks are trained on the basis of the same experimental data used for the physics-based model calibration. The methodology used for the calibration of the model, including the selection of the optimal number of neurons, is discussed in detail in Section 5.

4. Selection of the Model Input Variables

Selecting the most appropriate input parameters as independent variables for all the previously mentioned approaches is a key-aspect to obtain accurate and robust models.

In this paper, optimal sets of input parameters have to be defined for:

- The power-law correlations or the ANNs that evaluate the calibration parameters of the baseline physics-based model and of the ANN physics-based model, respectively.
- The direct semi-empirical model that estimates MFB50, PFP, BMEP.
- The direct ANN model that estimates MFB50, PFP, BMEP.

In general, the input parameters are related to the engine operating conditions, and can be either directly measured quantities or derived quantities. The directly measured parameters (such as engine speed N, the start of injection of the pilot/main pulses SOI_{pil}/SOI_{main}, the total injection quantity of the pilot pulses $q_{pil,tot}$, the injection quantity of the main pulse q_{main}, the total injection quantity q_{tot}, the intake/exhaust pressure and temperature $p_{IMF},T_{IMF}/p_{EMF},T_{EMF}$, the common rail pressure p_f, the trapped air mass m_{air} and the EGR valve opening position $Valve_{EGR}$) are the most commonly used input variables for control-oriented applications, since they are readily available for the ECU without the need of any additional sub-model. However, derived parameters (such as intake density ρ_{IMF}, in-cylinder gas density and temperature at the start of each injection pulse $\rho_{SOI,pil}$, $\rho_{SOI,main}$, $T_{SOI,pil}$, $T_{SOI,main}$, the start of combustion of the pilot and main pulses SOC_{pil}/SOC_{main}, in-cylinder gas density and temperature at the start of combustion for each injection pulse $\rho_{SOC,pil}$, $\rho_{SOC,main}$, $T_{SOC,pil}$, $T_{SOC,main}$, the pressure difference between the exhaust and intake manifold $\Delta p_{exh\text{-}int}$) or even outcomes of sub-models (such as PFP, MFB50) have also been used as input variables in models reported in the literature, since their correlation with the dependent variables is sometimes greater than that of the directly measured parameters. In this study, all these kinds of parameters were initially considered as candidate input variables to develop the models.

Table 3 reports the list of the directly measured input parameters (1st column), of the generated parameters (from the directly measured ones or from the combustion model) which can be used as input parameters (2nd column), and of the dependent output variables which have been considered in the present investigation. A total of 46 parameters have been included. It should be noted that some of the dependent variables are the calibration parameters of the physics-based model, which needs to be estimated for the model application (3rd column), while MFB50, PFP, and BMEP are the combustion metrics that are adopted for combustion control applications, i.e., the final output variables of all the four modeling approaches that have been investigated in this study (4th column).

The procedure that was identified to select the optimal set of input parameters is based on the joint use of the Pearson correlation and partial correlation coefficients (see [42–45] for details concerning the use of these coefficients), and is reported in detail in Appendix A for the paper readability purposes. However, a synthetic description is reported hereafter.

Basically, the procedure involves 3 steps:

1. First, the Pearson correlation coefficient is calculated for all the possible combinations of the 46 variables reported in Table 3 (the results are reported in Appendix A in Table A1). This coefficient measures the linear dependence between two variables, x_i and x_j, and is defined as follows:

$$\rho\left(x_i, x_j\right) = \frac{cov\left(x_i, x_j\right)}{\sqrt{var(x_i) \cdot var(x_j)}} \tag{23}$$

where $cov(x_i, x_j)$ is the covariance between x_i and x_j, while $var(x_i)$ and $var(x_j)$ are the variances of x_i and x_j, respectively. The Pearson correlation coefficient varies between -1 and $+1$. After the analysis of the Pearson correlation coefficients, an initial candidate set of input variables is identified for each output quantity that has to be estimated (see Table A2 in Appendix A).

2. In the second step, the partial correlation coefficient is evaluated between the dependent variables and each input variable of the initial candidate set, when removing the effect of the remaining input variables of the same set. It should be recalled that the partial correlation coefficient is able to measure the linear dependence between two variables, x_i and x_j, when the effect of the other input variables variable ($z_1, z_2, \ldots, z_i$) is removed:

$$\rho\left(x_i, x_j; z_1, z_2, \ldots, z_i\right) = \frac{\rho\left(x_i, x_j; z_1, z_2, \ldots, z_{i-1}\right) - \rho\left(x_j, z_i; z_1, z_2, \ldots, z_{i-1}\right) \cdot \rho\left(x_i, z_i; z_1, z_2, \ldots, z_{i-1}\right)}{\sqrt{\left[1 - \rho\left(x_j, z_i; z_1, z_2, \ldots, z_{i-1}\right)^2\right] \cdot \left[1 - \rho(x_i, z_i; z_1, z_2, \ldots, z_{i-1})^2\right]}} \tag{24}$$

The partial correlation coefficient is more robust than the Pearson coefficient. However, the number of parameters should not be too high, otherwise, the consistency of the method decreases. Also, the partial correlation coefficient varies between -1 and $+1$. As reported in Appendix A (see Table A3), the simultaneous analysis of the Pearson and of the partial correlation coefficients has allowed the least and most robust correlation variables of the initial candidate set to be identified for each dependent output variable. Therefore, some variables were excluded from the initial candidate set identified in step 1 (since they showed low values of the Pearson and partial correlation coefficients), while other variables were kept since they showed high values of both the Pearson and partial correlation coefficients.

3. The remaining correlation variables of the initial candidate set which showed intermediate values of the Pearson and partial correlation coefficients were analyzed on the basis of a sensitivity analysis, in order to identify those that had to be kept and those that had to be excluded. A power law function, such as that reported in Equation (21), was used to model each output dependent parameter, and all the possible combinations of input variables were identified by evaluating, for each combination, the related model fitting precision, which was quantified by a correlation coefficient R_{adj}:

$$R_{adj} = \sqrt{1 - \frac{n_{exp} - 1}{n_{exp} - k}(1 - R^2)} \tag{25}$$

where n_{exp} is the total number of experimental data and k is the total number of free coefficients. The adjusted correlation coefficient R_{adj} takes into account the number of available experimental tests used for the model fitting, in relation to the number of free parameters of the model. In the end, the combinations which led to the best trade-off between the prediction accuracy and the number of input parameters were selected. Details on the sensitivity analysis have not been reported in this paper for the sake of brevity.

It should be noted that the proposed approach, i.e., the preliminary selection of the input variables through Pearson and partial correlation coefficients, and the subsequent sensitivity analysis carried out on a reduced set of variables, is more computationally efficient than a pure sensitivity analysis approach which includes the entire set of input variables.

In fact, the number of possible combinations that have to be investigated in the sensitivity analysis can be calculated through the following formula:

$$N_c = 2^j - 1 \tag{26}$$

where N_c is the number of possible combinations for the sensitivity analysis and j is the number of independent variables.

Table 3. List of the input and dependent parameters.

Input Parameters (Directly Measured Quantities)	Generated Quantities Which Can Be Used as Input Parameters	Dependent Output Variables (RAF + Calibration Parameters of the Physics-Based Model)	Dependent Output Variables (Combustion Metrics for Use for Model-Based Control)
N	X_{rEGR}	RAF	MFB50
p_f	O_2	τ_{pil}	PFP
SOI_{pil}	ρ_{IMF}	τ_{main}	BMEP
SOI_{main}	$\rho_{SOI,pil}$	K_{pil}	
$q_{tot,pil}$	$\rho_{SOI,main}$	$K_{1,main}$	
q_{main}	$\rho_{SOC,pil}$	$K_{2,main}$	
q_{tot}	$\rho_{SOC,main}$	$Q_{ht,glob}$	
p_{IMF}	$T_{SOI,pil}$	$Q_{f,evap}$	
T_{IMF}	$T_{SOI,main}$	Δp_{IMF}	
p_{EMF}	$T_{SOC,pil}$	n	
T_{EMF}	$T_{SOC,main}$	n'	
m_{air}	$\Delta p_{exh-int}$	PMEP	
$Valve_{EGR}$	SOC_{pil}	FMEP	
	SOC_{main}		
	IMEP360		
	IMEP720		
	Q_{netmax}		

According to Equation (26), N_c increases to a great extent as j increases. Therefore, the combined use of the Pearson correlation and partial correlation analysis is a very useful and efficient way of decreasing the computational effort of the sensitivity analysis, since it allows the poorly correlated variables to be excluded a-priori, or the highly correlated variables to be selected, so that the sensitivity analysis (which is highly time consuming, since it requires model fitting of each possible combination), is only carried out for a reduced set of variables.

5. Results and Discussion

The improvement in the prediction performance of the physics-based model (compared to the previous version reported in [12]), when adopting the new methodology for the input parameter selection described in Section 4, is reported first. Subsequently, the performances of the four investigated models described in Section 3 are compared, in terms of accuracy at steady-state conditions and over WHTC, and in terms of required computational time. The main advantages and drawbacks of each kind of approach are also discussed.

5.1. Improvement to the Baseline Physics-Based Combustion Model

The performance of the baseline physics-based combustion model has been improved, with respect to the previous version reported in other papers [12]. The set of input variables for the model correlations has been revised, on the basis of the Pearson correlation and partial correlation analysis, according to the procedure described in the previous section.

The final list of selected input variables for each dependent variable and the corresponding fitting precision are shown in Table 4. It should be noted that some correlations are improved with respect to pure power-law functions.

Table 4. Final list of the selected input variables, the related correlations, for each tuning parameter of the physics-based model and the corresponding fitting precision in terms of correlation coefficient (R^2), root mean squared error (RMSE) and relative RMSE ($RMSE_{rel}$).

Dependent Variable	Final List of the Independent Variables	Fitted Correlation	R^2, RMSE, RMSErel
RAF	$N, q_{tot}, P_{IMF}, Valve_{EGR}$	$RAF = 43.172 \times N^{-0.0349} \times qtot^{-0.9530} \times P_{IMF}^{0.8989} \times Valve_{EGR}^{-0.2163}$	0.98, 0.135, 5.8%
τ_{pil}	$N, SOI_{pil}, \rho_{SOI,pil}, T_{SOI,pil}$	Power-law function : $\tau_{pil} = 8.177 \times 10^6 \times N^{0.5816} \times SOI_{pil}^{-2.828}$ $\times \rho_{SOI,pil}^{-1.043} \times T_{SOI,pil}^{0.1445}$ Improved correlation : $\tau_{pil} = 0.8688 \times N^{0.4900}$ $\times \left(365 - SOI_{pil}\right)^{0.2647} \times \rho_{SOI,pil}^{-0.9870} \times T_{SOI,pil}^{0.0468}$	0.906, 1.06, 26.9% 0.908, 1.04, 25.6%
τ_{main}	$N, p_f, q_{tot}, \rho_{SOI,pil}, \rho_{SOI,main}$	$\tau_{main} = 0.1862 \times N^{2.2085} \times p_f^{-0.7866} \times q_{tot}^{0.2847} \times \rho_{SOI,pil}^{-4.6496} \times \rho_{SOI,main}^{1.5852}$	0.886, 0.339, 47.9%
K_{pil}	$N, p_f, SOI_{pil}, q_{tot,pil}, \rho_{SOI,pil}, T_{SOI,pil}$	$K_{pil} = 1.2273 \times 10^6 \times N^{-1.0983}$ $\times p_f^{0.4252} \times SOI_{pil}^{3.5487} \times q_{tot,pil}^{-0.4497} \times \rho_{SOI,pil}^{-0.6886} \times T_{SOI,pil}^{-4.2469}$	0.365, 0.0894, 40.4%
$K_{1,main}$	$N, SOI_{main}, q_{tot}, p_{IMF}, T_{IMF}, \rho_{SOI,main}$	$K_{1,main} = 1.0106 \times 10^{-13} \times N^{-0.1030} \times SOI_{main}^{6.7329} \times q_{tot}^{-0.3414} \times p_{IMF}^{1.0216}$ $\times T_{IMF}^{-1.7314} \times \rho_{SOI,main}^{-0.2703}$	0.833, 0.0026, 5.4%
$K_{2,main}$	$N, p_f, q_{main}, \rho_{SOI,main}, T_{SOI,main}$	Power – law function : $K_{2,main} = 85.0613 \times N^{-0.2778} \times p_f^{0.3397} \times q_{main}^{0.3948}$ $\times \rho_{SOI,main}^{-1.2203} \times T_{SOI,main}^{-0.5323}$ Improved correlation : $K_{2,main} = 197.5434 \times N^{-3.2244 \times 10^{-4}}$ $\times p_f^{3.6373 \times 10^{-4}} \times q_{main}^{4.793 \times 10^{-4}} \times \rho_{SOI,main}^{-0.0015} \times T_{SOI,main}^{-7.6126 \times 10^{-4}} - 1957.1$	0.634, 0.0311, 61.5% 0.663, 0.0299, 55.2%
$Q_{f,evap}$	$N, p_f, SOI_{pil}, q_{tot}, m_{air}, \rho_{SOI,pil}$	$Q_{f,evap} = 2.016 \times 10^4 \times N^{0.4478} \times p_f^{-0.2891} \times SOI_{pil}^{-3.499} \times q_{tot}^{-0.1624}$ $\times m_{air}^{-0.5281} \times \rho_{SOI,pil}^{0.0807}$	0.532, 0.00145, 15.5%
n	$N, SOI_{pil}, p_{IMF}, T_{IMF}, Valve_{EGR}, \rho_{SOI,pil}, T_{SOC,main}$	$n = 2.89 \times 10^5 \times N^{-0.013} \times SOI_{pil}^{-2.103} \times p_{IMF}^{-0.154} \times T_{IMF}^{0.347} \times Valve_{EGR}^{0.0042}$ $\times \rho_{SOI,pil}^{0.168} \times T_{SOC,main}^{-0.340}$	0.654, 0.0053,0.39%
n'	$N, p_f, q_{tot}, p_{IMF}, Valve_{EGR}$	Power – law function : $n' = 2.2173 \times N^{-0.0647} \times p_f^{0.0218} \times q_{tot}^{-0.0613}$ $\times p_{IMF}^{0.0783} \times Valve_{EGR}^{-0.0025}$ Improved correlation : $n' = 5.6312 \times 10^{-4} \times N^{0.6839} \times p_f^{-0.2032}$ $\times q_{tot}^{0.6574} \times p_{IMF}^{-1.0419} \times Valve_{EGR}^{0.0418} + 1.4812$	0.913, 0.01, 0.8% 0.923, 0.00936, 0.709%
$Q_{ht,glob}$	$N, p_f, q_{tot}, \rho_{SOI,pil}$	$Q_{ht,glob} = 0.4596 \times N^{-0.3909} \times p_f^{-0.1981} \times q_{tot}^{1.4489} \times \rho_{SOI,pil}^{-0.5557}$	0.99, 0.0143, 11.5%
Δp_{IMF}	N, q_{tot}, p_{IMF}	Power – law function : $\Delta p_{IMF} = 0.0929 \times N^{0.032} \times q_{tot}^{-0.0319} \times p_{IMF}^{1.15}$ Improved correlation : $\Delta p_{IMF} = 0.12254 \times \left[1 - exp\left(\frac{-N+538.806}{131.94}\right)\right]$ $\times q_{tot}^{-0.044} \times p_{IMF}^{1.1932}$	0.966, 0.00733, 4.6% 0.967, 0.00723, 4.37%
$PMEP$	$N, p_{IMF}, Valve_{EGR}, \Delta p_{exh-int}$	$PMEP = 9.5393 \times 10^{-4} \times N^{0.8798} \times p_{IMF}^{0.3571} \times Valve_{EGR}^{0.0010} \times \Delta p_{exh-int}^{0.5164}$	0.994, 0.0204, 14.4%
$FMEP$	$N, SOI_{pil}, q_{tot}, p_{IMF}$	$FMEP = 2.4730 \times 10^{-5} \times N^{0.5640} \times SOI_{pil}^{1.1170} \times q_{tot}^{-0.0502} \times p_{IMF}^{0.6530}$	0.939, 0.0978, 7.91%

Figures 4 and 5 report the accuracy of the original physics-based model with the baseline calibration (presented in [12]) and the revised model which embeds the improved correlations, concerning the estimation of BMEP, PFP, and MFB50, for the steady-state tests reported in Figure 2.

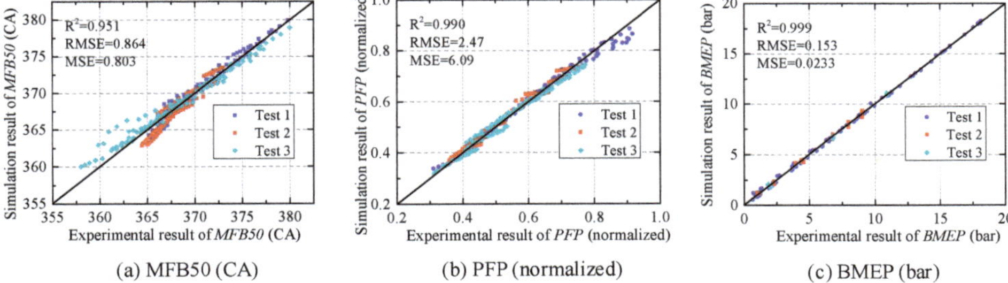

Figure 4. Predicted vs. experimental values of MFB50, PFP and BMEP for the original physics-based model [12], considering the steady-state conditions.

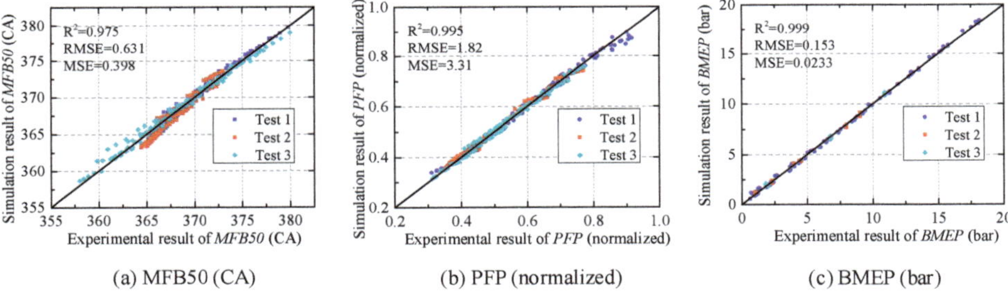

Figure 5. Predicted vs. experimental values of MFB50, PFP and BMEP for the improved physics-based model, considering the steady-state conditions.

It can be seen that, although the accuracy in the estimation of BMEP is somewhat similar, an improvement in the estimation of PFP, and MFB50 can be obtained, since the related RMSE values decrease from 2.47 bar to 1.82 bar and from 0.86 deg to 0.63 deg, respectively.

5.2. Direct Semi-Empirical Models of MFB50, PFP, and BMEP

The procedure used to identify the input parameters of the physics-based model, which was described in Section 4, was also adopted to select the input variables of the direct semi-empirical models. The correlations identified for MFB50, PFP, and BMEP are reported in Table 5.

Table 5. Final list of the selected independent variables, and the related correlations, for MFB50, PFP, and BMEP and the corresponding fitting precision.

Dependent Variable	Final List of the Independent Variables	Fitted Correlation	R^2, RMSE, RMSErel
MFB50	$N, p_f, SOI_{main}, q_{tot}, p_{IMF},$ $T_{IMF}, Valve_{EGR}$	$MFB50 = 0.5125 \times N^{0.0335} \times p_f^{-0.0241} \times SOI_{main}^{1.0036} \times q_{tot}^{0.0102} \times p_{IMF}^{0.0230}$ $\times T_{IMF}^{0.0965} \times Valve_{EGR}^{-0.0038}$	0.974, 0.633, 0.17%
PFP	$N, p_f, SOI_{main}, q_{tot}, p_{IMF},$ $T_{IMF}, \rho_{SOI,main}$	$PFP = 1.125 \times 10^6 \times N^{-0.386}$ $\times p_f^{0.275} \times (365 - SOI_{main})^{0.147} \times q_{tot}^{0.1625} \times p_{IMF}^{1.116} \times T_{IMF}^{-1.341} \times \rho_{SOI,main}^{-0.610}$	0.983, 3.03, 3.63%
BMEP	$N, p_f, SOI_{main}, q_{tot}, p_{IMF},$ $T_{IMF}, Valve_{EGR}, \rho_{SOI,main}$	$BMEP = 5.7727 \times 10^6 \quad \times N^{-0.2182} \times p_f^{0.2620} \times SOI_{main}^{-3.2588} \times q_{tot}^{1.1765}$ $\times p_{IMF}^{-0.5545} \times T_{IMF}^{0.0131} \times Valve_{EGR}^{0.0304} \times \rho_{SOI,main}^{0.3577}$	0.996, 0.257, 14.9%

5.3. ANN-Based Models

Artificial neural networks have a great capacity to capture complex nonlinear relationships between variables through relatively simple mathematical operations, and require very little computational effort. Therefore, they are good candidates for control-oriented applications. In this section, ANNs have been used to:

- Replace the power law-based correlations of the physics-based models shown in Table 4. The resulting new physics-based model is referred to as "ANN physics-based model".
- Directly simulate the MFB50, PFP, and BMEP combustion metrics, analogously to the semi-empirical models shown in Table 5.

In both cases, the sets of input variables that were used for the ANNs are the same as those shown in Table 4 (physics-based model) and Table 5 (semi-empirical models). The experimental data used for ANN training is constituted by the 410 tests shown in Figure 2.

In general, the implementation of ANNs involves three steps: the selection of the input variables, the training phase, and the testing phase. The available experimental dataset is usually split randomly for training, testing and validation purposes. The main parameters that have to be selected are the net type, the number of inner layers and the number of neurons for each layer, the activation function, and the training algorithm.

As far as the network type is concerned, feed-forward networks are typically used in the literature, and have also been used in this paper. These types of networks contain an input layer, some hidden layers, and an output layer. The back-propagation training algorithm is generally used as a training algorithm for feed-forward neural networks [46]. The Bayesian regularization training function 'trainbr' is one of the best-known BP training algorithms; it can lead to very precise ANNs [46,47] and is able to provide good generalization for difficult, small or noisy datasets [48], even though its number of epochs is usually higher. In this work, the training dataset is not large and the network structure is quite simple, therefore a relatively large epoch number should not lead to an unacceptably long computation time. For these reasons, 'trainbr' was selected as the training algorithm.

In order to speed up the learning process for a high number of networks, the inputs are usually normalized before sending them to the input layer. In this work, the 'mapminmax' process function [49] has been used to normalize the inputs, so that all the values fall into the $[-1, 1]$ interval. Similarly, network outputs can also be associated with processing functions [49].

The activation function that was used in this work is the 'tansig' function. It was found, with reference to the number of hidden layers, that one layer was enough to simulate the parameters of the physics-based model and the combustion metrics (MFB50, PFP, BMEP). The use of one hidden layer for feed-forward neural networks is also quite common in the literature [47,50,51], especially when the number of training data is not very high.

The MSE minimization criterion was used for the training process, as suggested in [52]. The ANN performance is evaluated by comparing the network outputs with the experimental data through regression analysis [53]. The values of the mean-squared-error MSE and the determination coefficient R^2 are used to measure the performance of ANNs. Details of the ANN parameters developed on the MATLAB (R2016b, MathWorks, Natick, MA, USA) platform are shown in Table 6.

The optimal number of neurons in the hidden layer depends on the specific application, and it cannot, therefore, be identified a-priori. The ANN predictive performance is influenced to a great extent by the number of neurons. According to the literature, a too large number of neurons may lead to overfitting, while a too small number of neurons may make the ANN too simple to be able to capture the characteristics of a complex system (that is, underfitting [37]). Therefore, a trade-off must be made between the number of hidden layer neurons and the predictive performance of the ANN, and a sensitivity analysis is always needed in order to identify what the optimal number of neurons is for a specific application. In this section, a suitable number of neurons is selected for each of the dependent parameters that have to be estimated (i.e., those reported in Tables 4 and 5).

Table 6. Details of the artificial neural network (ANN) parameters developed on the MATLAB platform.

Parameter	Selected Option
Network type	Feed-forward back-propagation
Hidden layer	Single layer
Process function	mapminmax
Training algorithm	trainbr
Transfer function	tansig
Loss function	MSE
Performance	MSE, R^2

All the available steady-state experimental data (410 tests) were used to generate the ANNs. Overall, 80% of the data was used as training data, and the other 20% was used to test the trained ANNs. Considering the smallness of the training dataset, the training process features some uncertainties, because the initial value of the weights of the neural networks is generated randomly. Thus, when the training process of a given neural network is repeated, its performance may vary to a great extent, even when the same input training dataset is used. Moreover, the experimental dataset for each training process is randomly divided into training and testing datasets with a certain ratio. This can also influence the performance of the trained neural networks when the training process is repeated. However, it has been shown in the literature that these uncertainties decrease considerably, and may even be ignored, when the size of the training dataset is sufficiently large [54], but this was not the case in this study.

Therefore, 200 training and testing repetitions were carried out for each ANN, with random initial weights and random splitting of the experimental dataset into training and testing datasets. The determination coefficient R^2 and the mean square error MSE were used to measure the performance of the ANNs. The mean values of R^2 and MSE for 200 trials were calculated for the training and testing of each ANN.

The trade-off between the number of hidden layer neurons and the performance of the ANN was explored for each ANN that simulated the parameters of the physics-based model (in the "ANN physics-based model") and for each ANN that simulated the MFB50, PFP, and BMEP combustion metrics (in the "direct ANN model"). The results of the sensitivity analysis are shown in Appendix B.

The sensitivity analysis allowed the optimal number of hidden layer neurons to be identified for each parameter. The most suitable number of neurons for each parameter is listed in Table 7, along with the mean values of R^2 and MSE for the training and testing phases. In general, it can be seen that the most suitable neuron number is no higher than 7 in all cases.

Once the sensitivity analysis had been carried out, the best ANN out of the 200 training trials was selected on the basis of the criteria reported in Appendix B for each parameter.

After this stage, the following investigations were conducted

1. With reference to the physics-based model, the power law-based correlations of the calibration parameters were replaced by the corresponding ANNs, and the performance of the resulting model (i.e., the "ANN physics-based model") was evaluated. Table 8 shows a comparison between the accuracy of the ANN-based correlations and of the power-law based ones in the estimation of the parameters of the physics-based model. It can be observed that all the precisions based on ANNs are higher than those based on the power law correlations. ANNs have in fact been proved to have a powerful ability to catch the non-linear characteristics between input and output parameters

2. The performance of the ANNs that were used to directly simulate MFB50, PFP, and BMEP was evaluated

A detailed comparison of the performance of the four developed approaches, for the estimation of the MFB50, PFP, and BMEP combustion metrics, is reported in the next section.

Table 7. The most suitable number of neurons and the corresponding mean values of R^2 and MSE (over 200 training trials) for the ANNs that simulated the parameters of the physics-based model (a) and the combustion metrics (b).

(a) Physics-Based Model Parameters

Dependent Parameter	τ_{pil}	τ_{main}	K_{pil}	$K_{1,main}$	$K_{2,main}$	$Q_{ht,glob}$	$Q_{f,evap}$	Δp_{IMF}	n	n'	PMEP	FMEP
Suitable number of neurons	6	6	6	6	6	4	5	7	5	5	5	4
Mean value of the training and testing R^2	0.977, 0.978	0.981, 0.983	0.889, 0.794	0.975, 0.968	0.939, 0.945	0.998, 0.998	0.914, 0.899	0.992, 0.995	0.977, 0.975	0.989, 0.989	1.00, 1.00	0.988, 0.986
Mean value of the training and testing MSE	0.53, 0.57	0.038, 0.063	0.0026, 0.0034	2×10^{-6}, 2.6×10^{-6}	3.15×10^{-4}, 2.95×10^{-4}	8.1×10^{-5}, 9.6×10^{-5}	7.2×10^{-7}, 9.0×10^{-7}	2.4×10^{-4}, 2.2×10^{-4}	3.6×10^{-6}, 4.3×10^{-6}	2.6×10^{-5}, 2.6×10^{-5}	5.7×10^{-5}, 6.96×10^{-5}	3.9×10^{-3}, 3.9×10^{-3}

(b) Combustion Metrics

Dependent Parameter	MFB50	PFP	BMEP
Suitable number of neurons	5	6	5
Mean value of the training and testing R^2	0.999, 0.999	1.00, 1.00	1.00, 1.00
Mean value of the training and testing MSE	0.030, 0.043	0.27, 0.40	0.0007, 0.0011

Table 8. Comparison of the correlation coefficient of the ANNs and of the power-law empirical functions that estimate the tuning parameters of the physics-based model.

Dependent Parameter	τ_{pil}	τ_{main}	K_{pil}	$K_{1,main}$	$K_{2,main}$	$Q_{f,evap}$	$Q_{ht,glob}$	n	n'	Δp_{IMF}	*PMEP*	*FMEP*
Most suitable number of neurons	6	6	6	6	6	5	4	5	5	7	5	4
The training and testing R^2 for ANNs	0.977, 0.975	0.977, 0.974	0.878, 0.872	0.981, 0.973	0.941, 0.910	0.919, 0.917	0.998, 0.998	0.978, 0.974	0.989, 0.986	0.992, 0.991	1.00, 0.999	0.975, 0.978
R^2 for the power-law based functions	0.908	0.886	0.365	0.833	0.663	0.532	0.99	0.654	0.923	0.967	0.994	0.939

5.4. Comparison of the Four Different Models under Steady-State and Transient Conditions

In this section, the performance of the four different models has been tested under steady-state and transient operating conditions.

First, the performance of the developed models was compared under steady-state conditions. The results are shown in Figure 6.

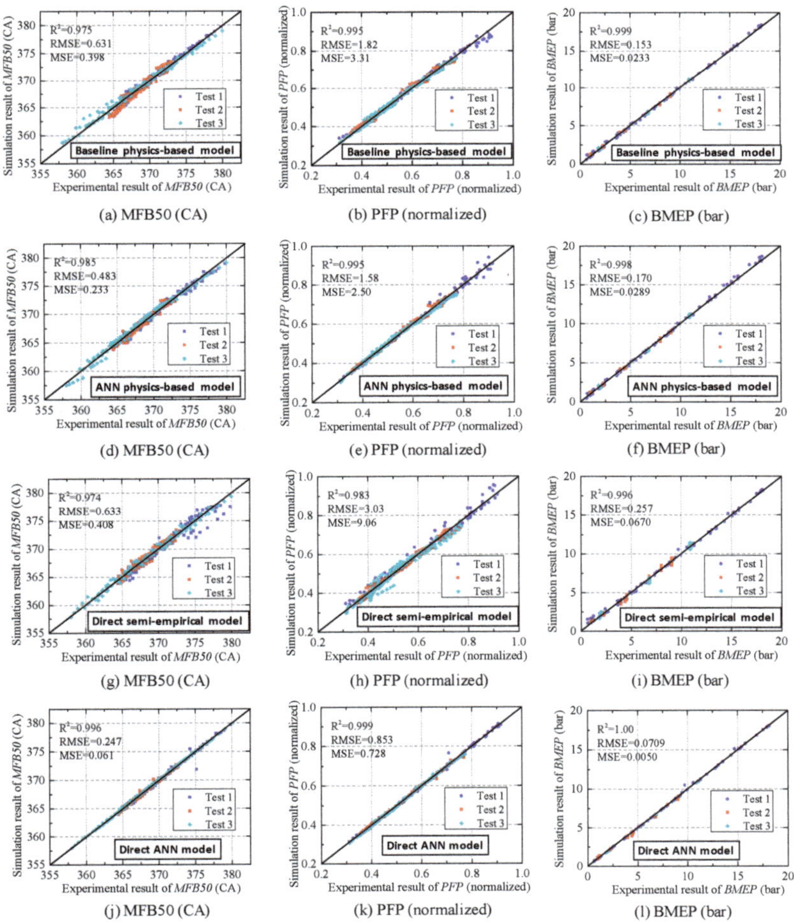

Figure 6. (a–l) Comparison of the performance of the developed models under steady-state operating conditions.

It can be seen, from the previous charts, that the accuracy of the ANN physics-based model is higher than that of the baseline physics-based model concerning MFB50 and PFP. This indicates that the ANNs are more powerful than the power-function-based empirical functions in catching the nonlinear characteristics between the input and dependent parameters of the model for these two metrics under steady-state conditions. Overall, the accuracy of the direct ANN model is the best, while the accuracy of the semi-empirical model is the lowest. However, the absolute values of the error of the semi-empirical model are still acceptable, considering that this kind of model ignores the detailed simulation of the in-cylinder physical process and requires a limited number of free parameters to be tuned (as a consequence, it also requires a limited number of experimental calibration tests). This suggests that the semi-empirical model, based on power functions, is also suitable for control-oriented combustion control applications, since its computational effort is much less than that of the physics-based model.

The performance of the developed models was then evaluated and compared under transient operating conditions over WHTC. The results are shown in Figures 7–9.

Table 9 reports a summary of the accuracy of the models over WHTC and under steady-state operating conditions.

Table 9. Comparison of the performance of the four different models under steady-state and transient operating conditions.

Model Type	Steady-State Condition (R^2 and RMSE)			WHTC Transient Condition (RMSE)		
	MFB50	**PFP**	**BMEP**	**MFB50**	**PFP**	**BMEP**
Baseline physics-based model	0.975 0.63 deg	0.995 1.8 bar	0.998 0.15 bar	1.2 deg	10.5 bar	0.7 bar
ANN physics-based model	0.984 0.48 deg	0.995 1.6 bar	0.998 0.17 bar	1.2 deg	13.8 bar	0.8 bar
Direct semi-empirical model	0.974 0.63 deg	0.983 3.0 bar	0.996 0.26 bar	1.4 deg	11.7 bar	0.8 bar
Direct ANN model	0.996 0.25 deg	0.999 0.85 bar	1.0 0.071 bar	1.1 deg	9.6 bar	0.7 bar

It can be seen from the graphs of Figures 6–9 and from Table 9 that the accuracy of the direct ANN model is the best among all the different approaches, not only for steady-state operation (RMSE equal to 0.25 deg, 0.85 bar and 0.071 bar for MFB50, PFP, and BMEP, respectively), but also for transient operation over WHTC (RMSE of 1.1 deg, 9.6 bar and 0.7 bar for MFB50, PFP, and BMEP, respectively).

The accuracy of the baseline physics-based model under transient operating conditions is slightly lower than that of the direct ANN model in terms of RMSE (1.2 deg vs. 1.1 deg for MFB50, 10.5 bar vs. 9.6 bar for PFP, 0.7 bar vs. 0.7 bar for BMEP), but is higher than that of the ANN physics-based model (1.2 deg vs. 1.2 deg for MFB50, 10.5 bar vs. 13.8 bar for PFP, 0.7 bar vs. 0.8 bar for BMEP) and of the semi-empirical model (1.2 deg vs. 1.4 deg for MFB50, 10.5 bar vs. 11.7 bar for PFP, 0.7 bar vs. 0.8 bar for BMEP).

A remarkable result is the deterioration of the accuracy of the ANN physics-based model (compared to the baseline physics-based model) which occurs for transient operation, while this approach leads to more accurate results for steady-state conditions. Therefore, this approach lacks robustness, compared to the baseline physics-based model.

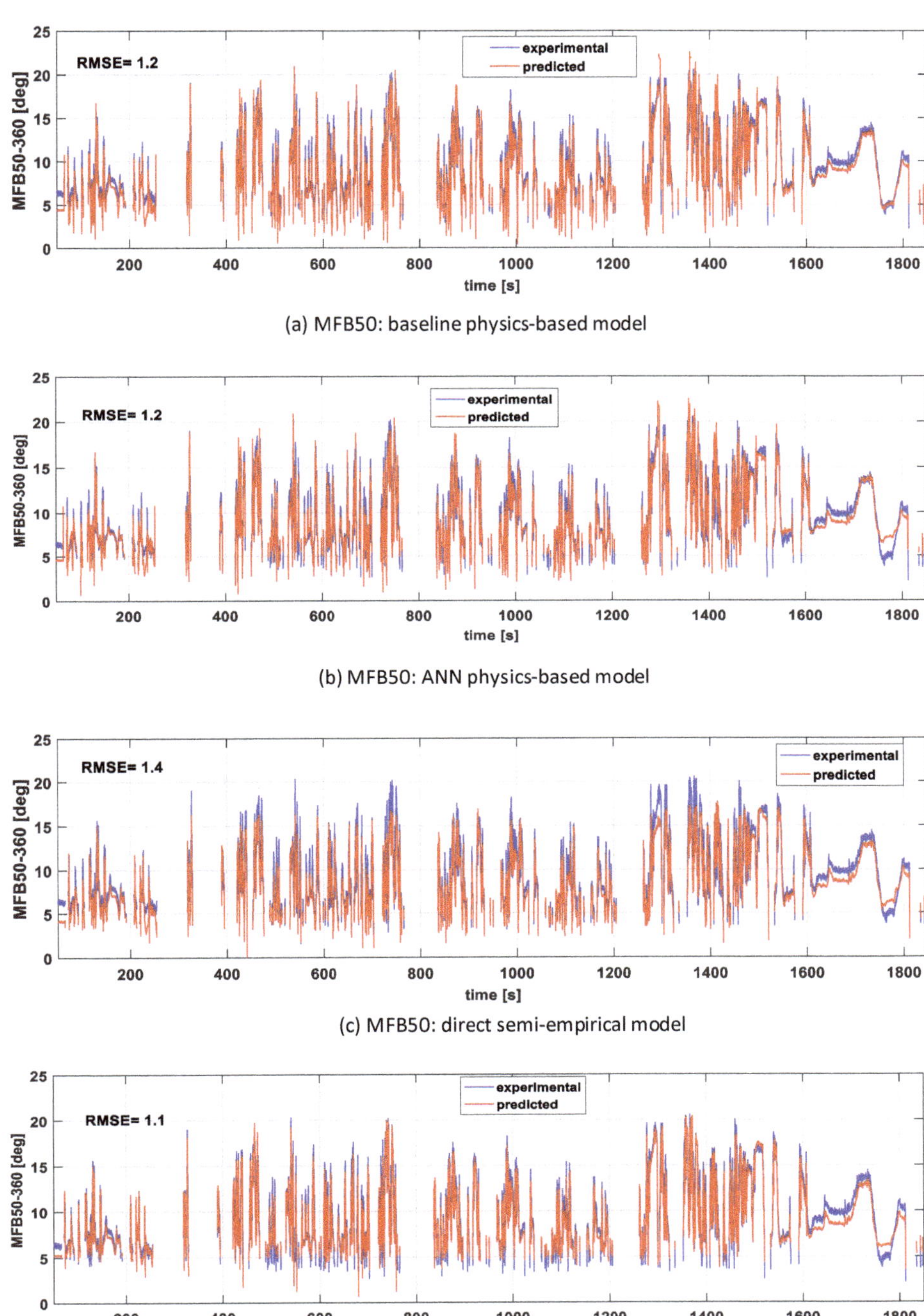

(a) MFB50: baseline physics-based model

(b) MFB50: ANN physics-based model

(c) MFB50: direct semi-empirical model

(d) MFB50: direct ANN model

Figure 7. (**a–d**) Comparison of the performance of the developed models over WHTC: MFB50.

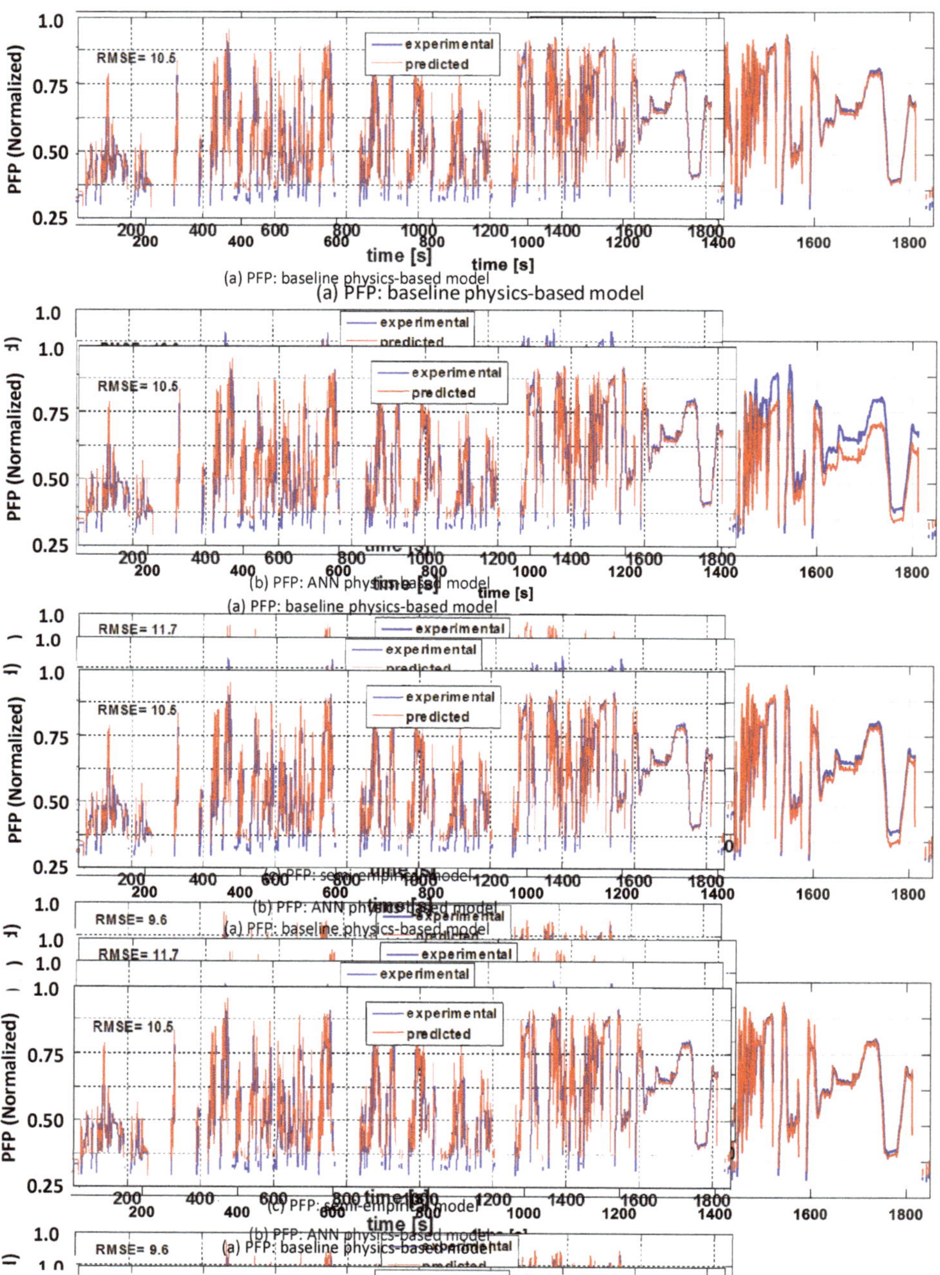

Figure 8. (**a–d**) Comparison of the performance of the developed models over WHTC: PFP.

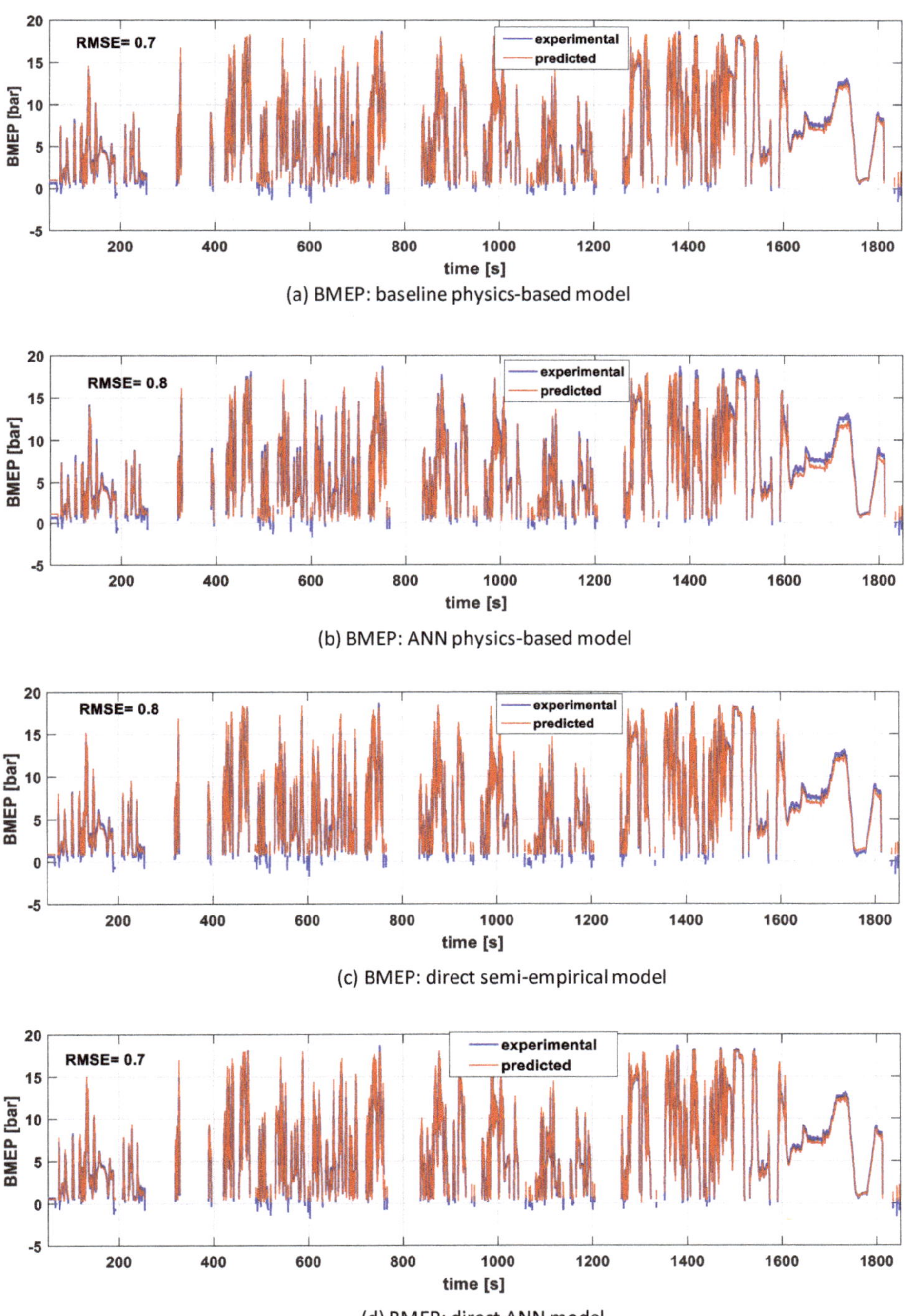

Figure 9. (**a–d**) Comparison of the performance of the developed models over WHTC: BMEP.

Finally, the semi-empirical model shows a lower level of accuracy than the baseline physics-based model and the direct ANN model. However, the accuracy deterioration can still be considered acceptable, considering the simple structure of this model and the low calibration effort that is required, as discussed above.

Overall, the direct ANN model can be considered as the best candidate for control-oriented applications, since:

- it is much simpler in structure and theory and easier to build than the physics-based model;
- it does not require any detailed knowledge or modeling of the physical and chemical processes;
- it is robust not only under steady-state conditions, but also for transient operation;
- its accuracy is high even when a not so large dataset of experimental tests (410) is used for training;
- it features the best accuracy under steady-state and transient operating conditions;
- the required computational time is less than that of the physics-based model (see the next section).

The main drawback of ANNs is that they are a black-box type models, so they do not provide detailed information about the physical and chemical processes which occur inside the combustion chamber (e.g., heat release trend, in-cylinder pressure, temperatures …).

5.5. Analysis of the Computational Time

The computational time of all the models (excluding the ANN physics-based model, which was found to be the least robust approach) was estimated when they were run on the ETAS ES910 device. The results are reported in Table 10.

Table 10. Comparison of the computational time for the different models.

Model	Calculated Quantity	Average Computational Time, per Iteration, on the ETAS ES910 Device. The Reported Computational Time Values are Cumulative.
Baseline physics-based model	MFB50	≈200 µs
	PFP	≈350 µs
	BMEP	≈350 µs
Direct semi-empirical model	MFB50	<50 µs
	PFP	<50 µs
	BMEP	<50 µs
Direct ANN model	MFB50	<50 µs
	PFP	<50 µs
	BMEP	<50 µs

It can be seen that the direct ANN direct model and the semi-empirical model feature the lowest computational times, so that they can be considered the most suitable approaches to be implemented on an ECU for onboard control. The physics-based model requires the highest computational time. However, it is also compatible with real-time combustion control applications, since it is much lower than the typical engine cycle time. On the basis of the results reported in this study, a model-based combustion controller will be developed in the near future, and will be tested on the engine through the ETAS ES910 rapid prototyping device.

6. Conclusions

A comparison of the performance of physics-based, semi-empirical and artificial neural network (ANN)-based models has been carried out in this paper to estimate the main combustion metrics for an FPT Euro VI 3.0 L F1C diesel engine. The models were developed for combustion control-oriented applications. The combustion metrics that are predicted by the models are MFB50, PFP, and BMEP. These metrics are interesting for control-oriented applications, since they are closely related to the thermal efficiency and to the pollutant formation process of the engine. Four different types of modeling approaches have been investigated. The first approach is based on a previously developed low-throughput mean-value physics-based model, which is capable of simulating the heat release rate, the in-cylinder pressure, and the related metrics on the basis of the accumulated fuel mass principle. In this approach, the tuning parameters of the model are estimated on the basis of physically-consistent correlations, which are modeled using power-law functions. The second investigated approach is based on the joint use of artificial neural networks (ANNs) and physical modeling: instead of using the power-law functions, ANNs have been used to identify the parameters of the physics-based model. The third investigated approach is based on a direct estimation of MFB50, PFP, and BMEP on the basis of semi-empirical correlations. The fourth approach is based on the use of ANNs to directly estimate MFB50, PFP, and BMEP. The performance of the four models was evaluated under steady-state conditions and transient operation over WHTC.

Moreover, a new methodology, which is based on the sequential use of the Pearson correlation and partial correlation analysis, has been developed to identify the optimal set of input model parameters in order to a priori identify the correlation variables that should be excluded and the most robust correlation variables.

The main results can be summarized as follows:

(1) The new methodology set up for the identification of the model input parameters has allowed an improvement to be obtained in the calibration of the baseline physics-based model. Although the accuracy in the estimation of BMEP is quite similar, an improvement in the estimation of PFP, and MFB50 can be obtained, since the related RMSE values decrease from 2.47 bar to 1.82 bar and from 0.86 deg to 0.63 deg, respectively, at steady-state conditions. Moreover, it allows a saving in the model calibration computational time to be achieved.

(2) The direct ANN model has shown the best accuracy in the estimation of MFB50, PFP, and BMEP under both steady-state conditions (RMSE = 0.25 deg, 0.85 bar and 0.071 bar, respectively) and transient operation over the WHTC (RMSE = 1.1 deg, 9.6 bar and 0.7 bar, respectively); the accuracy of the baseline physics-based model is slightly worse than that of the direct ANN model (RMSE = 0.63 deg, 1.8 bar, 0.15 bar under steady-state conditions, RMSE = 1.2 deg, 10.5 bar, 0.8 bar over WHTC).

(3) The accuracy of the ANN physics-based model is higher than that of the baseline physics-based model under steady-state operations (RMSE = 0.48 deg, 1.6 bar, 0.17 bar), but it shows a marked deterioration for transient operation (RMSE = 1.2 deg, 13.8 bar, 0.8 bar), and this approach, therefore, seems to lack robustness.

(4) The accuracy of the semi-empirical model is worse than that of the physics-based model and of the direct ANN model under steady-state conditions (RMSE = 0.63 deg, 3 bar, 0.26 bar) and over the WHTC (RMSE = 1.4 deg, 11.7 bar, 0.8 bar). However, the accuracy can still be considered acceptable, in view of the simple mathematical structure of this kind of model, and the low number of tuning parameters that are necessary (and therefore of experimental data needed for calibration).

(5) The computational time required for the direct ANN model and semi-empirical model is less than 50 µs on an ETAS ES910 rapid prototyping device, while that of the baseline physics-based model is of the order of 350 µs.

Overall, the direct ANN model can be considered the best candidate for control-oriented applications, since:

- it is much simpler in structure and theory and easier to build than the physics-based model;
- it does not require any detailed knowledge or modeling of the physical and chemical processes;

- it is robust not only under steady-state conditions, but also in transient operation;
- its accuracy is high, even when a not so large dataset of experimental tests (410) is used for training;
- it features the best accuracy under steady-state and transient operating conditions
- the required computational time is lower than that of the physics-based model.

The main drawback of this kind of model is that it is of the black-box type. Thus, it does not provide detailed information about the physical and chemical processes that occur inside the combustion chamber, such as the heat release trend, in-cylinder pressure, or temperatures.

Author Contributions: The authors contributed equally to the preparation of the paper. Conceptualization, S.H., R.F., S.d.; Methodology, S.H., R.F., S.d.; Software, S.H.; Formal Analysis, S.H., R.F., S.d.; Data Curation, A.M. (Andrea Manelli), L.V., S.d.; Writing-Original Draft, S.H., R.F.; Writing-Review and Editing, R.F., M.M., A.M. (Antonio Mittica); Supervision, R.F., S.d., A.M. (Antonio Mittica), H.W., Y.W.

Funding: FPT is kindly acknowledged for its support in the activities. This work was also supported by the China Scholarship Council (Grant No. 201706680011) and Marine Low-Speed Engine Project-Phase I (CDG01-KT0302). The authors would like to thank Omar Marello for his efforts in discussing and carrying out this research.

Conflicts of Interest: The authors declare no conflict of interest.

Abbreviations

ABDC	After Bottom Dead Center
AF_{st}	stoichiometric air-to-fuel ratio
ANN	artificial neural network
BMEP	Brake Mean Effective Pressure (bar)
CFD	Computer Fluid-Dynamics
DoE	Design of Experiment
ECU	Engine Control Unit
EGR	Exhaust Gas Recirculation
EOC	End of Combustion
EOI	End of Injection
EVO	Exhaust Valve Opening
FMEP	Friction Mean Effective Pressure (bar)
H_L	lower heating value of the fuel
HCCI	Homogeneous Charge Compression Ignition
ICE	Internal Combustion Engines
IMEP360	gross Indicated Mean Effective Pressure (bar)
IMEP720	net Indicated Mean Effective Pressure (bar)
IVC	Intake Valve Closing
K	combustion rate coefficient
m	mass
m_{air}	trapped air mass
m_{EGR}	trapped EGR mass
$m_{f,inj}$	injected fuel mass (mg per cycle/cylinder)
$\dot{m}_{f,inj}$	fuel injection rate
MFB50	crank angle at which 50% of the fuel mass fraction has burned (deg)
n	polytropic coefficient for the compression phase
n'	polytropic coefficient for the expansion phase
N	engine rotational speed (1/min)
O_2	intake charge oxygen concentration (%)
p	pressure (bar)
PCCI	Premixed Charge Compression Ignition
p_{EMF}	exhaust manifold pressure (bar abs)
p_f	injection pressure (bar)
PFP	peak firing pressure

p_{IMF}	intake manifold pressure (bar abs)
PMEP	Pumping Mean Effective Pressure (bar)
q	injected fuel volume quantity (mm^3)
Q_{ch}	chemical heat release
$Q_{f,evap}$	fuel evaporation heat
q_{main}	total injected fuel volume quantity of the main pulse per cycle/cylinder
q_{tot}	total injected fuel volume quantity per cycle/cylinder
$q_{tot,pil}$	total injected fuel volume quantity of the pilot pulses per cycle/cylinder
Q_{fuel}	chemical energy associated with the injected fuel
$Q_{ht,glob}$	global heat transfer of the charge with the walls
Q_{net}	net heat release
R^2	squared correlation coefficient
RAF	relative air-to-fuel ratio
RMS	root mean square
RMSE	root mean square error
SOC	start of combustion
SOI	electric start of injection
SVP	support vector machine
t	time
T	temperature (K)
T_{EMF}	exhaust manifold temperature
T_{IMF}	intake manifold temperature
T_{SOC}	charge temperature at start of combustion
T_{SOI}	charge temperature at start of injection
V	volume
V2X	Vehicle-to-Everything
$Valve_{EGR}$	opening position of the high pressure EGR valve
VGT	Variable Geometry Turbine
WHTC	Worldwide Harmonized Heavy-duty Transient Cycle
$X_{r,EGR}$	EGR rate

Greek Symbols

$\Delta p_{exh\text{-}int}$	difference between the exhaust and intake manifold pressure
γ	isentropic coefficient
ρ	density
ρ_{IMF}	density in the inlet manifold
ρ_{SOC}	charge density at the start of combustion
ρ_{SOI}	charge density at the start of injection
τ	ignition delay coefficient

Appendix A Correlation Analysis for the Selection of the Model Input Variables

In order to make the correlation analysis more synthetic, the considered variables have been associated with indexing numbers (IN), as shown in Table A1.

For the sake of clarity, Figure A1 reports the temporal sequence of the calculation of the different variables used in the physics-based model.

The Pearson correlation coefficient was calculated, using Matlab R2016b, in order to identify the correlations between the variables identified in Table A1. The results are reported in Figure A2.

Table A1. Collected variables and corresponding index number (IN).

Variable Name	IN	Variable Name	IN	Variable Name	IN	Variable Name	IN
N	1	$Valve_{EGR}$	13	$T_{SOC,main}$	25	Q_{netmax}	37
p_f	2	X_{rEGR}	14	SOC_{pil}	26	Δp_{IMF}	38
SOI_{pil}	3	RAF	15	SOC_{main}	27	n	39
SOI_{main}	4	O_2	16	$\Delta p_{exh-int}$	28	n'	40
$q_{tot,pil}$	5	ρ_{IMF}	17	τ_{pil}	29	PFP	41
q_{main}	6	$\rho_{SOI,pil}$	18	τ_{main}	30	IMEP360	42
q_{tot}	7	$\rho_{SOI,main}$	19	K_{pil}	31	PMEP	43
p_{IMF}	8	$\rho_{SOC,pil}$	20	$K_{1,main}$	32	FMEP	44
T_{IMF}	9	$\rho_{SOC,main}$	21	$K_{2,main}$	33	IMEP720	45
p_{EMF}	10	$T_{SOI,pil}$	22	MFB50	34	BMEP	46
T_{EMF}	11	$T_{SOI,main}$	23	$Q_{ht,glob}$	35		
m_{air}	12	$T_{SOC,pil}$	24	$Q_{f,evap}$	36		

The numbers in the rows and columns in Figure A2 are the indexing numbers of the variables shown in Table A1. The background color changes from green to white and from white to red when the Pearson coefficient value varies in the $(-1, 0)$ range or in the $(0, 1)$ range, respectively. The linear relationship between two variables is considered strong when their Pearson coefficient value is in the $(-1.0, -0.5)$ or $(0.5, 1)$ ranges, moderate when it is in the $(-0.5, -0.3)$ or $(0.3, 0.5)$ ranges, weak when it is in the $(-0.3, -0.1)$ or $(0.1, 0.3)$ ranges, and non-existent or very weak when from it is in the $(-0.1, 0.1)$ range [43].

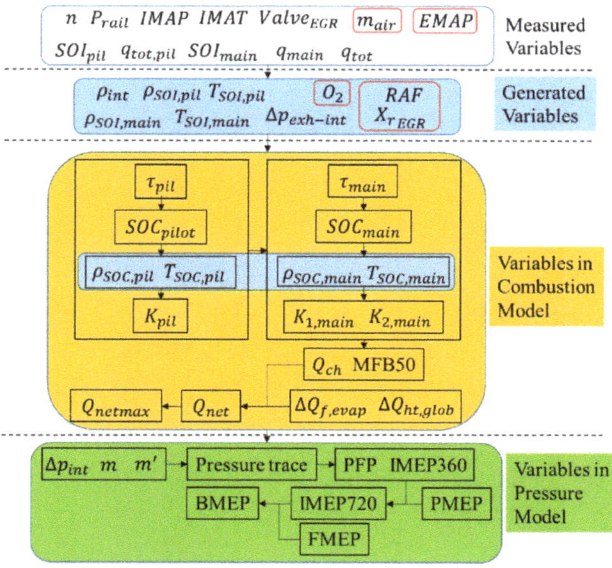

Figure A1. Temporal calculation sequence of the physics-based model variables.

The Pearson coefficient can provide useful information for the selection of:

- the input parameters of the correlations which are used to estimate the calibration parameters of the physics-based models (i.e., those reported in Table 3), or the input parameters of the ANNs that are used to estimate the same calibration parameters;
- the input parameters of the semi-empirical correlations that are used to directly estimate MFB50, PFP, BMEP;
- the input parameters of the ANNs that are used to directly estimate MFB50, PFP, BMEP.

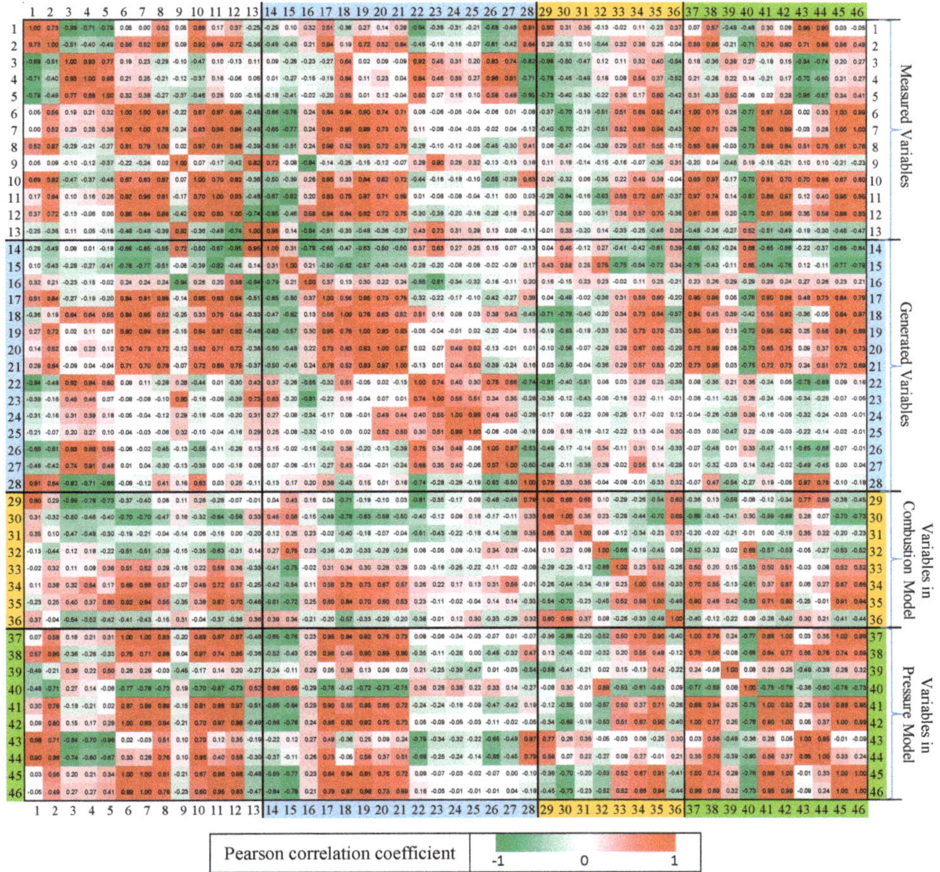

Figure A2. Pearson correlation coefficients for all the possible combinations of the variables reported in Table A1.

As can be seen in Figure A2, the groups of variables [IN 3, IN 4], [IN 6, IN 7], [IN 13, IN 14], [IN 17, IN 19], [IN 20, IN 21], [IN 24, IN 25] and [IN 26, IN 27], namely $[SOI_{pil}, SOI_{main}]$ $[q_{main}, q_{tot}]$, $[Valve_{EGR}, X_{r,EGR}]$, $[p_{IMF}, p_{SOImain}]$, $[p_{SOC,pil}, p_{SOC,main}]$, $[T_{SOC,pil}, T_{SOC,main}]$ and $[SOC_{pil}, SOC_{main}]$, are very closely correlated to each other, since their crossed Pearson coefficient is almost equal to 1. Therefore, during the identification of the input variables of the models, it is sufficient to select only one of them as representative of each variable group. Moreover, it should be noted that the variables which are calculated at an earlier stage are expected to be closely correlated to the variables that are calculated in a subsequent stage (see Figure A1).

The following rules were identified to select the set of input variables to use in the correlations of the physics-based model or in the semi-empirical and ANN-based models, on the basis of the information derived from the Pearson correlation analysis:

- the temporal sequences of the variables in the model have to be taken into account, i.e., variables cannot be used as the independent variables if they are calculated or occur after the dependent variable, according to the scheme in Figure A1;
- the directly measured engine operation variables (namely N, p_f, SOI, fuel injection quantity, p_{IMF}, T_{IMF}, $Valve_{EGR}$) should be considered as independent variable candidates with top priority, because they are the primary cause of the changes in engine operation. Moreover, these variables are usually very precise because they are measured directly in real-time. The generated parameters (ρ_{SOI}, T_{SOI}), based on SOI, should be considered with higher priority than those based on SOC (ρ_{SOC}, T_{SOC}), because of the uncertainty chain for the estimation of SOC on the basis of SOI;
- the dependent variables (such as τ_{pil}, τ_{main}, SOC, K_{pil}, $K_{1,main}$, $K_{2,main}$, $Q_{f,evap}$, $Q_{ht,glob}$, n, n', Δp_{IMF}, PMEP, FMEP) should be considered with a lower priority as correlation parameters, because they are usually less precise than the directly measured parameters. The $X_{r,EGR}$, RAF, O_2, ρ_{SOC} and T_{SOC} generated parameters should also be considered with a lower priority because of their relatively low precision;
- some variable groups (e.g., [SOI_{pil}, SOI_{main}] [q_{main}, q_{tot}], [$Valve_{EGR}$, $X_{r,EGR}$], [p_{IMF}, m_{air}, ρ_{IMF}], [$\rho_{SOC,pil}$, $\rho_{SOC,main}$], [$T_{SOC,pil}$, $T_{SOC,main}$] and [SOC_{pil}, SOC_{main}]) are constituted by two variables which are closely correlated to each other. Therefore, only one variable should be selected as representative of each group;
- the total number of independent variable candidates should not be too high (in this study a maximum of 15 variables was chosen for the estimation of each dependent one), in order to ensure that the partial coefficient order is not too large and to make the model more robust to input uncertainty.

Some considerations should also been made concerning the m_{air} and p_{EMF} variables. Although m_{air} is measured directly, it is usually not a precise or stable variable, since the accuracy of the engine air mass flow sensor is not very high (~5%), and its use, therefore, needs to be evaluated carefully. Moreover, many medium and heavy-duty engines are not generally equipped with air mass flow sensors and the exhaust manifold pressure is generally not measured in production engines, so its use as an input variable for the model should also be evaluated carefully (in this study it was avoided).

On the basis of the rules explained above, an analysis of the Pearson correlation coefficients was carried out to select independent variable candidates for each dependent variable. A summary is reported in Table A2.

It should be recalled that some of the dependent variables are the calibration parameters of the physics-based model (see Table 3), and they need to be modeled through correlations or ANNs, while other dependent variables (i.e., MFB50, PFP, BMEP) are the direct outcome of the semi-empirical model or of the direct ANN model, i.e., the metrics to be used for combustion control applications.

After the previous step, the partial correlation coefficient between each input variable of the initial candidate set reported in Table A2 and the corresponding dependent variable was calculated. The results are reported in Table A3, where the previously estimated Pearson correlation coefficients are also reported. The simultaneous analysis of the Pearson correlation and of the partial correlation coefficients is very useful, since it allows the least and most robust correlation variables to be identified.

Table A2. Initial candidate set of the input variables for each output parameter, identified on the basis of the Pearson correlation analysis.

Dependent Variable	Initial Candidate Set of the Independent Variables Selected Using Pearson Correlation Analysis
RAF	$N, p_f, SOI_{pil}, q_{tot,pil}, q_{tot}, p_{IMF}, T_{IMF}, X_{rEGR}, \rho_{SOI,pil}, \rho_{SOI,main}$
τ_{pil}	$N, p_f, SOI_{pil}, q_{tot,pil}, p_{IMF}, T_{IMF}, Valve_{EGR}, RAF, \rho_{SOI,pil}, T_{SOI,pil}$
τ_{main}	$N, p_f, SOI_{pil}, q_{tot,pil}, q_{tot}, T_{IMF}, m_{air}, X_{rEGR}, RAF, \rho_{SOI,pil}, \rho_{SOI,main}, \rho_{SOC,pil}, T_{SOI,pil}, \Delta p_{exh-int}, \tau_{pil}$
K_{pil}	$N, p_f, SOI_{pil}, q_{tot,pil}, p_{IMF}, T_{IMF}, Valve_{EGR}, \rho_{SOI,pil}, T_{SOI,pil}, \tau_{pil}, \Delta p_{exh-int}$
$K_{1,main}$	$N, p_f, SOI_{main}, q_{tot,pil}, q_{tot}, p_{IMF}, T_{IMF}, Valve_{EGR}, RAF, \rho_{SOI,main}, \rho_{SOC,pil}$
$K_{2,main}$	$N, p_f, SOI_{pil}, q_{tot,pil}, q_{tot}, T_{IMF}, m_{air}, X_{rEGR}, RAF, \rho_{SOI,pil}, \rho_{SOI,main}, T_{SOI,main}, K_{1,main}$
$Q_{ht,glob}$	$N, p_f, SOI_{pil}, q_{tot,pil}, q_{tot}, T_{IMF}, m_{air}, X_{rEGR}, RAF, \rho_{SOI,pil}, \rho_{SOC,pil}, \tau_{main}, K_{2,main}$
$Q_{f,evap}$	$N, p_f, SOI_{pil}, q_{tot,pil}, q_{tot}, T_{IMF}, m_{air}, \rho_{SOI,pil}, \rho_{SOI,main}, RAF, Valve_{EGR}, T_{SOI,pil}, \Delta p_{exh-int}$
Δp_{IMF}	$N, p_f, SOI_{pil}, q_{tot,pil}, q_{main}, p_{IMF}, X_{rEGR}, \rho_{SOI,pil}, \rho_{SOI,main}, T_{SOI,pil}, \rho_{SOC,main}, \Delta p_{exh-int}$
n	$N, p_f, SOI_{pil}, q_{tot,pil}, q_{tot}, p_{IMF}, T_{IMF}, Valve_{EGR}, \rho_{SOI,pil}, \rho_{SOI,main}, T_{SOI,pil}, T_{SOI,main}, T_{SOC,main}, \Delta p_{exh-int}$
n'	$N, p_f, q_{tot,pil}, q_{main}, SOI_{pil}, T_{IMF}, Valve_{EGR}, RAF, \rho_{IMF}, \rho_{SOI,main}, T_{SOI,pil}, \Delta p_{exh-int}, K_{1,main}, K_{2,main}$
$MFB50$	$N, p_f, SOI_{main}, q_{tot,pil}, q_{main}, T_{IMF}, X_{rEGR}, RAF, \rho_{IMF}, \rho_{SOI,pil}, \rho_{SOI,main}, T_{SOI,pil}$
PFP	$N, p_f, q_{tot,pil}, q_{main}, SOI_{main}, T_{IMF}, X_{rEGR}, RAF, \rho_{IMF}, \rho_{SOI,pil}, \rho_{SOI,main}, T_{SOI,main}$
$PMEP$	$N, p_f, q_{tot,pil}, SOI_{pil}, p_{IMF}, T_{IMF}, Valve_{EGR}, \tau_{pil}, T_{SOI,pil}, \Delta p_{exh-int}$
$FMEP$	$N, p_f, SOI_{pil}, q_{tot,pil}, q_{main}, p_{IMF}, T_{IMF}, Valve_{EGR}, \Delta p_{IMF}, PFP, \rho_{SOI,mian}, T_{SOI,pil}, \Delta p_{exh-int}$
$BMEP$	$N, p_f, q_{tot,pil}, q_{tot}, SOI_{main}, T_{IMF}, m_{air}, X_{rEGR}, RAF, \rho_{SOI,pil}, \rho_{SOI,main}$

In fact, input variables that are characterized by low Pearson and partial correlation coefficients can be removed from the candidate list, while input variables which are characterized by a high value of the Pearson correlation and the partial correlation coefficient at the same time can be selected as very robust correlation variables. Particular attention should also be paid to the input variables which, although showing a strong correlation with the dependent variables, are characterized by a limited precision (since they are derived parameters or measured parameters with low accuracy sensors). On the basis of the previous considerations, the least robust correlation variables are highlighted in Table A3 with a gray background, while the highly correlated variables are marked with an orange background, and finally the highly correlated variables characterized by a low precision are marked with a blue background.

It was found that some input variables (i.e., those marked with a white background in Table A3) are characterized by an intermediate value of the correlation coefficients. In order to understand whether they should have been included or excluded from the input candidate list, a sensitivity analysis was carried out, considering all the possible combinations of these variables and estimating the accuracy of the related models, as reported in Section 4.

As reported in the same section, it should be noted that the proposed approach, i.e., the preliminary selection of the input variables through Pearson and partial correlation coefficients, and a subsequent sensitivity analysis carried out on a reduced set of variables, is more computationally efficient than a pure sensitivity analysis approach which includes the entire set of input variables.

Table A3. Pearson and partial correlation coefficients for the input variables of the initial candidate set.

Initial Candidate of the Independent Variables

Dependent Variable: RAF

	N	p_f	SOI_{pil}	$q_{tot,pil}$	q_{tot}	p_{IMF}	T_{IMF}	X_{rEGR}	$\rho_{SOI,pil}$	$\rho_{SOI,main}$
Pearson coefficient	0.10	−0.43	−0.28	−0.41	−0.77	−0.51	−0.08	0.31	−0.62	−0.57
Partial coefficient	−0.46	−0.56	−0.47	−0.48	−0.29	−0.24	0.19	−0.50	0.42	0.21

Dependent Variable: τ_{pil}

	N	p_f	SOI_{pil}	$q_{tot,pil}$	p_{IMF}	T_{IMF}	$Valve_{EGR}$	RAF	$\rho_{SOI,pil}$	$T_{SOI,pil}$
Pearson coefficient	0.80	0.29	−0.89	−0.73	0.06	0.11	−0.01	0.43	−0.71	−0.81
Partial coefficient	0.44	−0.37	−0.13	−0.06	−0.15	0.13	−0.16	0.12	−0.05	0.44

Dependent Variable: τ_{main}

	N	p_f	SOI_{pil}	$q_{tot,pil}$	q_{tot}	T_{IMF}	m_{air}	X_{rEGR}	RAF	$\rho_{SOI,pil}$	$\rho_{SOI,main}$	$\rho_{SOC,pil}$	$T_{SOI,pil}$	$\Delta p_{exh-int}$	τ_{pil}
Pearson coefficient	0.31	−0.32	−0.50	−0.40	−0.70	0.18	−0.58	0.45	0.58	−0.78	−0.63	−0.58	−0.40	0.33	0.68
Partial coefficient	0.26	−0.47	−0.27	0.03	−0.18	−0.01	−0.05	−0.01	−0.13	0.34	−0.29	0.06	0.00	0.08	0.26

Dependent Variable: K_{pil}

	N	p_f	SOI_{pil}	$q_{tot,pil}$	p_{IMF}	T_{IMF}	$Valve_{EGR}$	$\rho_{SOI,pil}$	$T_{SOI,pil}$	τ_{pil}	$\Delta p_{exh-int}$
Pearson coefficient	0.35	0.10	−0.47	−0.30	−0.04	−0.14	0.25	−0.40	−0.51	0.65	0.35
Partial coefficient	−0.45	0.36	0.02	−0.01	0.13	−0.15	−0.07	−0.13	0.10	0.76	−0.16

Dependent Variable: $K_{1,main}$

	N	p_f	$q_{tot,pil}$	q_{tot}	SOI_{main}	p_{IMF}	T_{IMF}	$Valve_{EGR}$	RAF	$\rho_{SOI,main}$	$\rho_{SOC,main}$
Pearson coefficient	−0.13	−0.44	−0.22	−0.51	0.18	−0.39	−0.15	0.14	0.75	−0.33	−0.36
Partial coefficient	−0.34	0.13	−0.20	0.04	−0.44	−0.19	0.15	−0.26	−0.12	−0.49	0.69

Dependent Variable: $K_{2,main}$

	N	p_f	SOI_{pil}	$q_{tot,pil}$	q_{tot}	T_{IMF}	m_{air}	X_{rEGR}	RAF	$\rho_{SOI,pil}$	$\rho_{SOI,main}$	$T_{SOI,main}$	$K_{1,main}$
Pearson coefficient	−0.02	0.32	0.11	0.36	0.52	−0.16	0.36	−0.41	−0.75	0.34	0.30	−0.16	−0.68
Partial coefficient	0.01	−0.06	−0.09	−0.17	−0.20	0.13	−0.06	−0.50	0.21	−0.27	−0.10	−0.27	0.01

Dependent Variable: $Q_{ht,glob}$

	N	p_f	SOI_{pil}	$q_{tot,pil}$	q_{tot}	T_{IMF}	m_{air}	RAF	X_{rEGR}	$\rho_{SOI,pil}$	$\rho_{SOC,pil}$	τ_{main}	$K_{2,main}$
Pearson coefficient	−0.23	0.25	0.40	0.60	0.94	−0.35	0.70	−0.72	−0.61	0.84	0.60	−0.70	0.52
Partial coefficient	0.33	−0.48	0.40	−0.04	0.93	−0.44	−0.08	0.34	0.14	−0.43	−0.24	−0.42	0.09

Dependent Variable: $Q_{f,evap}$

	N	p_f	SOI_{pil}	$q_{tot,pil}$	q_{tot}	T_{IMF}	m_{air}	RAF	$Valve_{EGR}$	$\rho_{SOI,pil}$	$\rho_{SOI,main}$	$T_{SOI,pil}$	$\Delta p_{exh-int}$
Pearson coefficient	0.37	−0.04	−0.54	−0.42	−0.43	0.31	−0.36	0.34	0.36	−0.57	−0.33	−0.38	0.32
Partial coefficient	0.07	−0.14	−0.31	0.04	−0.01	−0.07	−0.27	0.03	−0.03	0.26	0.09	−0.06	0.07

Dependent Variable: Δp_{IMF}

	N	p_f	SOI_{pil}	$q_{tot,pil}$	q_{main}	p_{IMF}	X_{rEGR}	$\rho_{SOI,pil}$	$\rho_{SOI,main}$	$T_{SOI,pil}$	$\rho_{SOC,main}$	$\Delta p_{exh-int}$
Pearson coefficient	0.57	0.86	−0.36	−0.33	0.75	0.98	−0.52	0.45	0.90	−0.35	0.95	0.47
Partial coefficient	−0.09	−0.13	0.06	−0.03	−0.02	0.42	−0.02	−0.09	0.16	−0.12	−0.15	0.18

Dependent Variable: n

	N	p_f	SOI_{pil}	$q_{tot,pil}$	q_{tot}	p_{IMF}	T_{IMF}	$Valve_{EGR}$	$\rho_{SOI,pil}$	$\rho_{SOI,main}$	$T_{SOI,pil}$	$T_{SOI,main}$	$T_{SOC,main}$	$\Delta p_{exh-int}$
Pearson coefficient	−0.49	−0.21	0.39	0.35	0.29	−0.03	−0.45	−0.27	0.39	0.13	0.21	−0.25	−0.47	−0.54
Partial coefficient	−0.45	0.09	−0.15	−0.28	0.24	−0.19	0.19	−0.01	−0.01	0.13	−0.05	0.03	−0.42	−0.16

Dependent Variable: n'

	N	p_f	SOI_{pil}	$q_{tot,pil}$	q_{main}	T_{IMF}	X_{rEGR}	RAF	p_{IMF}	$\rho_{SOI,main}$	$T_{SOI,pil}$	$\Delta p_{exh-int}$	$K_{1,main}$	$K_{2,main}$
Pearson coefficient	−0.48	−0.71	0.27	−0.06	−0.77	0.19	0.68	0.65	−0.76	−0.72	0.36	−0.27	0.69	−0.53
Partial coefficient	−0.69	−0.13	−0.04	−0.24	−0.36	0.05	0.04	−0.04	0.07	0.05	−0.03	0.25	0.52	0.20

Dependent Variable: MFB50

	N	p_f	SOI_{main}	$q_{tot,pil}$	q_{main}	T_{IMF}	X_{rEGR}	RAF	p_{IMF}	$\rho_{SOI,pil}$	$\rho_{SOI,main}$	$T_{SOI,pil}$
Pearson coefficient	0.11	0.38	0.54	0.17	0.69	0.57	−0.42	−0.54	0.58	0.73	0.73	0.26
Partial coefficient	0.76	−0.64	0.84	0.07	0.35	0.13	−0.10	−0.10	0.07	−0.10	0.16	0.13

Dependent Variable: PFP

	N	p_f	SOI_{main}	$q_{tot,pil}$	q_{main}	T_{IMF}	X_{rEGR}	RAF	p_{IMF}	$\rho_{SOI,pil}$	$\rho_{SOI,main}$	$T_{SOI,main}$
Pearson coefficient	0.30	0.76	−0.21	0.02	0.87	0.89	−0.65	−0.64	0.90	0.55	0.85	−0.24
Partial coefficient	−0.68	0.59	−0.69	0.21	0.36	−0.32	−0.10	−0.13	0.35	−0.20	−0.25	0.30

Table A3. *Cont.*

Dependent Variable		Initial Candidate of the Independent Variables												
PMEP	Independent variable	N	p_f	$q_{tot,pil}$	SOI_{pil}	p_{IMF}	T_{IMF}	$Valve_{EGR}$	$T_{SOI,pil}$	τ_{pil}	$\Delta p_{exh-int}$			
	Pearson coefficient	0.96	0.71	−0.86	−0.84	0.51	0.10	−0.19	−0.78	0.77	0.97			
	Partial coefficient	0.57	−0.05	−0.58	0.25	0.32	0.09	0.04	−0.21	−0.05	0.84			
FMEP	Independent variable	N	p_f	SOI_{pil}	$q_{tot,pil}$	q_{main}	p_{IMF}	T_{IMF}	$Valve_{EGR}$	Δp_{IMF}	PFP	$\rho_{SOI,main}$	$T_{SOI,pil}$	$\Delta p_{exh-int}$
	Pearson coefficient	0.90	0.86	−0.74	−0.67	0.33	0.76	0.10	−0.30	0.56	0.56	0.56	−0.68	0.79
	Partial coefficient	0.28	0.08	−0.03	0.18	−0.25	0.15	0.15	−0.18	−0.28	−0.00	0.25	−0.04	−0.07
BMEP	Independent variable	N	p_f	SOI_{main}	$q_{tot,pil}$	q_{tot}	T_{IMF}	m_{air}	X_{rEGR}	RAF		$\rho_{SOI,main}$		
	Pearson coefficient	−0.05	0.49	0.27	0.41	1.00	−0.23	0.83	−0.64	−0.78	0.87	0.88		
	Partial coefficient	−0.78	0.51	−0.64	0.43	0.97	−0.01	−0.07	−0.11	−0.43	0.20	0.61		

Appendix B Training and Selection of the ANNs

Figures A3–A6 report the mean values of R^2 and MSE for 200 training trials of the ANNs that simulate the parameters of the physics-based model (Figures A3–A5) and the combustion metrics (Figure A6).

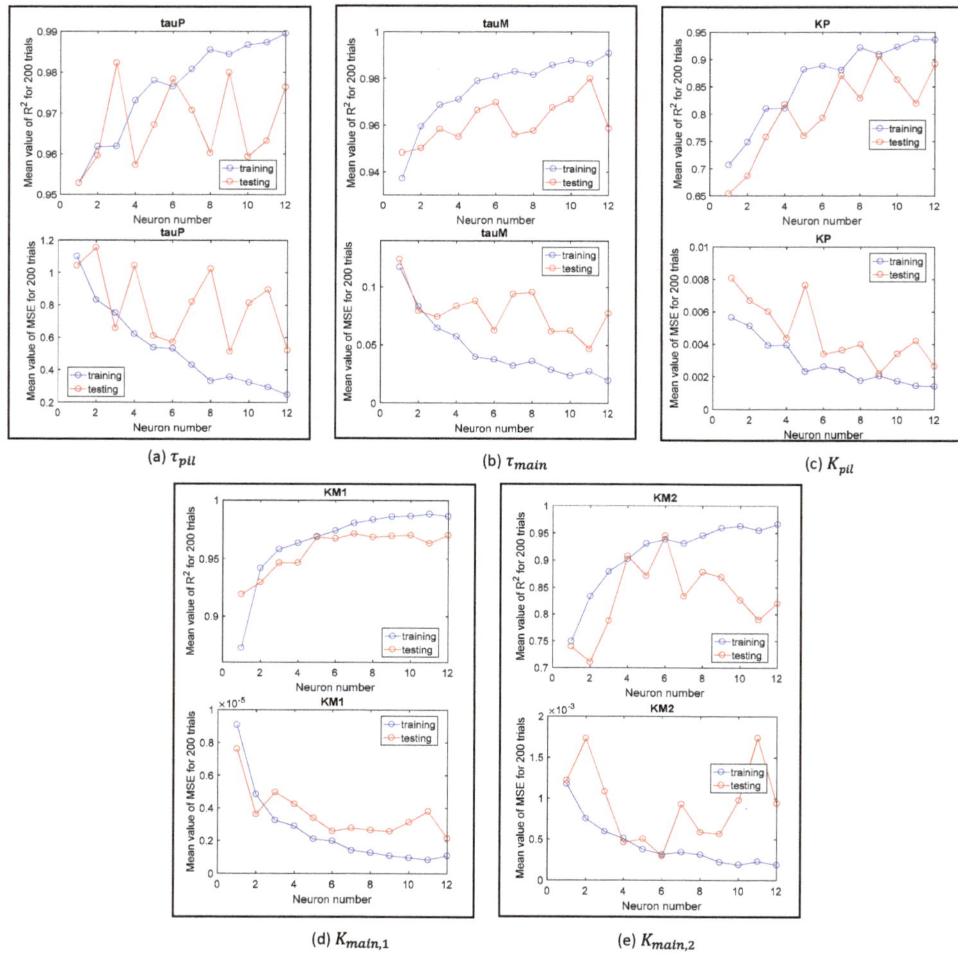

Figure A3. (a–e) Mean value of R^2 and MSE over 200 training trials for the ANNs that simulate the different parameters of the physics-based model (heat release simulation), as a function of the number of neurons.

It can be noted, from the charts, that the training precision is generally higher than the testing precision, for both R^2 and MSE, in all cases. It is obvious that the training precision will always increase when the number of neurons increases, since the neural network is tested over the same dataset as that used for training. Moreover, it can be seen that the training precision increases slightly as the number of neurons increases when the number of neurons is above a certain value. As far as the testing precision is concerned, although some fluctuations occur, a rising trend of R^2 and a decreasing trend of MSE are also in general observed at the beginning when the number of neurons increases, and the trends then tend to become stable or even reverse. Reversing of the trend indicates overfitting. The precision of both the training and testing is quite low when the number of neurons is low, and this means that a too small number of neurons cannot capture the data characteristics accurately (underfitting). An interesting phenomenon can be seen with reference to the trends of the training precision of MFB50, PFP, and BMEP (Figure A6), since they fluctuate less than the other parameters of the physics-based model (Figures A3–A5). This can be ascribed to the fact that the calibration precision of the ANNs that simulate MFB50, PFP, and BMEP is higher than that of the other parameters.

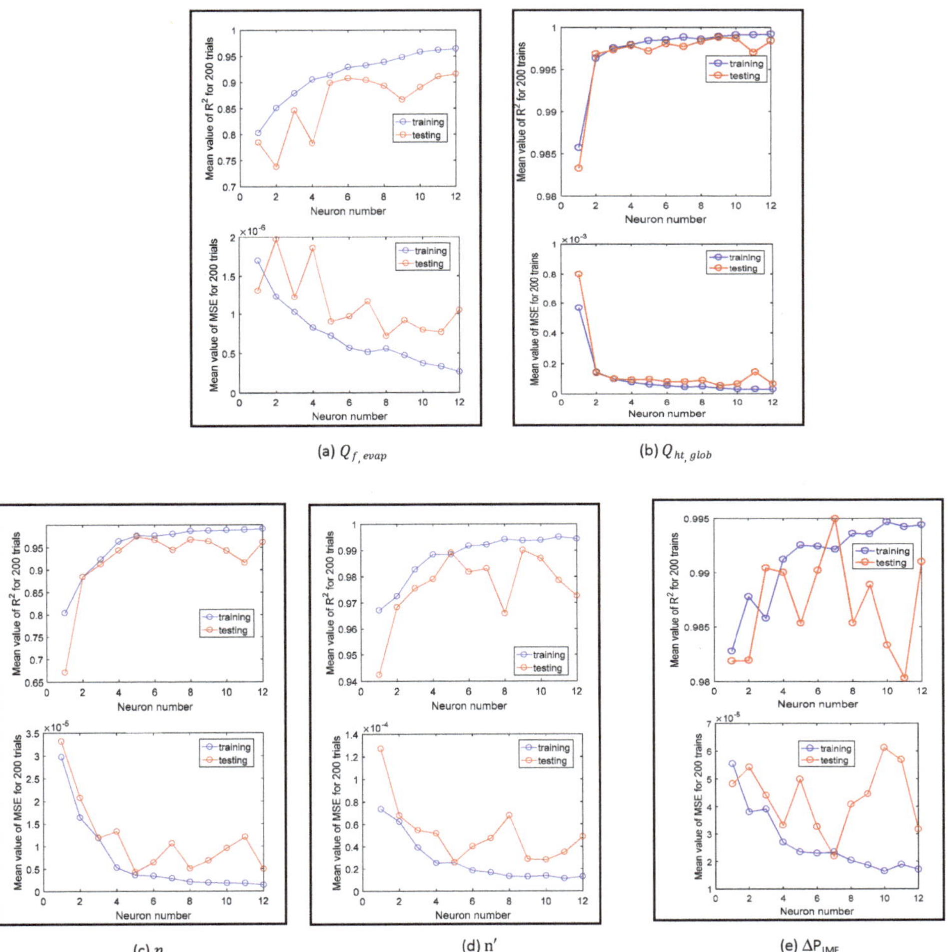

Figure A4. (**a–e**) Mean value of R^2 and MSE over 200 training trials for the ANNs that simulate the different parameters of the physics-based model (pressure simulation), as a function of the number of neurons.

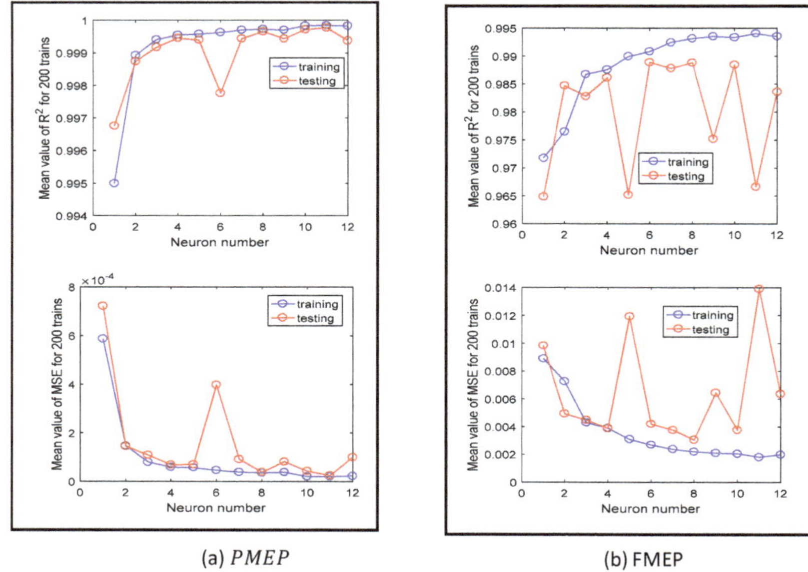

(a) *PMEP* (b) FMEP

Figure A5. (**a**,**b**) Mean value of R^2 and MSE over 200 training trials for the ANNs that simulate the different parameters of the physics-based model (BMEP simulation), as a function of the number of neurons.

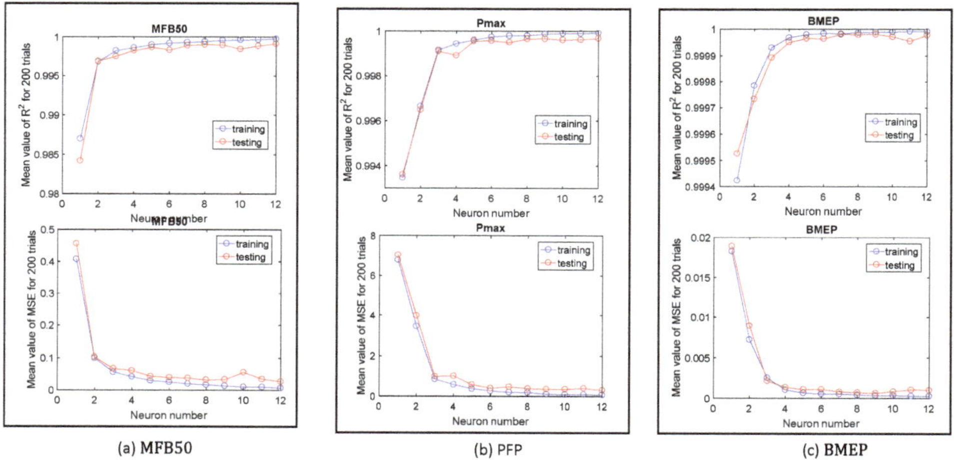

(a) MFB50 (b) PFP (c) BMEP

Figure A6. (**a**–**c**) Mean value of R^2 and MSE over 200 training trials for the ANNs that simulate MFB50 (**a**), PFP (**b**) and BMEP (**c**) combustion metrics, as a function of the number of neurons.

Once the sensitivity analysis had been carried out, the best ANN out of the 200 trials was selected for each parameter. The automatic selection algorithm is shown hereafter (Algorithm A1):

Algorithm A1. Automatic selection

```
    for i = 1:200
if i = 1
ANN_best = ANN_i;
R2_best = [ R2(i,1), R2 (i,2), R2 (i,3)];
MSE_best = [ MSE (i,1), MSE (i,2), MSE (i,3)];
else
if R2(i,3) > R2_best(3) && (R2 (i,2) > R2_best(2) || R2 (i,1) > R2_best(1)) && abs(R2 (i,1)−R2(i,2)) < 0.10*
R2_best(3)
ANN_best = ANN_i;
R2_best = [ R2(i,1), R2 (i,2), R2 (i,3)];
MSE_best = [ MSE (i,1), MSE (i,2), MSE (i,3)];
end
end
end
```

Where, i is the trial number; ANN_best is the best ANN selected by this algorithm; ANN_i is the trained ANN for i[th] trial; R2(i,1), R2 (i,2) and R2 (i,3) are the R^2 precisions for training, testing and total data validation, respectively; MSE (i,1), MSE (i,2) and MSE (i,3) are the MSE precisions for training, testing and total data validation, respectively; abs is the absolute value function.

Three conditions are used to decide whether to update the ANN_best variable for this algorithm, namely R2(i,3) > R2_best(3), (R2 (i,2) > R2_best(2) || R2 (i,1) > R2_best(1)) and abs(R2 (i,1) − R2(i,2)) < 0.10* R2_best(3), respectively. ANN_best is updated only when all the three conditions are true. The first condition, i.e., R2(i,3) > R2_best(3), means that the precision of the total data validation becomes better; the second one, i.e., (R2 (i,2) > R2_best(2) || R2 (i,1) > R2_best(1)), means that the precision of the training data or testing data becomes better; the third one, i.e., abs(R2 (i,1)−R2(i,2)) < 0.10* R2_best(3), means that the difference between the training and testing precisions is smaller than a given level. The latter condition is used to ensure that the testing and training precisions are not so different, and this can prevent the need to select a trained ANN which may present overfitting or underfitting.

Finally the best ANN out of the 200 trials is selected for each parameter, and the fitting results are compared with experimental data, as shown in Figures A7–A10. It can be observed that both the training and testing precisions are quite high (all the R^2 values are higher than 0.85) and there is not much difference between them. This indicates that all the selected ANNs present quite good fitting and prediction performances.

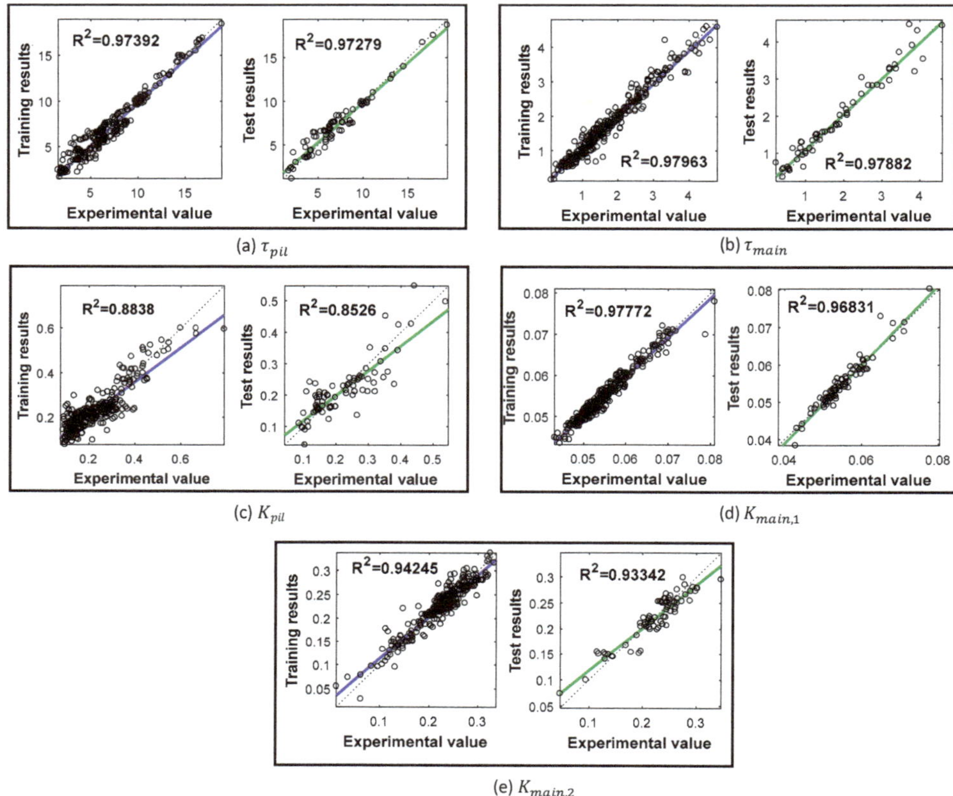

(a) τ_{pil}

(b) τ_{main}

(c) K_{pil}

(d) $K_{main,1}$

(e) $K_{main,2}$

Figure A7. (**a–e**) Comparison of the experimental and ANN fitting results for the physics-based model parameters (heat release simulation).

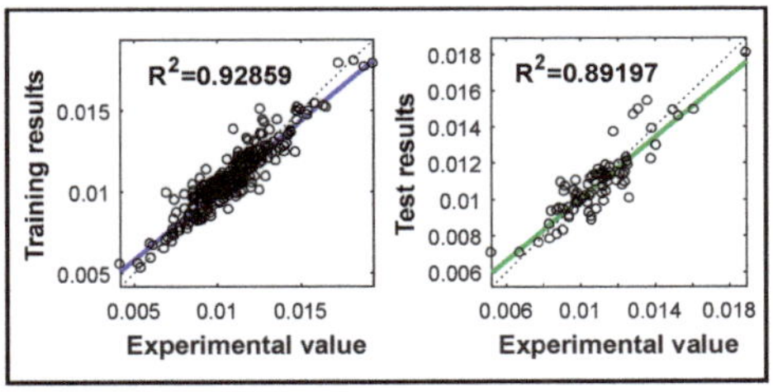

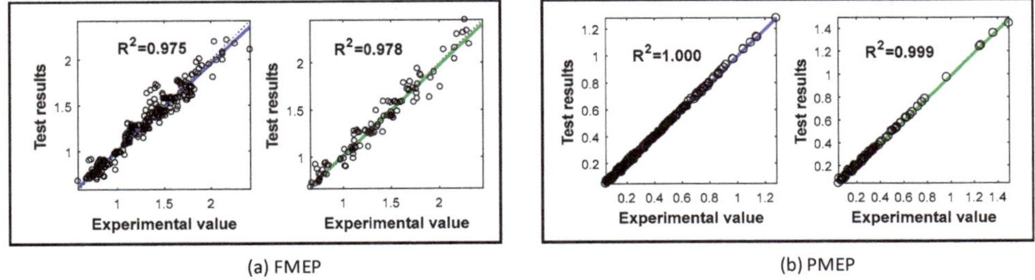

Figure A8. (**a–e**) Comparison of the experimental and ANN fitting results for the physics-based model parameters (pressure simulation).

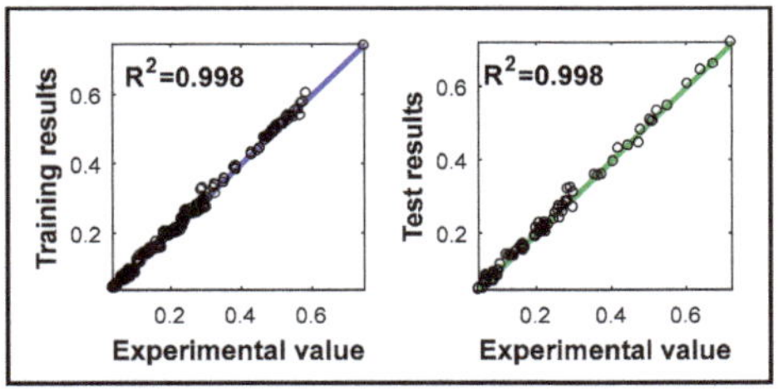

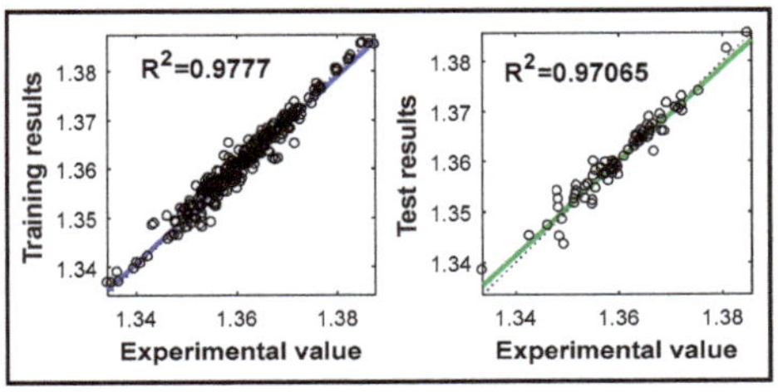

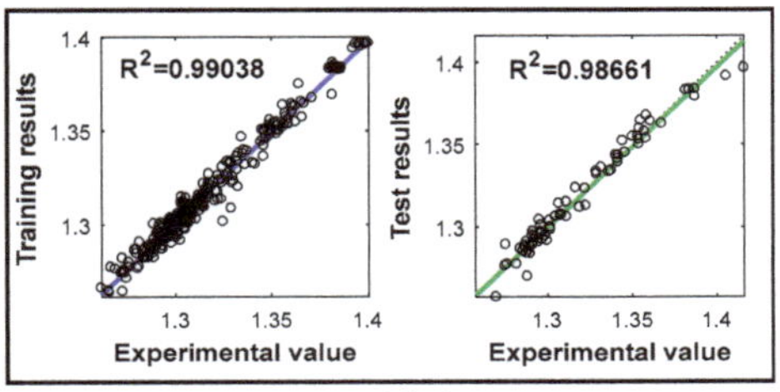

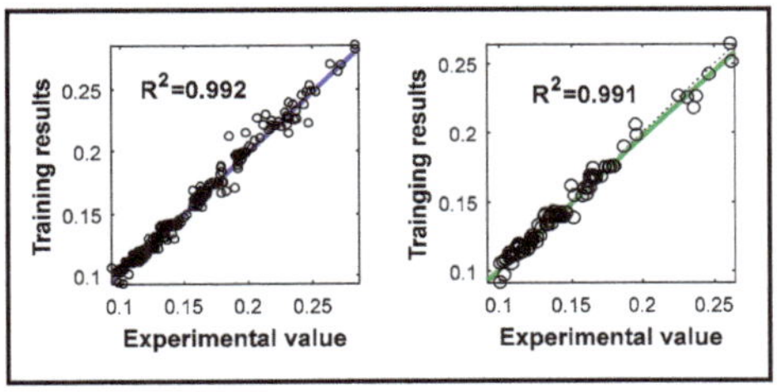

Figure A9. (**a,b**) Comparison of the experimental and ANN fitting results for the physics-based model parameters (BMEP simulation).

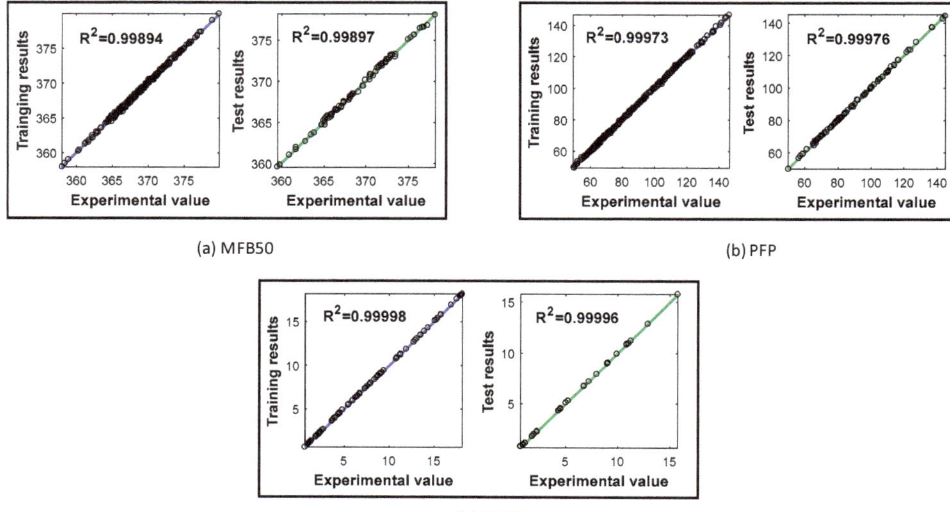

(a) MFB50

(b) PFP

(c) BMEP

Figure A10. Comparison of the experimental and ANN fitting results for the direct estimation of MFB50 (**a**), PFP (**b**) and BMEP (**c**).

References

1. Payri, F.; Olmeda, P.; Martín, J.; García, A. A complete 0D thermodynamic predictive model for direct injection diesel engines. *Appl. Energy* **2011**, *88*, 4632–4641. [CrossRef]
2. Maroteaux, F.; Saad, C.; Aubertin, F. Development and validation of double and single Wiebe function for multi-injection mode Diesel engine combustion modelling for hardware-in-the-loop applications. *Energy Convers. Manag.* **2015**, *105*, 630–641. [CrossRef]
3. Hu, S.; Wang, H.; Niu, X.; Li, X.; Wang, Y. Automatic calibration algorithm of 0-D combustion model applied to DICI diesel engine. *Appl. Therm. Eng.* **2018**, *130*, 331–342. [CrossRef]
4. Ferrari, A.; Mittica, A.; Pizzo, P.; Jin, Z. PID Controller Modelling and Optimization in Cr Systems with Standard and Reduced Accumulators. *Int. J. Automot. Technol.* **2018**, *19*, 771–781. [CrossRef]
5. Ferrari, A.; Mittica, A.; Pizzo, P.; Wu, X.; Zhou, H. New methodology for the identification of the leakage paths and guidelines for the design of common rail injectors with reduced leakage. *J. Eng. Gas Turbines Power* **2018**, *140*, 022801. [CrossRef]
6. Ferrari, A.; Mittica, A.; Paolicelli, F.; Pizzo, P. Hydraulic Characterization of Solenoid-actuated Injectors for Diesel Engine Common Rail Systems. *Energy Procedia* **2016**, *101*, 878–885. [CrossRef]
7. Catania, A.E.; Ferrari, A.; Mittica, A.; Spessa, E. Common Rail without Accumulator: Development, Theoretical-Experimental Analysis and Performance Enhancement at DI-HCCI Level of a New Generation FIS. *SAE Tech. Paper Ser.* **2007**. [CrossRef]
8. Catania, A.; Ferrari, A.; Mittica, A. High-pressure rotary pump performance in multi-jet common rail systems. In Proceedings of the 8th Biennial ASME Conference on Engineering Systems Design and Analysis; ESDA2006, Engineering Systems Design and Analysis, Fatigue and Fracture, Heat Transfer, Internal Combustion Engines, Manufacturing, and Technology and Society, Torino, Italy, 4–7 July 2006; Volume 4, pp. 557–565. [CrossRef]
9. Baratta, M.; Finesso, R.; Misul, D.; Spessa, E. Comparison between Internal and External EGR Performance on a heavy Duty Diesel Engine by Means of a Refined 1D Fluid-Dynamic Engine Model. *SAE Int. J. Engines* **2015**, *8*, 1977–1992. [CrossRef]

10. D'Ambrosio, S.; Gaia, F.; Iemmolo, D.; Mancarella, A.; Salamone, N.; Vitolo, R.; Hardy, G. Performance and Emission Comparison between a Conventional Euro VI Diesel Engine and an Optimized PCCI Version and Effect of EGR Cooler Fouling on PCCI Combustion. *SAE Tech. Paper Ser.* **2018**. [CrossRef]

11. Finesso, R.; Marello, O.; Misul, D.; Spessa, E.; Violante, M.; Yang, Y.; Hardy, G.; Maier, C. Development and Assessment of Pressure-Based and Model-Based Techniques for the MFB50 Control of a Euro VI 3.0L Diesel Engine. *SAE Int. J. Engines* **2017**, *10*, 1538–1555. [CrossRef]

12. Finesso, R.; Marello, O.; Spessa, E.; Yang, Y.; Hardy, G. Model-Based Control of BMEP and NOx Emissions in a Euro VI 3.0L Diesel Engine. *SAE Int. J. Engines* **2017**, *10*, 2288–2304. [CrossRef]

13. Finesso, R.; Spessa, E.; Yang, Y.; Conte, G.; Merlino, G. *Neural-Network Based Approach for Real-Time Control of BMEP and MFB50 in A Euro 6 Diesel Engine*; SAE Technical Paper 2017-24-0068; SAE International: Warrendale, PA, USA, 2018. [CrossRef]

14. Finesso, R.; Hardy, G.; Mancarella, A.; Marello, O.; Mittica, A.; Spessa, E. Real-Time Simulation of Torque and Nitrogen Oxide Emissions in an 11.0 L Heavy-Duty Diesel Engine for Model-Based Combustion Control. *Energies* **2019**, *12*, 460. [CrossRef]

15. Hu, S.; Wang, H.; Sun, Y.; Wang, Y. Zero-Dimensional Prediction Combustion Modelling of a Turbocharging Diesel Engine. *Trans. CSICE* **2016**, *34*, 311–318. [CrossRef]

16. Catania, A.E.; Finesso, R.; Spessa, E. Predictive zero-dimensional combustion model for DI diesel engine feed-forward control. *Energy Convers. Manag.* **2011**, *52*, 3159–3175. [CrossRef]

17. Finesso, R.; Spessa, E.; Yang, Y. Development and Validation of a Real-Time Model for the Simulation of the Heat Release Rate, In-Cylinder Pressure and Pollutant Emissions in Diesel Engines. *SAE Int. J. Engines* **2016**, *9*, 322–341. [CrossRef]

18. Orthaber, G.C.; Chmela, F.G. *Rate of Heat Release Prediction for Direct Injection Diesel Engines Based on Purely Mixing Controlled Combustion*; SAE Technical Paper 1999-01-0186; SAE International: Warrendale, PA, USA, 2018. [CrossRef]

19. Egnell, R. *A Simple Approach to Studying the Relation between Fuel Rate Heat Release Rate and NO Formation in Diesel Engines*; SAE Technical Paper 1999-01-3548; SAE International: Warrendale, PA, USA, 2018. [CrossRef]

20. Ericson, C.; Westerberg, B. *Modelling Diesel Engine Combustion and NOx Formation for Model Based Control and Simulation of Engine and Exhaust Aftertreatment Systems*; SAE Technical Paper 2006-01-0687; SAE International: Warrendale, PA, USA, 2018. [CrossRef]

21. Finesso, R.; Spessa, E.; Yang, Y.; Alfieri, V.; Conte, G. HRR and MFB50 Estimation in a Euro 6 Diesel Engine by Means of Control-Oriented Predictive Models. *SAE Int. J. Engines* **2015**, *8*, 1055–1068. [CrossRef]

22. Hu, S.; Wang, H.; Yang, C.; Wang, Y. Burnt fraction sensitivity analysis and 0-D modelling of common rail diesel engine using Wiebe function. *Appl. Therm. Eng.* **2017**, *115*, 170–177. [CrossRef]

23. Finesso, R.; Spessa, E.; Yang, Y. Fast estimation of combustion metrics in DI diesel engines for control-oriented applications. *Energy Convers. Manag.* **2016**, *112*, 254–273. [CrossRef]

24. Roy, S.; Ghosh, A.; Das, A.K.; Banerjee, R. A comparative study of GEP and an ANN strategy to model engine performance and emission characteristics of a CRDI assisted single cylinder diesel engine under CNG dual-fuel operation. *J. Nat. Gas Sci. Eng.* **2014**, *21*, 814–828. [CrossRef]

25. Brusca, S.; Lanzafame, R.; Messina, M. *A Combustion Model for ICE by Means of Neural Network*; SAE Technology Paper 2005-01-2110; SAE International: Warrendale, PA, USA, 2005. [CrossRef]

26. Yusri, I.; Majeed, A.A.; Mamat, R.; Ghazali, M.; Awad, O.I.; Azmi, W. A review on the application of response surface method and artificial neural network in engine performance and exhaust emissions characteristics in alternative fuel. *Renew. Sustain. Energy Rev.* **2018**, *90*, 665–686. [CrossRef]

27. Niu, X.; Yang, C.; Wang, H.; Wang, Y. Investigation of ANN and SVM based on limited samples for performance and emissions prediction of a CRDI-assisted marine diesel engine. *Appl. Therm. Eng.* **2017**, *111*, 1353–1364. [CrossRef]

28. Turkson, R.F.; Yan, F.; Ali, M.K.A.; Hu, J. Artificial neural network applications in the calibration of spark-ignition engines: An overview. *Eng. Sci. Technol. Int. J.* **2016**, *19*, 1346–1359. [CrossRef]

29. Yap, W.K.; Karri, V. ANN virtual sensors for emissions prediction and control. *Appl. Energy* **2011**, *88*, 4505–4516. [CrossRef]

30. Shi, Y.; Yu, D.-L.; Tian, Y.; Shi, Y. Air–fuel ratio prediction and NMPC for SI engines with modified Volterra model and RBF network. *Eng. Appl. Artif. Intell.* **2015**, *45*, 313–324. [CrossRef]

31. Kshirsagar, C.M.; Anand, R. Artificial neural network applied forecast on a parametric study of Calophyllum inophyllum methyl ester-diesel engine out responses. *Appl. Energy* **2017**, *189*, 555–567. [CrossRef]
32. Xu, K.; Xie, M.; Tang, L.; Ho, S. Application of neural networks in forecasting engine systems reliability. *Appl. Soft Comput.* **2003**, *2*, 255–268. [CrossRef]
33. Pai, P.S.; Rao, B.S. Artificial Neural Network based prediction of performance and emission characteristics of a variable compression ratio CI engine using WCO as a biodiesel at different injection timings. *Appl. Energy* **2011**, *88*, 2344–2354.
34. Kiani, M.K.D.; Ghobadian, B.; Tavakoli, T.; Nikbakht, A.; Najafi, G. Application of artificial neural networks for the prediction of performance and exhaust emissions in SI engine using ethanol- gasoline blends. *Energy* **2010**, *35*, 65–69. [CrossRef]
35. Javed, S.; Murthy, Y.S.; Baig, R.U.; Rao, D.P. Development of ANN model for prediction of performance and emission characteristics of hydrogen dual fueled diesel engine with Jatropha Methyl Ester biodiesel blends. *J. Nat. Gas Sci. Eng.* **2015**, *26*, 549–557. [CrossRef]
36. Bahri, B.; Shahbakhti, M.; Kannan, K.; Aziz, A.A. Identification of ringing operation for low temperature combustion engines. *Appl. Energy* **2016**, *171*, 142–152. [CrossRef]
37. Lawrence, S.; Giles, C. Overfitting and neural networks: Conjugate gradient and backpropagation. In Proceedings of the IEEE-INNS-ENNS International Joint Conference on Neural Networks. IJCNN 2000. Neural Computing: New Challenges and Perspectives for the New Millennium, Como, Italy, 27 June 2000; Volume 1, pp. 114–119.
38. Finesso, R.; Hardy, G.; Maino, C.; Marello, O.; Spessa, E. A New Control-Oriented Semi-Empirical Approach to Predict Engine-Out NOx Emissions in a Euro VI 3.0 L Diesel Engine. *Energies* **2017**, *10*, 1978. [CrossRef]
39. Heywood, J. *Internal Combustion Engine Fundamentals*; McGraw-Hill Intern: Columbus, OH, USA, 1988.
40. Chen, S.K.; Flynn, P.F. *Development of a Single Cylinder Compression Ignition Research Engine*; SAE Technical SAE Technical Paper 650733; SAE International: Warrendale, PA, USA, 2018. [CrossRef]
41. Catania, A.; Finesso, R.; Spessa, E. *Real-Time Calculation of EGR Rate and Intake Charge Oxygen Concentration for Misfire Detection in Diesel Engines*; SAE Technical SAE Technical Pape r2011-24-0149; SAE International: Warrendale, PA, USA, 2018. [CrossRef]
42. Pearson, K. Contributions to the mathematical theory of evolution. *Philos. Trans. R. Soc. Lond. A* **1894**, *185*, 71–110. [CrossRef]
43. Wilson, L.T. Pearson Product-Moment Correlation. Available online: https://explorable.com/pearson-product-moment-correlation?gid=1586 (accessed on 3 June 2018).
44. Melissa, A.S.; Raghuraj, K.R.; Lakshminarayanan, S. Partial correlation metric based classifier for food product characterization. *J. Food Eng.* **2009**, *90*, 146–152. [CrossRef]
45. Marrelec, G.; Krainik, A.; Duffau, H.; Pélégrini-Issac, M.; Lehéricy, S.; Doyon, J.; Benali, H. Partial correlation for functional brain interactivity investigation in functional MRI. *NeuroImage* **2006**, *32*, 228–237. [CrossRef]
46. Rezaei, J.; Shahbakhti, M.; Bahri, B.; Aziz, A.A. Performance prediction of HCCI engines with oxygenated fuels using artificial neural networks. *Appl. Energy* **2015**, *138*, 460–473. [CrossRef]
47. Najafi, G.; Ghobadian, B.; Tavakoli, T.; Buttsworth, D.; Yusaf, T.; Faizollahnejad, M. Performance and exhaust emissions of a gasoline engine with ethanol blended gasoline fuels using artificial neural network. *Appl. Energy* **2009**, *86*, 630–639. [CrossRef]
48. MATLAB Documentation. *Matlab User Guide*; The MathWorks, Inc.: Natick, MA, USA, 2016.
49. Beale, M.H.; Hagan, M.T.; Demuth, H.B. *Neural Network Toolbox™ User's Guide*; The MathWorks, Inc.: Natick, MA, USA, 2018.
50. Ismail, H.M.; Ng, H.K.; Queck, C.W.; Gan, S. Artificial neural networks modelling of engine-out responses for a light-duty diesel engine fuelled with biodiesel blends. *Appl. Energy* **2012**, *92*, 769–777. [CrossRef]
51. Yusaf, T.F.; Buttsworth, D.; Saleh, K.H.; Yousif, B.; Yousif, B. CNG-diesel engine performance and exhaust emission analysis with the aid of artificial neural network. *Appl. Energy* **2010**, *87*, 1661–1669. [CrossRef]
52. Roy, S.; Banerjee, R.; Bose, P.K. Performance and exhaust emissions prediction of a CRDI assisted single cylinder diesel engine coupled with EGR using artificial neural network. *Appl. Energy* **2014**, *119*, 330–340. [CrossRef]

53. Sayin, C.; Ertunc, H.M.; Hosoz, M.; Kilicaslan, I.; Canakci, M. Performance and exhaust emissions of a gasoline engine using artificial neural network. *Appl. Therm. Eng.* **2007**, *27*, 46–54. [CrossRef]
54. Baum, E.B.; Haussler, D. What Size Net Gives Valid Generalization? *Adv. Neural Inf. Process. Syst.* **1989**, *1*, 81–90. [CrossRef]

Article

Implementation and Assessment of a Model-Based Controller of Torque and Nitrogen Oxide Emissions in an 11 L Heavy-Duty Diesel Engine

Fabio Cococcetta [1], Roberto Finesso [2,*], Gilles Hardy [1], Omar Marello [2] and Ezio Spessa [2]

[1] FPT Motorenforschung AG, Schlossgasse 2, 9320 Arbon, Switzerland; fabio.cococcetta@cnhind.com (F.C.); gilles.hardy@cnhind.com (G.H.)

[2] Department of Energy, Politecnico di Torino, Corso Duca degli Abruzzi 24, 10129 Torino, Italy; omar.marello@polito.it (O.M.); ezio.spessa@polito.it (E.S.)

* Correspondence: roberto.finesso@polito.it; Tel.: +39-011-090-4493

Received: 31 October 2019; Accepted: 6 December 2019; Published: 10 December 2019

Abstract: A previously developed model-based controller of torque and nitrogen oxides emissions has been implemented and assessed on a heavy-duty 11 L FPT prototype Cursor 11 diesel engine. The implementation has been realized by means of a rapid prototyping device, which has allowed the standard functions of the engine control unit to be by-passed. The activity was carried out within the IMPERIUM H2020 EU Project, which is aimed at reducing the consumption of fuel and urea in heavy-duty trucks up to 20%, while maintaining the compliance with the legal emission limits. In particular, the developed controller is able to achieve desired targets of brake mean effective pressure (BMEP) (or brake torque) and engine-out nitrogen oxides emissions. To this aim, the controller adjusts the fuel quantity and the start of injection of the main pulse in real-time. The controller is based on a previously developed low-throughput combustion model, which estimates the heat release rate, the in-cylinder pressure, the BMEP (or torque) and the engine-out nitrogen oxide emissions. The controller has been assessed at both steady-state and transient operations, through rapid prototyping tests at the engine test bench and on the road.

Keywords: torque; nitrogen oxide emissions; model-based control; engines

---

1. Introduction

Fuel consumption reduction has become a key aspect in the mobility sector, in order to reduce operating costs and face the impact of global warming. This is especially true for heavy-duty trucks, which typically perform long haul driving missions.

Recent technologies have been developed to reduce fuel consumption and include engine downsizing [1], alternative fuels [2], innovative combustion concepts [3,4], advanced high-pressure injection systems [5–10], model-based control [11–15], advanced aftertreatment systems [16], kinetic and thermal energy recovery [17], vehicle lightweight [18], electrification of the powertrain [19], and vehicle connectivity [20].

With reference to the previous solutions, model-based control will be a key-technology in the near future. In fact, this technology is suitable to be coupled with the emerging V2X systems, and will also be supported by the implementation of more and more powerful engine control units, as well as by the installation of new sensors on the vehicle.

The V2X communication technology is particularly attractive since it will allow the realization of intelligent transport systems, in order to reduce fuel consumption also on the basis of real traffic information. For example, in [21], the authors proposed a dynamic programming-based optimal speed planning algorithm for heavy-duty vehicles, which embeds a look-ahead function using traffic information. In that study, the authors showed that a significant decrease in fuel consumption can be obtained.

With reference to combustion control, model-based algorithms are attractive since they can be used to realize a real-time optimization of the engine performance. In addition, it should be noted that model-based controllers require a lower calibration effort than standard map-based systems. Model-based combustion control is expected to provide significant advantages for diesel engines, which are typically managed by a large number of control parameters.

Although the market share of diesel vehicles is decreasing in the last year for passenger car applications, it is expected that light-duty and heavy-duty vehicles will still be equipped with diesel engines in the next decades. With reference to the latter aspect, a three-year EU H2020 collaborative research project (IMPERIUM) started in 2016, with the aim of achieving fuel consumption and urea reduction of about 20% in heavy-duty trucks, while guaranteeing the fulfillment of the legal emission limits at the same time. Within the Euro VI regulation, heavy-duty engines are tested on the dynamometer over the WHSC (worldwide harmonized steady cycle) and the WHTC (worldwide harmonized transient cycle). The current legal limits for tailpipe NOx (nitrogen oxide) emissions are equal to 0.40 g/kWh over the WHSC and of 0.46 g/kWh over the WHTC. In addition, vehicle emissions are also measured directly on the road, by means of PEMS (portable emissions measurement system). The conformity factor concerning tailpipe NOx emissions is currently equal to 1.5.

The IMPERIUM consortium includes several partners from the academia and industry (see [13–22]). At the end of the project, three heavy-duty demonstrators were realized, in order to show the benefits achieved by means of the developed technologies. These benefits are realized through:

- The direct optimization of the control strategy for the main powertrain components (e.g., engine, transmission), in order to maximize their performance.
- The development of a model-based global powertrain energy manager supervisor, which optimizes the utilization of the different energy sources depending on the current driving situation.
- The implementation and use of "look-ahead" systems, such as eHorizon, in order to realize long-term optimization strategies.

Within the IMPERIUM project, a real-time model-based controller of BMEP (Brake Mean Effective Pressure) and engine-out NOx emissions was realized by the authors and applied to an 11 L FPT prototype Cursor 11 heavy-duty diesel engine. The controller is capable of controlling BMEP (brake mean effective pressure), or torque, and NOx emissions by acting on the quantity of fuel injected in the combustion chamber and on the start of injection of the main pulse. The baseline version of the controller was presented and tested offline in a previous paper [13].

The present investigation is focused on the implementation of the controller on the same engine through rapid prototyping, and on the subsequent assessment at steady-state conditions and in transient operation by means of tests carried out at the engine test bench and on the road.

2. Experimental Setup and Engine Conditions

The engine adopted for the present activity is an FPT prototype 6-cylinder Cursor 11 diesel engine. The engine features a displacement of 11.1 dm^3 and 4 valves per cylinder. Moreover, it is equipped with a VGT (Variable Geometry Turbine) turbine and a high pressure common rail injection system. The maximum power and torque of the engine are 353 kW and 2300 Nm, respectively.

The rapid prototyping tests on the engine were realized at a highly dynamic test bench in FPT Motorenforschung AG, Arbon.

The schemes and detailed description of the engine and of the test bench, including the main measurement sensors and related accuracy, are reported in [13] and have not been shown here for the sake of brevity.

The developed controller was tested on the engine by using the ETAS ES910 (ETAS, Stuttgart, Germany) rapid prototyping device, which is equipped with a Freescale PowerQUICC™ III MPC8548 processor with 800 MHz clock (double precision floating point unit), and features 512 Mbyte of DDR2 RAM (400 MHz clock) and 64 Mbyte of Flash Memory.

With reference to the development and calibration of the real-time combustion model that is used in the model-based controller, a set of about 2200 steady-state points was used (see [13]), which includes a full engine map carried out with or without pilot injection/EGR (Exhaust Gas Recirculation), sweeps of SOI_{main} (main injection timing) and p_f (injection pressure) at fixed key-points, and sweeps of EGR and VGT position. The details of these tests and of the model calibration can be found in [13]. With reference to the input variables used to assess the combustion model, three cases were considered in that study:

- Case 1: All the input variables were derived from test bench sensors (best case).
- Case 2: The available ECU (Engine Control Unit) variables were used as model inputs, except for the intake O_2 concentration, which was obtained from test bench sensors.
- Case 3: Only the available ECU variables were used as model inputs.

It should be noted that Case 3 is the most significant one, when the controller is implemented onboard.

The experimental tests which are considered in this paper, instead, have been used for the assessment and validation of the controller, and include:

1. Selected steady-state points on the engine map (Figure 1). These tests were used to perform a preliminary check of the controller functionality when implemented in the engine through rapid prototyping. In particular, for these tests, targets of BMEP and NOx for the controller were set according to the baseline engine map. The steady-state points were chosen in order to explore a wide range of the engine map, but keeping at the same time a safety margin from the full-load curve.

2. Load ramp tests carried out at several engine speed conditions, with and without EGR, in which the controller was activated or deactivated. The EGR valve position was controlled in the engine to realize the nominal EGR levels according to a look-up table. For a given ramp test, different NOx targets (nominal, ±20%, −40%) were set when the controller was activated, in order to fully assess its functionality. The summary of the ramp tests is reported in Table 1, while Figure 2 reports the time histories of the accelerator pedal position for the ramp tests carried out under baseline engine operating conditions (i.e., without controller active). As can be seen, two types of load ramp tests were considered, i.e., variations from 0% to 60%–70% of maximum load with intermediate steps, and variations from 0% to 60% of maximum load with different ramp durations.

Table 1. Summary of the load ramp tests.

Ramp Test	Engine Speed	Load (Accelerator Pedal Position)	EGR	Controller Activation	NOx Target When Controller ON
Ramp test 1	N = 800 rpm	0%–60% with intermediate steps	ON/OFF	ON/OFF	Nominal/+20% − 20%/−40%
Ramp test 2	N = 1300 rpm	0%–60% with intermediate steps	ON/OFF	ON/OFF	Nominal/+20% − 20%/−40%
Ramp test 3	N = 1900 rpm	0%–60% with intermediate steps	ON/OFF	ON/OFF	Nominal/+20% − 20%/−40%
Ramp test 4	N = 1100 rpm	0%–60% with different ramp durations	ON/OFF	ON/OFF	Nominal/+20% − 20%/−40%
Ramp test 5	N = 1500 rpm	0%–60% with different ramp durations	ON/OFF	ON/OFF	Nominal/+20% − 20%/−40%

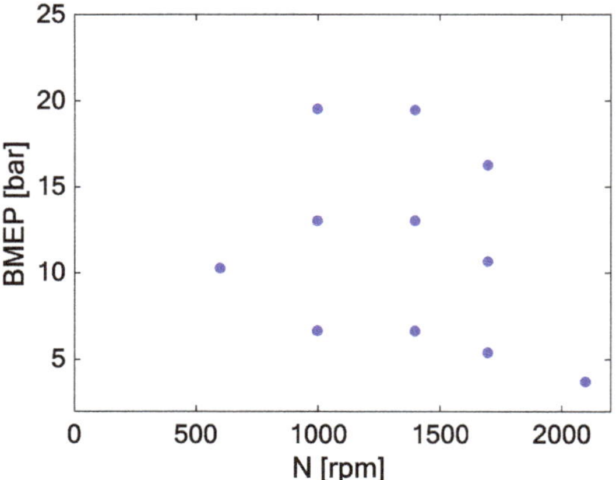

Figure 1. Summary of the steady-state tests used for the preliminary assessment of the model-based controller of BMEP (Brake Mean Effective Pressure) and engine-out NOx (nitrogen oxide) emissions on the engine.

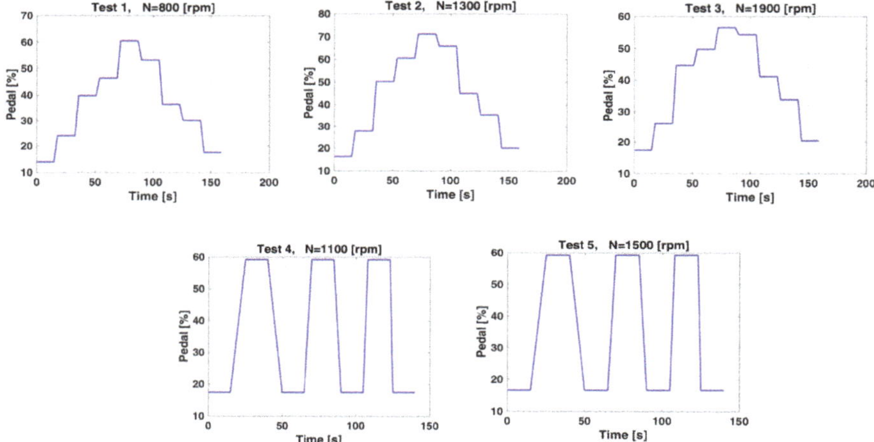

Figure 2. Accelerator pedal position vs. time for the baseline load ramp tests.

Conventional diesel fuel (according to EN 590 regulations) was used, which is characterized by a cetane number equal to 54 and a lower heating value of 43 MJ/kg.

3. Description of the Model-Based Controller of BMEP and NOx

The controller is based on a previously developed real-time combustion model, whose scheme is reported in Figure 3.

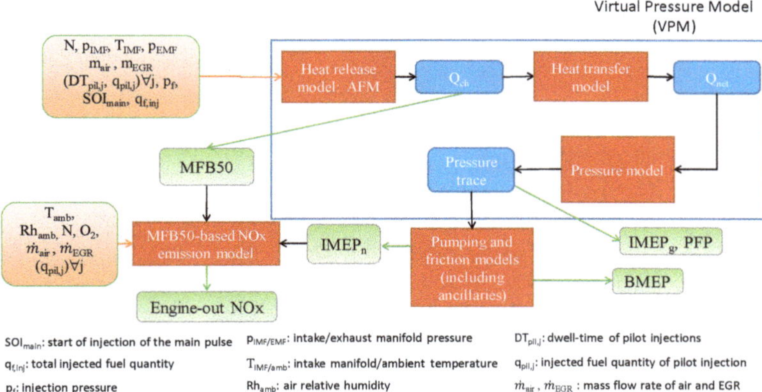

SOI$_{main}$: start of injection of the main pulse $p_{IMF/EMF}$: intake/exhaust manifold pressure   DT$_{pil,j}$: dwell-time of pilot injections

$q_{f,inj}$: total injected fuel quantity $T_{IMF/amb}$: intake manifold/ambient temperature $q_{pil,j}$: injected fuel quantity of pilot injection

p_f: injection pressure Rh_{amb}: air relative humidity $\dot{m}_{air}$, $\dot{m}_{EGR}$: mass flow rate of air and EGR

Figure 3. Scheme of the real-time combustion model.

The combustion model is capable of simulating the following quantities:

1. Released chemical energy (Q_{ch}): The model is based on an improved version of the accumulated fuel mass approach [23–28].

2. Net energy of the charge (Q_{net}): It is calculated as the difference between the released chemical energy and the heat exchanged by the charge with the walls.

3. In-cylinder pressure: A single-zone thermodynamic model [29] is applied, in which the net energy of the charge is used as input in order to calculate the in-cylinder pressure during the combustion phase. The in-cylinder pressure during the compression and expansion phases is obtained assuming polytropic processes. The PFP (peak firing pressure) and IMEP$_g$ (gross indicated mean effective pressure) metrics are then evaluated on the basis of the in-cylinder pressure [25].

4. Pumping and friction losses: A pumping model, which is based on the intake and exhaust manifold pressure levels, is used to estimate PMEP (pumping mean effective pressure), while FMEP (friction mean effective pressure) is evaluated by adopting the Chen–Flynn model [30]. The evaluation of PMEP and FMEP allows IMEP$_n$ (net indicated mean effective pressure) and BMEP (brake mean effective pressure) to be estimated.

5. Engine-out NOx emissions: The approach described in [13,31] was adopted. In particular, NOx emissions are estimated as the sum of two terms, i.e., the nominal NOx level, which is emitted by the engine when it operates with the baseline calibration map, and a NOx deviation term. The latter term is a function of the deviations of the intake oxygen concentration (δO_2) and MFB50 (δMFB50) with respect to the values that occur with the baseline engine calibration map.

The previously described model has been used to develop the model-based controller of BMEP and engine-out NOx emissions. The scheme of the controller is provided in Figure 4.

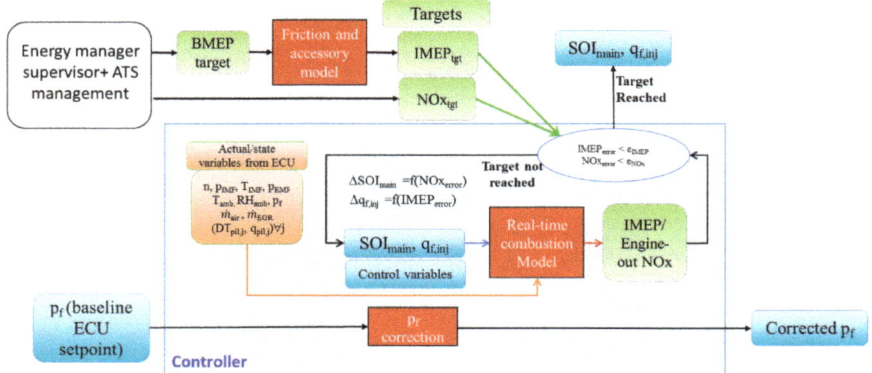

Figure 4. Scheme of the model-based controller of BMEP and engine-out NOx emissions.

The controller is included in a more comprehensive energy manager supervisor of the vehicle, which optimizes the operation of the different energy sources on the basis of the current driving conditions. This supervisor includes a look-ahead function, which exploits a dynamic eHorizon system, in order to realize a long-term optimization strategy. From the scheme reported in Figure 4, it can be seen that the BMEP/NOx controller receives from the vehicle energy supervisor the targets of engine-out NOx emissions (NOx_{tgt}) and indicated mean effective pressure ($IMEP_{tgt}$), and corrects the injected fuel quantity ($q_{f,inj}$) and the start of injection of the main pulse (SOI_{main}) in order to achieve the desired targets. To this aim, an iterative procedure is implemented, in which the correction of SOI_{main} is based on the NOx error, while the correction of $q_{f,inj}$ is based on the IMEP error. The iterative procedure is completed within a single engine firing (120 degrees of crank angle), so that a cycle-by-cycle control is realized. A correction of the injection pressure p_f, with respect to the baseline value which derives from the ECU maps, is also implemented. This correction has been adopted in order to reduce the soot penalties which occur when SOI_{main} is delayed, as a consequence of a low NOx target request.

The core of the controller is a real-time mean value combustion model, that was presented and assessed in [13] for the same engine considered in the present study.

The controller structure is described in detail in [13], in which the correction schemes for SOI_{main} and $q_{f,inj}$ are reported.

Finally, several features were implemented in the controller in order to improve its robustness, safety sensitivity in case of failure in the estimation of the input variables. These features include:

- the adoption of upper/lower saturation for all the input quantities;
- the adoption of upper/lower saturation of SOI_{main} and $q_{f,inj}$, in order to avoid the exceeding of the peak firing pressure and exhaust gas temperature limits, as well as to avoid excessive penalizations in terms of BSFC;
- the adoption of an output buffer for SOI_{main} and $q_{f,inj}$, in order to remove high frequency oscillations deriving from input variable noise.

Moreover, the controller is deactivated when the values of engine speed and injected fuel quantity are below a user-defined threshold. This feature was added in order to improve functionality robustness during the engine start and when a transition occurs from positive torque to cut-off conditions and vice versa.

The developed controller has been implemented on the real engine through rapid prototyping (RP). The rapid prototyping setup is shown in Figure 5.

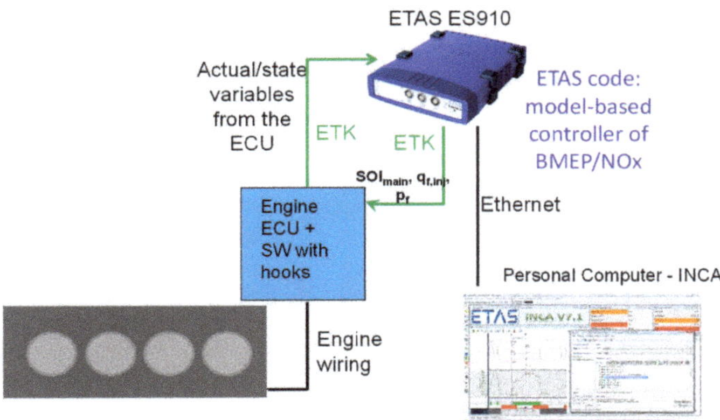

Figure 5. Scheme of the rapid prototyping setup for the BMEP/NOx controller.

It can be seen in the figure that the ETAS RP device receives the needed input signals for the BMEP/NOx controller (i.e., the actual/state variables) from the ECU via ETK, which is an ethernet-based communication interface. The algorithm then evaluates the values of $q_{f,inj}$ and SOI_{main} that allow the desired targets of BMEP and NOx to be achieved (see Figure 3), and writes these values in the ECU via ETK. Therefore, the standard values which derive from the engine maps are by-passed. This has been made possible by the implementation of a modified ECU software, which includes several hooks for the variables to be by-passed. The injection pressure p_f is also sent to the ECU in by-pass, since a correction with respect to the baseline ECU value is applied, in order to compensate for the soot increase that would occur when delaying the start of injection.

4. Results and Discussion

4.1. Real-Time Combustion Model

A detailed analysis of the accuracy of the real-time combustion model which was used in the controller was reported in [13]. Table 2 reports the RMSE (root mean square error) for the prediction of BMEP and NOx at steady-state conditions.

Table 2. Values of RMSE (Root Means Square Error) of BMEP and NOx for the real-time combustion model at steady-state conditions [13].

Input Type	RMSE NOx	RMSE BMEP
Test bench sensors	80 ppm	0.28 bar
Engine sensors except for intake O_2	80 ppm	0.32 bar
Engine sensors	111 ppm	0.32 bar

It can be seen in the table that the BMEP prediction is not affected significantly by the type of input signals, while the NOx prediction accuracy deteriorates when using the ECU signals, as a consequence of a less precise estimation of the EGR rate than that obtained using test bench sensors.

4.2. Controller of BMEP and NOx: Assessment at Steady-State Operation

The controller was first tested at steady-state conditions for the engine points reported in Figure 1. The input variables derived from the engine sensors were used in the controller.

Figure 6 shows, for the different operating conditions, a comparison between the target values (blue bars) and the measured values (orange bars) of NOx emissions (Figure 6a) and BMEP (Figure 6b). The NOx emissions were measured using the test bench gas analyzer. The absolute values of NOx emissions were not reported for confidentiality reasons. The reported points are ordered from low power to high power.

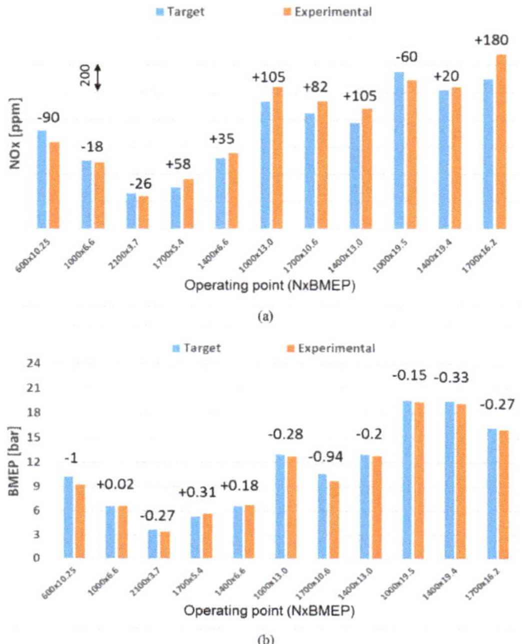

Figure 6. Comparison between the target values and the measured values of NOx emissions (**a**) and BMEP (**b**) for the rapid prototyping tests at steady-state operation. The absolute errors are also reported above each bar.

It can be seen that the errors are in line with the combustion model accuracy shown in Table 2.

The root mean square error of the relative NOx error is of the order of 7.8%. With reference to the operation of the controller, it was verified that the average number of required iterations to achieve convergence is of the order of 6. It should also be noted that a maximum number of 8 iterations is set in the controller, in order to improve robustness and avoid time overrun.

4.3. Controller of BMEP and NOx, Assessment in Transient Operation: NOx Control

The controller has then been assessed over the transient tests reported in Table 1. With reference to the NOx control, considering that several conditions were investigated for each single ramp test (controller 'ON' with nominal target, controller 'ON' with NOx target deviations of ±20% and −40%), the detailed results of only two ramp tests (i.e., ramp test 3 and ramp test 4 in Table 1) have been reported in this section for the sake of brevity. These two ramp tests have been selected since they are characterized by different time histories of the engine load. In particular, with reference to ramp test 3, a load variation from 0% to 75% of the maximum torque was realized at N = 1900 rpm, with intermediate load steps. With reference to ramp test 4, instead, ramp load variations from 0% to 60% of the maximum torque (and vice-versa) were realized at N = 1100 rpm, considering different ramp durations. The time histories of the load are reported in Figure 2.

With reference to ramp test 3, Figure 7 reports the time histories of the measured NOx emissions (Figure 7a,c) and of SOI_{main} (Figure 7b,d) for the cases in which EGR is closed and with nominal EGR level, considering different NOx targets. The absolute values of NOx emissions and of SOI_{main} were not reported for confidentiality reasons. The nominal NOx target levels were identified on the basis of a look-up table, function of engine speed and load, which was derived on the basis of the measured NOx emissions at steady-state operation over the full engine map, with the baseline configuration of the ECU variables. In particular, in each chart four different cases were considered, i.e.:

- Engine operation with controller enabled and nominal NOx target (blue lines).
- Engine operation with controller enabled and NOx target increased of 20% with respect to the nominal one (magenta lines).
- Engine operation with controller enabled and NOx target decreased of 20% with respect to the nominal one (green lines).
- Engine operation with controller enabled and NOx target decreased of 40% with respect to the nominal one (black lines).

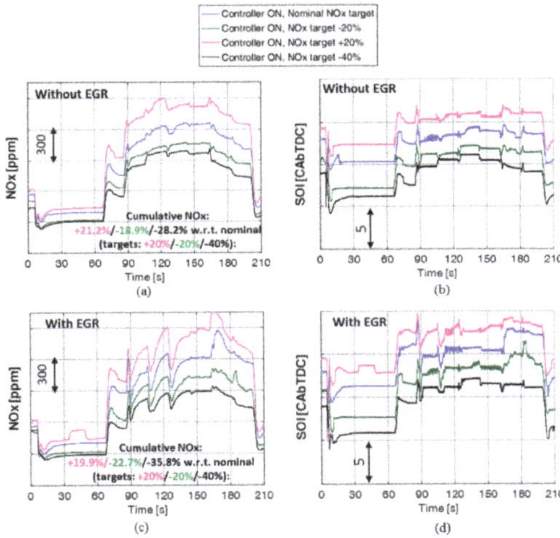

Figure 7. Comparison of the time histories of the measured NOx emissions (**a**,**c**) and of SOI_{main} (**b**,**d**) for the ramp test 3 without EGR (Exhaust Gas Recirculation) (**a**,**b**) and with EGR (**c**,**d**). The origin of the y-axis is the same for all the related figures.

For the cases in which NOx target deviations of +20% and −20%/−40% were set, the percentage difference of the measured cumulative NOx emissions (ppm), with respect to the baseline case with nominal NOx target have also been reported. The NOx trends shown in the charts were measured by means of the engine NOx sensor. In all the charts, the origin of the y-axis is the same for all the related figures.

In general, it can be seen from the charts how the controller progressively advances/delays the values of SOI_{main} in order to increase/decrease the engine-out NOx levels.

It can also be noted that the percentage differences of the cumulative NOx emissions, with respect to the baseline case with nominal NOx target, are in line with the targets, except for the case in which a NOx target of −40% is requested when EGR is closed (Figure 7a). In this case, in fact, the percentage difference of the cumulative NOx emissions is of the order of −28.2%. The main reason for this behavior is due to the fact that, for the considered case, the lower SOI_{main} safety boundary is often achieved. This can be seen in Figure 8, which shows the time histories of SOI_{main} for the ramp test 3 without EGR (a) and with EGR (b). In the figure, the actuated SOI_{main} is indicated with black lines, the reference SOI_{main} from ECU is reported with red lines, and the safety SOI_{main} boundaries are reported with green lines.

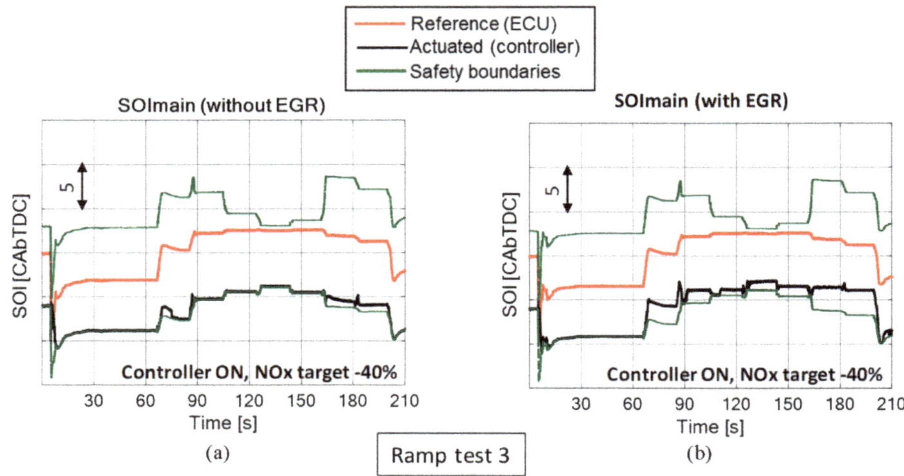

Figure 8. Time histories of SOI_{main} for the ramp test 3 without EGR (**a**) and with EGR (**b**).

It can be seen from Figure 8 that the lower SOI_{main} safety boundary is often achieved, especially for the case in which EGR is not used (Figure 9a). As a consequence, in these conditions, the desired NOx target cannot be achieved.

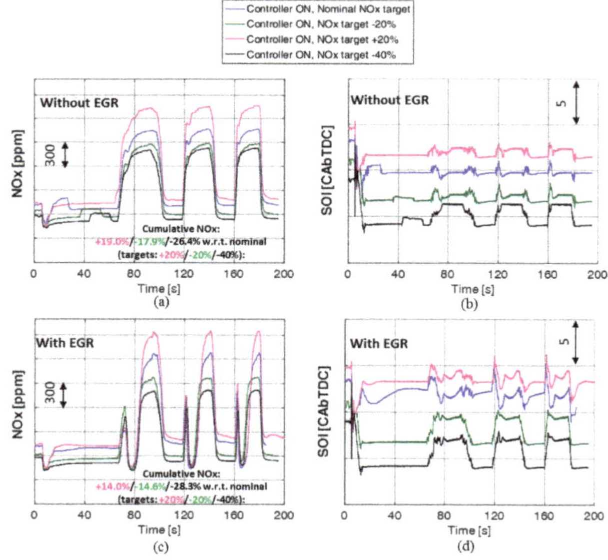

Figure 9. Comparison of the time histories of the measured NOx emissions (**a**,**c**) and of SOI_{main} (**b**,**d**) for the ramp test 4 without EGR (**a**,**b**) and with EGR (**c**,**d**). The origin of the y-axis is the same for all the related figures.

The results concerning the ramp test 4 are instead reported in Figure 9. Similarly to Figures 7 and 9 reports the time histories of the measured NOx levels (Figure 9a,c) and of SOI_{main} (Figure 9b,d) for the condition in which EGR is closed (Figure 9a,b) or with nominal EGR level (Figure 9c,d), considering different NOx targets. For the cases in which NOx target deviations of +20% and −20%/−40% were set, the percentage difference of the cumulative NOx emissions (ppm), with respect to the baseline case with nominal NOx target have also been reported.

Also, in this case, it can be seen how the controller progressively advances/delays the values of SOI_{main} in order to increase/decrease the engine-out NOx levels.

The percentage differences of the cumulative NOx emissions, with respect to the baseline case with nominal NOx target, are in line with the target requests also for this ramp test, except for the case in which a NOx target of −40% with respect to the nominal one, is requested. With reference to the case without EGR, similarly to the previous case, the main reason of this behavior is due to the fact that the lower SOI_{main} safety boundary is sometimes achieved when a very low NOx target is requested. With reference to the case with EGR, the lower SOI_{main} safety boundaries are not achieved. Therefore, the error is due to inaccuracies in the model. However, a significant contribution to these inaccuracies may be related to a difficult estimation of EGR in transient operation. This effect will be discussed in Section 4.5.

A summary is finally reported of the performance of the controller over the entire set of transient conditions, with specific reference to the NOx control. In particular, a cumulative NOx index was estimated by integrating the instantaneous NOx concentration for all the investigated tests, and the relative differences in the values of the cumulative NOx index, with respect to the baseline case in which the controller is activated with nominal NOx target, were calculated. The results are reported in Table 3. The values indicated with bold underlined text in Table 3 refer to test conditions in which the boundaries of SOI_{main} are achieved for a significant portion of the test, while the values indicated with underlined text indicate the test conditions in which the boundaries of SOI_{main} are achieved, but for a limited portion of the test.

Table 3. Relative differences in the values of the cumulative NOx index, with respect to the baseline case in which the controller is activated with nominal NOx target.

Input Type	Controller ON, Nominal NOx Target	Controller ON, NOx Target +20%	Controller ON, NOx Target −20%	Controller ON, NOx Target −40%
Ramp test 1, EGR OFF	Reference	+17.44	−15.77	**<u>−26.62</u>**
Ramp test 1, EGR ON	Reference	**+13.29%**	−19.48%	<u>−30.53%</u>
Ramp test 2, EGR OFF	Reference	+21.33%	−16.26	**<u>−24.25%</u>**
Ramp test 2, EGR ON	Reference	+13.28%	−21.31%	<u>−31.83%</u>
Ramp test 3, EGR OFF	Reference	+21.17%	−18.86%	**<u>−28.24%</u>**
Ramp test 3, EGR ON	Reference	+19.85%	−22.71%	<u>−35.81%</u>
Ramp test 4, EGR OFF	Reference	+18.99%	−17.89%	**<u>−26.41%</u>**
Ramp test 4, EGR ON	Reference	+14.03	−14.56%	<u>−28.35%</u>
Ramp test 5, EGR OFF	Reference	+20.44%	−18.51%	**<u>−25.44%</u>**
Ramp test 5, EGR ON	Reference	+16.39%	−22.07%	<u>−33.62%</u>

It can be noted in the table how, in general, the controller is effective in reducing/increasing the cumulative NOx emissions, with respect to the reference case, even for the conditions in which EGR is enabled. It was in fact observed, for these cases, that larger deviations occur between the instantaneous NOx target levels and the measured NOx emissions, as a consequence of a difficult EGR estimation in dynamic condition that is given to the controller as input.

4.4. Controller of BMEP and NOx, Assessment in Transient Operation: BMEP Control

In this section, the performance of the controller has been investigated concerning the BMEP control. In particular, Figures 10 and 11 report a comparison of the target and measured BMEP levels for all the transient tests reported in Table 1.

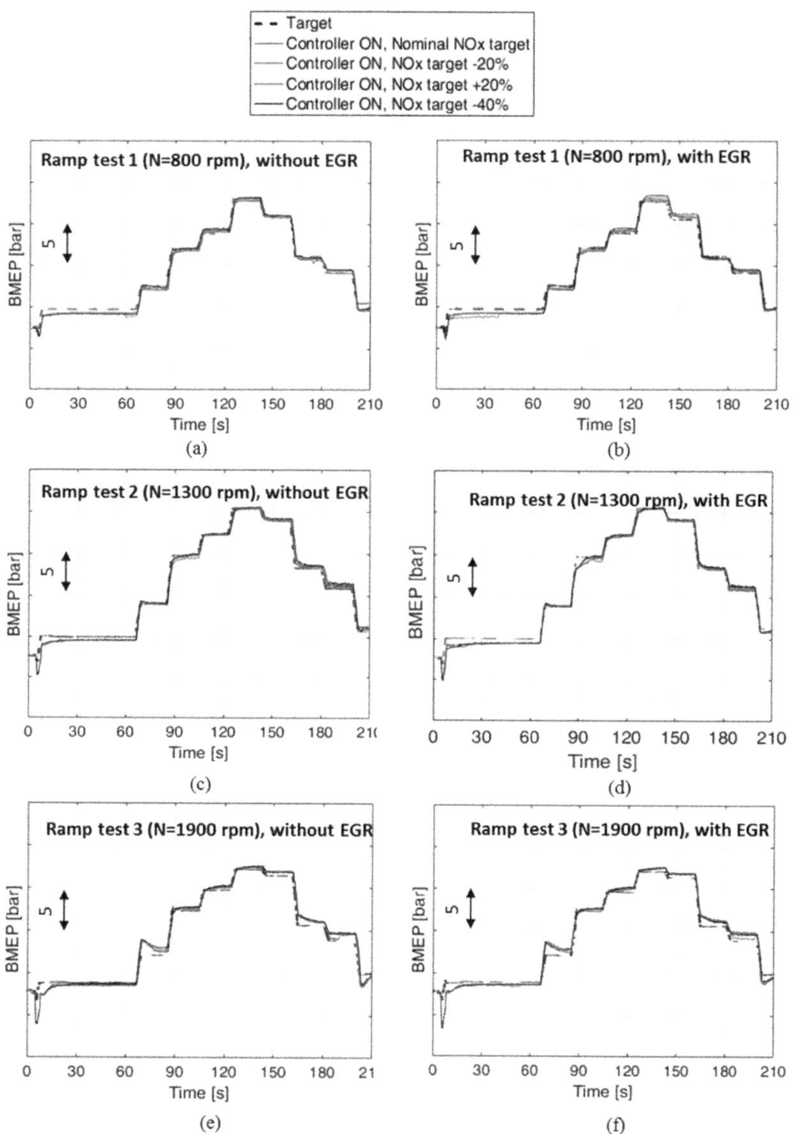

Figure 10. Time histories of the target and measured levels of BMEP for the ramp tests 1 (**a**,**b**), 2 (**c**,**d**) and 3 (**e**,**f**).

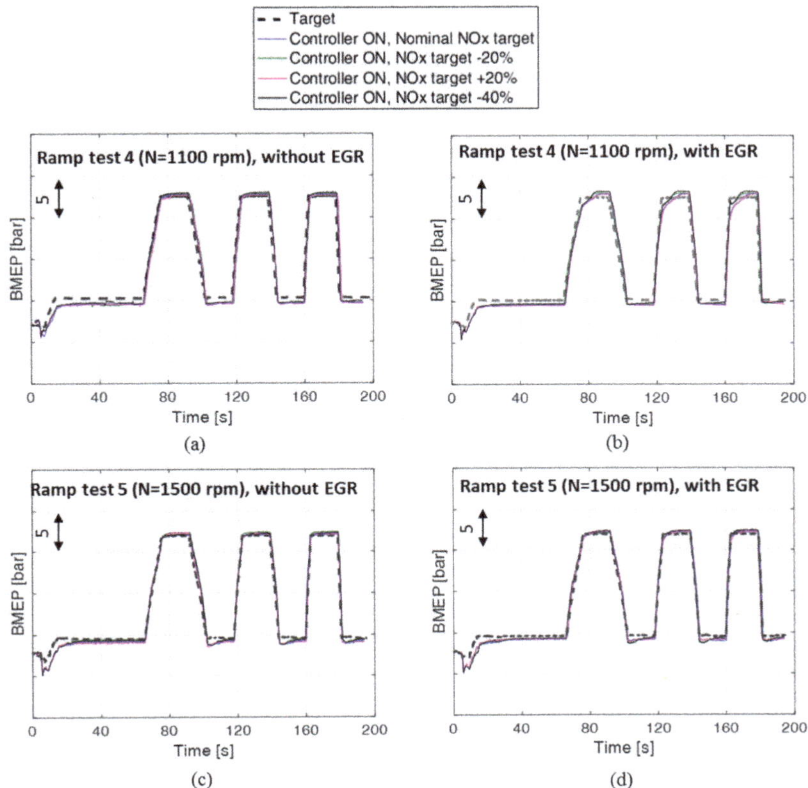

Figure 11. Time histories of the target and measured levels of BMEP for the ramp tests 4 (**a,b**) and 5 (**c,d**).

In general, the performance of the controller is very good for all the investigated tests. It can, in fact, be seen that the desired targets of BMEP are in general achieved, regardless of the NOx target and of the activation/deactivation of EGR.

Some deviations can sometimes be detected, such as those shown in Figure 10e,f during the time interval between 60 and 90 s. It was verified that these deviations derive from a combination of several effects, such as inaccuracy of the combustion model, inaccuracy of the injector maps embedded in the ECU, and inaccuracy of the EGR rate estimated by the ECU.

It can also be observed in Figure 11b that lower BMEP values are realized, with respect to the target, for the ramp test 4 with EGR at the end of the increasing load ramps. This behavior is likely to be due to a too low level of the intake O_2 concentration, which leads to a non-optimal combustion.

The general behavior of the controller, however, is acceptable, since the injected fuel quantity is increased/decreased, with respect to the baseline level, in order to compensate for the different engine thermal efficiency when SOI_{main} is delayed/anticipated in order to achieve the desired NOx targets. This can be seen in Figure 12a, which reports, as an example, the time histories of the measured injected fuel quantity for the ramp test 5 without EGR. It can be seen in the figure how the injected quantities are different for the different cases, and the realized BMEP levels are in line with the targets (see Figure 11c). For the sake of completeness, Figure 12b reports the time histories of the measured NOx levels, as well as the percentage difference of the cumulative NOx emissions (ppm), with respect to the baseline case with nominal NOx target.

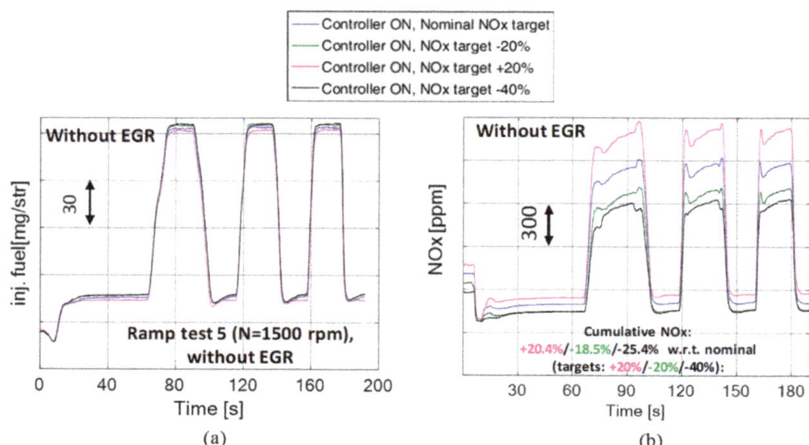

Figure 12. Time histories of the measured injected fuel quantity (**a**) and NOx emissions (**b**) for the ramp test 5 without EGR.

4.5. Preliminary Tests on the Road: NOx Control

Preliminary tests of the controller of BMEP and NOx have also been carried out on the road after the engine installation on the vehicle demonstrator, in view of the final assessment of the IMPERIUM project. These tests covered a run from the city of Genova to the city of Asti (and vice-versa), and have demonstrated the basic functionality of the controller, since no crashes occurred in the software. Some preliminary results are shown in this section.

During the tests, the controller was de-activated when the engine ran in warm-up mode. However, the combustion model was running in background in these cases, and it was able to simulate the NOx emissions (thus working as a virtual sensor) using, as input, the control variables which were set by the ECU. Some results are shown in Figure 13, which reports the predicted (blue line) and measured (red line) engine-out NOx emissions along a time interval between t = 500 s and t = 1200 s over the Genova–Asti run, when the engine was operating in warm up mode. Figure 13a shows the instantaneous emissions, while Figure 13b shows the cumulative NOx index. Figure 13c,d report a detail of the results in the time interval between t = 500 s and t = 600 s. It can be seen that the accuracy of the model is very high, also considering that it was not calibrated in this working mode. It should also be remarked that the mission had a duration of 1.5 h and the reported time interval is representative of the accuracy of the model over the whole mission. Moreover, in general, the trend of the NOx emissions measured by the engine sensor shows a smoother and little delayed behavior with respect to the trend calculated by the combustion model. This could be due to the fact that controller estimates a "cylinder out" NOx concentration for each combustion firing, while the engine sensor measures the concentration in the exhaust manifold, after that mixing phenomena occur.

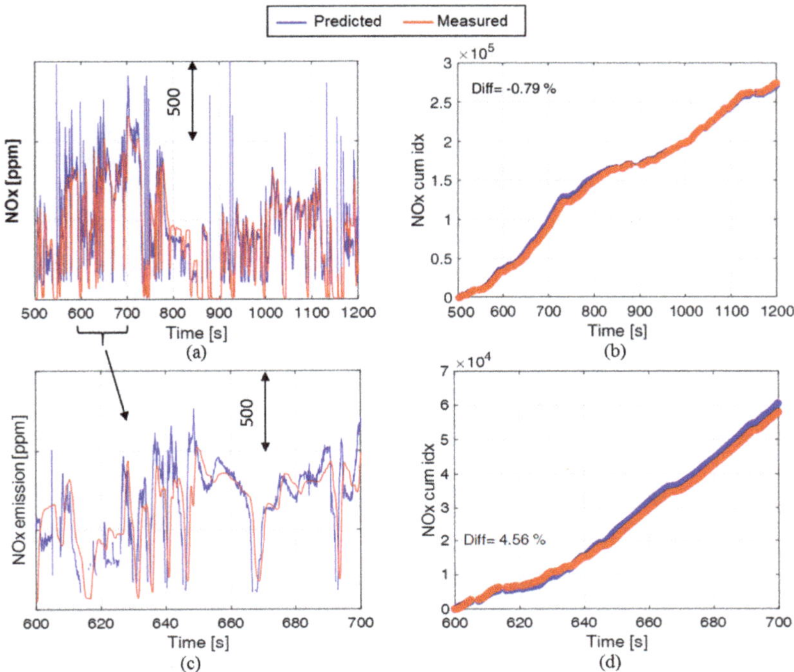

Figure 13. Predicted and measured engine-out NOx emissions along an interval of the Genova–Asti run, when the engine run in warm up mode and the controller operated as virtual NOx sensor. (**a**,**c**) instantaneous NOx emissions; (**b**,**d**) NOx cumulative index.

Figure 14 shows a time interval of the Asti–Genova run, in which the controller was activated (at t = 50 s) and the engine operated in normal mode.

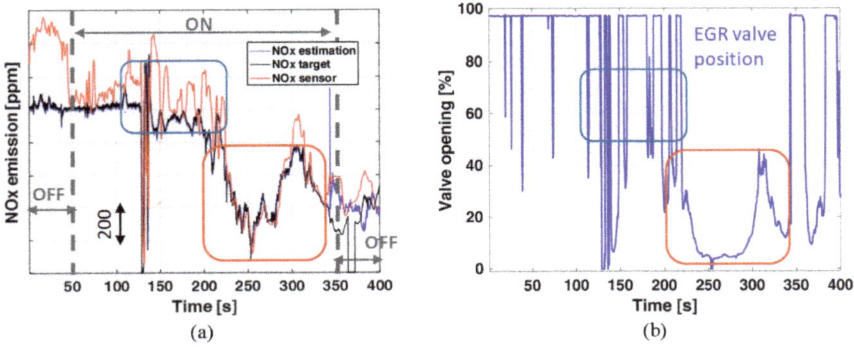

Figure 14. (**a**) Predicted, measured, and target engine-out NOx emissions along an interval of the Asti–Genova run, when the engine ran in normal mode and the controller was activated; (**b**) time history of the EGR valve position.

Figure 14a reports the time histories of the requested NOx emissions (black line), of the measured NOx emissions (red line) and of the NOx emissions estimated by the controller (blue line). The NOx emissions estimated by the controller should always be coincident with the requested levels, except for the cases in which the safety boundary level of SOI_{main} is achieved. Figure 14b, instead, reports the time history of the EGR valve position.

In the first part of the trace reported in Figure 14a, (indicated with 'OFF'), the by-pass hooks were disabled and therefore NOx target was not realized, since the values of SOI_{main} and $q_{f,inj}$ elaborated by the controller were not sent to the ECU. Subsequently, the by-pass hooks were enabled and NOx concentration quickly achieved the target. Moreover, considering the part of mission in which the optimizer is working, the results indicate that when the EGR valve tends to be closed, the accuracy in the instantaneous control of NOx emissions is very high, while it deteriorates when the EGR valve is activated.

This deviation is likely to be due to the fact that the controller requires the intake oxygen concentration as input, and in the present application this quantity was estimated on the basis of the EGR and air mass levels that are estimated by the air path sub-models embedded in the ECU. These models are accurate at steady-state operation, but were not yet calibrated in transient operation, and therefore, they cannot capture the highly transient phenomena which occur in the intake manifold in these conditions. A suitable calibration of the air path sub-models embedded in the ECU in dynamic conditions, or the installation of an intake O_2 sensor may lead to a significant improvement in the performance of the controller when EGR is adopted.

However, by calculating the cumulative emissions, a relatively small error between the desired and measured NOx levels is detected, even considering the interval in which EGR is activated.

5. Conclusions

In this paper, a previously developed model-based controller of BMEP (brake mean effective pressure), or torque, and nitrogen oxides emissions (NOx) has been implemented and assessed on an 11 L FPT prototype heavy-duty Cursor 11 diesel engine. The controller acts on the injected fuel quantity ($q_{f,inj}$) and on the start of injection of the main pulse (SOI_{main}) in order to achieve the desired targets.

The activity was carried out within the IMPERIUM H2020 EU Project.

The controller has been assessed at both steady-state and transient operations, through rapid prototyping tests carried out at the engine test bench. Moreover, preliminary tests were carried out on the road using the vehicle demonstrator.

The main results can be summarized as follows:

- Concerning the steady-state tests acquired at the engine test bench, the average values of the root mean square error (RMSE) are of the order of 100 ppm with reference to the control of NOx emissions, and of the order of 0.3 bar with reference to the control of BMEP. The root mean square error of the relative NOx error is of the order of 7.8%.

- Concerning the transient tests acquired at the engine test bench, it was found that the control of BMEP is very accurate for all the investigated tests. With reference to the control of NOx emissions, it was verified that the cumulative NOx emissions are in line with the target levels when different NOx targets are set, and this confirms the effectiveness of the proposed approach.

- By analyzing the data acquired at the engine test bench, it was verified that if a too low or too high target of NOx emissions is requested, the limitation in the actuated value of SOI_{main} is realized correctly. This limitation is designed for safety reasons (e.g., in order to avoid too high exhaust temperatures/peak firing pressure levels, or to avoid an excessive penalization in terms of engine thermal efficiency), so that the target cannot be achieved but the engine integrity is guaranteed.

- Concerning the road tests, the basic functionality of the controller and the NOx control accuracy observed at the engine test bench were confirmed. It was also verified that, when the EGR valve tends to be closed, the accuracy in the instantaneous control of NOx emissions is very high, while it deteriorates when the EGR valve is activated. This deviation is likely to be due to inaccuracies in the estimation of the intake oxygen concentration on the basis of the air-path models of EGR and air mass embedded in the ECU.

- The installation of an intake O_2 sensor may lead to a significant improvement in the performance of the controller, concerning the instantaneous control of NOx emissions, when EGR is adopted.
- During the on-road tests, it was verified that the controller could be used as an accurate virtual NOx sensor even when it was disabled over specific engine modes. This functionality can be very useful, especially in case the engine NOx sensor has not yet achieved warm-up conditions and cannot provide reliable measurements.

Author Contributions: The authors equally contributed to the deployment of the paper. Conceptualization, R.F., O.M. and E.S.; Methodology, R.F. and O.M.; Software, O.M.; Formal Analysis, R.F. and O.M; Data Curation, R.F. and O.M.; Writing—Original Draft Preparation, R.F. and O.M.; Writing—Review and Editing, R.F., O.M. and G.H.; Supervision, E.S.; Resources; F.C. and G.H.

Funding: The research leading to these results received funding from the European Union's Horizon 2020 research and innovation programme under grant agreement n°713,783 (IMPERIUM) and from the Swiss State Secretariat for Education, Research and Innovation (SERI) under contract n°16.0063 for the Swiss consortium members.

Conflicts of Interest: The authors declare no conflict of interest.

Abbreviations

ATS	after-treatment system
BMEP	Brake Mean Effective Pressure (bar)
CA	crank angle (deg)
DT	dwell-time
ECU	Engine Control Unit
EGR	Exhaust Gas Recirculation
FMEP	Friction Mean Effective Pressure (bar)
FPT	FPT Industrial
HCCI	Homogeneous Charge Compression Ignition
IMEP	Indicated Mean Effective Pressure (bar)
$IMEP_g$	gross Indicated Mean Effective Pressure (bar)
$IMEP_n$	net Indicated Mean Effective Pressure (bar)
IMPERIUM	IMplementation of Powertrain Control for Economic and Clean Real driving emIssion and fuel ConsUMption
m	mass
$\dot{m}_{air}$	mass flow rate of fresh air
$\dot{m}_{EGR}$	mass flow rate of EGR
MFB50	crank angle at which 50% of the fuel mass fraction has burned (deg)
N	engine rotational speed (1/min)
O_2	intake charge oxygen concentration (%)
p	pressure (bar)
PCCI	Premixed Charge Compression Ignition
p_{EMF}	exhaust manifold pressure (bar abs)
PEMS	Portable Emissions Measurement System
p_f	injection pressure (bar)
PFP	peak firing pressure
p_{IMF}	intake manifold pressure (bar abs)
PMEP	Pumping Mean Effective Pressure (bar)
q	injected fuel volume quantity (mm3)
Q_{ch}	chemical heat release
$q_{f,inj}$	total injected fuel volume quantity per cycle/cylinder
Q_{net}	net heat release

Rh_{amb}	ambient relative humidity
RMSE	root mean square error
SOI	electric start of injection
SOI_{main}	electric start of injection of the main pulse
t	time
T	temperature (K)
T_{amb}	ambient temperature
T_{IMF}	intake manifold temperature
V2X	vehicle-to-everything technology
VGT	Variable Geometry Turbine
VPM	Virtual Pressure Model
WHSC	Worldwide Harmonized Steady Cycle
WHTC	Worldwide Harmonized Transient Cycle

References

1. Xue, X.; Rutledge, J. *Potentials of Electrical Assist and Variable Geometry Turbocharging System for Heavy-Duty Diesel Engine Downsizing*; SAE Technical Paper 2017-01-1035; SAE International: Warrendale, PA, USA, 2017. [CrossRef]
2. Di Iorio, S.; Beatrice, C.; Guido, C.; Napolitano, P.; Vassallo, A.; Ciaravino, C. *Impact of Biodiesel on Particle Emissions and DPF Regeneration Management in a Euro5 Automotive Diesel Engine*; SAE Technical Paper 2012-01-0839; SAE International: Warrendale, PA, USA, 2012. [CrossRef]
3. D'Ambrosio, S.; Gaia, F.; Iemmolo, D.; Mancarella, A.; Salamone, N.; Vitolo, R.; Hardy, G. *Performance and Emission Comparison between a Conventional Euro VI Diesel Engine and an Optimized PCCI Version and Effect of EGR Cooler Fouling on PCCI Combustion*; SAE Technical Paper 2018-01-0221; SAE International: Warrendale, PA, USA, 2018. [CrossRef]
4. Andert, J.; Wick, M.; Lehrheuer, B.; Sohn, C.; Albin, T.; Pischinger, S. Autoregressive modeling of cycle-to-cycle correlations in homogeneous charge compression ignition combustion. *Int. J. Engine Res.* **2017**, *19*, 790–802. [CrossRef]
5. Ferrari, A.; Mittica, A.; Pizzo, P.; Jin, Z. PID Controller Modelling and Optimization in Cr Systems with Standard and Reduced Accumulators. *Int. J. Automot. Technol.* **2018**, *19*, 771–781. [CrossRef]
6. Catania, A.E.; Ferrari, A.; Mittica, A.; Spessa, E. *Common Rail without Accumulator: Development, Theoretical-Experimental Analysis and Performance Enhancement at DI-HCCI Level of a New Generation FIS*; SAE Technical Paper 2007-01-1258; SAE International: Warrendale, PA, USA, 2007. [CrossRef]
7. Ferrari, A.; Mittica, A.; Pizzo, P.; Wu, X.; Zhou, H. New methodology for the identification of the leakage paths and guidelines for the design of common rail injectors with reduced leakage. *J. Eng. Gas Turbines Power* **2018**, *140*, 022801. [CrossRef]
8. Ferrari, A.; Mittica, A.; Paolicelli, F.; Pizzo, P. Hydraulic Characterization of Solenoid-actuated Injectors for Diesel Engine Common Rail Systems. *Energy Procedia* **2016**, *101*, 878–885. [CrossRef]
9. Ferrari, A.; Manno, M.; Mittica, A. Cavitation analogy to gasdynamic shocks: Model conservativeness effects on the simulation of transient flows in high-pressure pipelines. *J. Fluids Eng.* **2008**, *130*, 031304. [CrossRef]
10. Catania, A.; Ferrari, A.; Mittica, A. High-pressure rotary pump performance in multi-jet common rail systems. ESDA2006, Engineering Systems Design and Analysis, Fatigue and Fracture, Heat Transfer, Internal Combustion Engines, Manufacturing and Technology and Society. In Proceedings of the 8th Biennial ASME Conference on Engineering Systems Design and Analysis, Torino, Italy, 4–7 July 2006; Volume 4, pp. 557–565. [CrossRef]
11. Finesso, R.; Marello, O.; Misul, D.; Spessa, E.; Violante, M.; Yang, Y.; Hardy, G.; Maier, C. Development and Assessment of Pressure-Based and Model-Based Techniques for the MFB50 Control of a Euro VI 3.0L Diesel Engine. *SAE Int. J. Engines* **2017**, *10*, 1538–1555. [CrossRef]
12. Finesso, R.; Marello, O.; Spessa, E.; Yang, Y.; Hardy, G. Model-Based Control of BMEP and NOx Emissions in a Euro VI 3.0L Diesel Engine. *SAE Int. J. Engines* **2017**, *10*, 2288–2304. [CrossRef]
13. Finesso, R.; Hardy, G.; Mancarella, A.; Marello, O.; Mittica, A.; Spessa, E. Real-Time Simulation of Torque and Nitrogen Oxide Emissions in an 11.0 L Heavy-Duty Diesel Engine for Model-Based Combustion Control. *Energies* **2019**, *12*, 460. [CrossRef]

14. Hu, S.; d'Ambrosio, S.; Finesso, R.; Manelli, A.; Marzano, M.R.; Mittica, A.; Ventura, L.; Wang, H.; Wang, Y. Comparison of Physics-Based, Semi-Empirical and Neural Network-Based Models for Model-Based Combustion Control in a 3.0 L Diesel Engine. *Energies* **2019**, *12*, 3423. [CrossRef]

15. Nuss, E.; Wick, M.; Andert, J.; De Schutter, J.; Diehl, M.; Abel, D.; Albin, T. Nonlinear model predictive control of a discrete-cycle gasoline-controlled auto ignition engine model: Simulative analysis. *Int. J. Engine Res.* **2019**, *20*, 1025–1036. [CrossRef]

16. Meda, L.; Shu, Y.; Romzek, M. *Heavy Duty Diesel After-Treatment System Analysis Based Design: Fluid, Thermal and Structural Considerations*; SAE Technical Paper 2009-01-0624; SAE International: Warrendale, PA, USA, 2009. [CrossRef]

17. Yamaguchi, T.; Aoyagi, Y.; Uchida, N.; Fukunaga, A.; Kobayashi, M.; Adachi, T.; Hashimoto, M. Fundamental Study of Waste Heat Recovery in the High Boosted 6-cylinder Heavy Duty Diesel Engine. *SAE Int. J. Mater. Manf.* **2015**, *8*, 209–226. [CrossRef]

18. Delogu, M.; Zanchi, L.; Dattilo, C.; Maltese, S.; Riccomagno, R.; Pierini, M. *Take-Home Messages from the Applications of Life Cycle Assessment on Lightweight Automotive Components*; SAE Technical Paper 2018-37-0029; SAE International: Warrendale, PA, USA, 2018. [CrossRef]

19. Finesso, R.; Misul, D.; Spessa, E.; Venditti, M. Optimal Design of Power-Split HEVs Based on Total Cost of Ownership and CO2 Emission Minimization. *Energies* **2018**, *11*, 1705. [CrossRef]

20. Talavera, E.; Díaz-Álvarez, A.; Jiménez, F.; Naranjo, J.E. Impact on Congestion and Fuel Consumption of a Cooperative Adaptive Cruise Control System with Lane-Level Position Estimation. *Energies* **2018**, *11*, 194. [CrossRef]

21. Johansson, I.; Jin, J.; Ma, X.; Pettersson, H. Look-ahead speed planning for heavy-duty vehicle platoons using traffic information. *Transp. Res. Procedia* **2017**, *22*, 561–569. [CrossRef]

22. Danninger, A.; Armengauda, E.; Milton, G.; Lützner, J.; Hakstege, B.; Zurlo, G.; Schöni, A.; Lindberg, J.; Krainer, F. IMplementation of Powertrain Control for Economic and Clean Real driving emIssion and fuel ConsUMption. In Proceedings of the 7th Transport Research Arena TRA 2018, Vienna, Austria, 16–19 April 2018.

23. Finesso, R.; Spessa, E.; Yang, Y.; Alfieri, V.; Conte, G. HRR and MFB50 Estimation in a Euro 6 Diesel Engine by Means of Control-Oriented Predictive Models. *SAE Int. J. Engines* **2015**, *8*, 1055–1068. [CrossRef]

24. Finesso, R.; Spessa, E.; Yang, Y. Development and Validation of a Real-Time Model for the Simulation of the Heat Release Rate, In-Cylinder Pressure and Pollutant Emissions in Diesel Engines. *SAE Int. J. Engines* **2016**, *9*, 322–341. [CrossRef]

25. Catania, A.E.; Finesso, R.; Spessa, E. Predictive zero-dimensional combustion model for DI diesel engine feed-forward control. *Energy Convers. Manag.* **2011**, *52*, 3159–3175. [CrossRef]

26. Orthaber, G.C.; Chmela, F.G. *Rate of Heat Release Prediction for Direct Injection Diesel Engines Based on Purely Mixing Controlled Combustion*; SAE Technical Paper 1999-01-0186; SAE International: Warrendale, PA, USA, 2018. [CrossRef]

27. Egnell, R. *A Simple Approach to Studying the Relation between Fuel Rate Heat Release Rate and NO Formation in Diesel Engines*; SAE Technical Paper 1999-01-3548; SAE International: Warrendale, PA, USA, 2018. [CrossRef]

28. Ericson, C.; Westerberg, B. *Modelling Diesel Engine Combustion and NOx Formation for Model Based Control and Simulation of Engine and Exhaust Aftertreatment Systems*; SAE Technical Paper 2006-01-0687; SAE International: Warrendale, PA, USA, 2018. [CrossRef]

29. Heywood, J. *Internal Combustion Engine Fundamentals*; McGraw-Hill Intern: Columbus, OH, USA, 1988.

30. Chen, S.K.; Flynn, P.F. *Development of a Single Cylinder Compression Ignition Research Engine*; SAE Technical SAE Technical Paper 650733; SAE International: Warrendale, PA, USA, 2018. [CrossRef]

31. Finesso, R.; Hardy, G.; Maino, C.; Marello, O.; Spessa, E. A New Control-Oriented Semi-Empirical Approach to Predict Engine-Out NOx Emissions in a Euro VI 3.0 L Diesel Engine. *Energies* **2017**, *10*, 1978. [CrossRef]

Article

A Control-Oriented Engine Torque Online Estimation Approach for Gasoline Engines Based on In-Cycle Crankshaft Speed Dynamics

Qiang Tong, Hui Xie *, Kang Song * and Dong Zou

State Key Laboratory of Engines, Tianjin University, Tianjin 300072, China; tongqiang@tju.edu.cn (Q.T.); zoudong@tju.edu.cn (D.Z.)
* Correspondence: xiehui@tju.edu.cn (H.X.); songkangtju@tju.edu.cn (K.S.)

Received: 27 October 2019; Accepted: 5 December 2019; Published: 9 December 2019

Abstract: Engine brake torque is a key feedback variable for the optimal torque split control of an engine–motor hybrid powertrain system. Due to the limitations in available sensors, however, engine torque is difficult to measure directly. For torque estimation, the unknown external load torque and the overlap of the expansion stroke between cylinders introduce a great disturbance to engine speed dynamics. This makes the conventional cycle average engine speed-based estimation approach unusable. In this article, an in-cycle crankshaft speed-based indicated torque estimation approach is proposed for a four-cylinder engine. First, a unique crankshaft angle window is selected for load torque estimation without the influence of combustion torque. Then, an in-cycle angle-domain crankshaft speed dynamic model is developed for engine indicated torque estimation. To account for the effects of model inaccuracy and unknown external disturbances, a "total disturbance" term is introduced. The total disturbance is then estimated by an adaptive observer using the engine's historical operating data. Finally, a real-time correction method for the friction torque is proposed in the fuel cut-off scenario. Combining the aforementioned torque estimators, the brake torque can be obtained. The proposed algorithm is implemented in an in-house developed multi-core engine control unit (ECU). Experimental validation results on an engine test bench show that the algorithm's execution time is about 3.2 ms, and the estimation error of the brake torque is within 5%. Therefore, the proposed method is a promising way to accurately estimate engine torque in real-time.

Keywords: engine torque estimation; GDI engines; extended state observer; online performance

---

1. Introduction

Engine–motor hybrid powertrain systems have been widely used in passenger vehicles [1] to meet increasingly strict emission legislation and improve fuel economy. Optimal torque split between the engine and torque is obviously essential to achieve the best overall efficiency for hybrid vehicles.

Due to the fundamental nature of internal combustion engines (ICEs), the torque's response is slower than the motor's and is usually difficult to measure directly [2,3]. The degradation in engine torque control performance will, in turn, have an adverse effect on the overall fuel economy of the hybrid powertrain systems [4]. This drives the need for real-time estimation of the engine torque, especially in the application of hybrid electric vehicles (HEVs) [5].

Various solutions were available for engine torque control in the past. The most straightforward method to measure the engine torque is by in-cylinder pressure sensors or torque sensors. These solutions, however, increase the hardware cost and create an issue of durability. For instance, the cylinder's pressure may suffer from its harsh thermal environment. These factors limit the application of this method in stock engines [6–9].

An alternative cost-effective solution without the need for additional sensors is to use a crankshaft instantaneous speed sensor, based on the causality between the engine torque and engine speed variation [10,11]. Theoretically, any changes in the engine brake torque can be sensed from the fluctuation of the crankshaft's instantaneous speed [12]. Three methodologies are commonly used in crankshaft speed-based torque estimation, consisting of black-box model-based estimation, frequency analysis-based mapping, and crankshaft dynamic model-based estimation [13]. The first solution is to use a black-box model (such as a neural network and nominal function) to describe the relationship between the engine torque and crankshaft instantaneous speed [14–18]. However, these methods need an amount of data to train the black-box model. Moreover, their model parameters vary with the operating conditions caused by the nonlinear nature of the engine speed dynamics. This makes a single model unsuitable for estimating engine torque under all operating conditions. To improve the estimation accuracy, a piece-wise linear model is a popular solution but comes at the cost of a heavy workload during calibration.

Frequency analysis-based mapping [19] can be used to estimate the indicated torque. After the crankshaft's instantaneous speed and indicated torque are processed by DFT (discrete Fourier transform), a significantly positive correlation can be observed between the two signals in the main harmonic order [20]. However, this requires complicated signal and computational processing and is unsuitable for online applications in the engine control unit (ECU). The third approach is to use crankshaft dynamic models, which can be expressed in the torque balance equation [21–25]. The crankshaft dynamics model can be divided into two categories: A rigid model and an elastic model [26]. The elastic crankshaft dynamics model has higher prediction accuracy and wider working conditions than the rigid model. However, the elastic model requires a large amount of calculation work, which limits its use in real-time applications [13].

In order to increase the adaptability and accuracy of the torque estimation algorithm, the intake process and combustion process are also considered in crankshaft speed modeling. Obviously, this makes the physical model too complicated to implement in an ECU without many model parameters for calibration [27]. Additionally, for the engine friction torque estimation, a look-up table approach is simple to implement and shows degraded estimation accuracy with the aging of the engine [28]. In addition, the unknown resistance torque from the gear box and the wheel makes the engine speed-based engine torque estimation more challenging. To sum up, it is clear that the on-board torque estimation algorithm is challenging, due to the dilemma between the estimation's accuracy and its feasibility in embedded system implementation [13].

Currently, the existing torque online estimation methods are primarily based on look-up tables calibrated offline. This is simple-to-straightforward to implement, but the estimation accuracy deteriorates as the engine ages. One contribution of the proposed algorithm is the ability to be adaptive to the aging of the engine.

In this paper, an engine brake torque estimation approach is proposed for a four-cylinder engine, consisting of the load torque estimator, indicated torque estimator, and a friction torque observer. In fact, the brake torque estimation is valid for both four-cylinder and three-cylinder engines. For engines with a cylinder number equal or less than four, there exists a unique crankshaft angle window, where there is no overlap of the combustion processes. For engines with more than four cylinders, such kind of crankshaft window does not exist, which affects the estimation of total gas torque. First, the load torque estimator is designed for the unique crankshaft angle window. Then, an in-cycle angle-domain crankshaft speed dynamic model is developed for engine-indicated torque estimation with a deviation from the model of a real plant lumped as the total disturbance for estimation. Finally, a real-time correction method for the friction torque is proposed for use in a fuel cut-off scenario. The proposed algorithm is implemented in a multicore ECU to testify its accuracy, computational time, and central processing unit (CPU) loads.

The rest of this article is organized as follows. In Section 2, the experiment setup is discussed briefly. Then, an engine torque observer algorithm is proposed in Section 3. The engine torque observer estimation results and embedded performance are discussed in Section 4. Finally, the conclusions of this study are shown in Section 5.

2. Experiment Setup

The experiment was conducted on a Greatwall Motor EC02 GDI (Gasoline Direct Injection) engine test bench (as shown in Figure 1) with a Horiba DYNAS3 LI 250 electric dynamometer. The schematic diagram of the test bench is shown in Figure 2. The engine has a firing order of 1-3-4-2, and all four cylinders are equipped with in-cylinder pressure sensors for the indicated torque estimation (the baseline for observer validation). The detailed engine specifications are tabulated in Table 1.

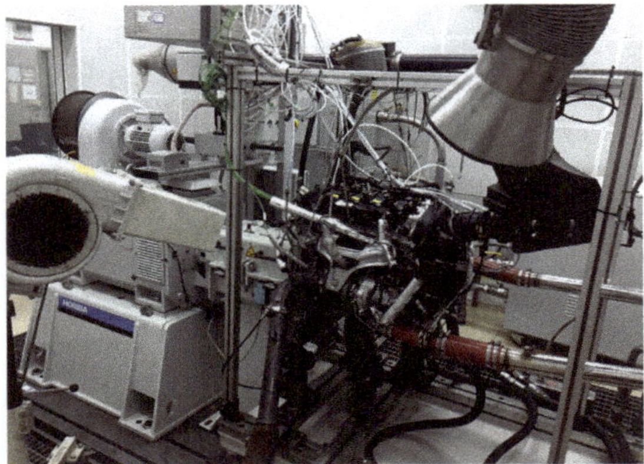

Figure 1. Experimental environment for the engine bench.

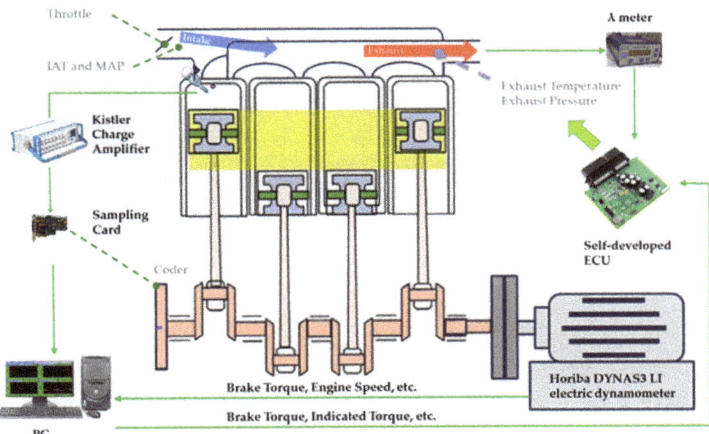

Figure 2. Schematic diagram of an engine experimental platform. ECU, engine control unit; IAT, intake air temperature; MAP, manifold absolute pressure.

Table 1. Engine specifications.

Variable	Value
Displacement (liter)	2.0 L
Cylinders	4
Compression ratio	9.6
Bore (mm)	82.5
Stroke (mm)	92
Connecting rod (mm)	144
Maximum torque (Nm)/speed (rpm)	385/1800
Rated power (kW)/speed (rpm)	165/5500
Intake mode	Naturally aspired

3. Engine Torque Observer Development

In this section, the derivation of the proposed engine torque estimation algorithm is discussed in detail, including engine dynamic model identification, the indicated torque observer [29], and the brake torque observer.

3.1. Engine Dynamic Model

The instantaneous rotational speed of the crankshaft is affected by the torque enforced by the crankshaft following Newton's law. In the crankshaft dynamics, a combination of the brake torque, the friction torque, the indicated torque, and the reciprocating inertia torque works on the crankshaft, causing engine instantaneous speed to fluctuate. There are two kinds of dynamic model in literature, the elastic model and the rigid-body model. The rigid-body model is simplified from the elastic model. Although the elastic model of the crankshaft has high accuracy, the calculation process needs to consume more computing resources. The rigid-body crankshaft dynamic model needs less computing resources, but it has a lower accuracy. This section proposes a rigid-body crankshaft model with a disturbance factor (1), which is used to compensate the error caused by dynamic model simplification process.

$$[J + \Delta \xi(\theta)]\ddot{\theta} = T_{ind} - T_r - T_{fric} - T_e \tag{1}$$

where J is the rotational inertia of the crankshaft, θ is the rotation angle of the crankshaft, $\Delta \xi(\theta)$ is the disturbance factor, $\ddot{\theta}$ is the angular acceleration of the crankshaft, T_{ind} is the indicated torque, T_r is the reciprocating inertia torque, T_{fric} is the friction torque, and T_e is the brake torque.

As seen in Equation (1), the angular acceleration is the second-order derivative of the crankshaft rotation angle ($\ddot{\theta}$), which is very noisy. So, a finite impulse response (FIR) filter is used to process the angular speed signal with a cut-off frequency of 300 Hz. A Kalman filter is used to calculate the angular acceleration.

3.1.1. Reciprocating Torque

The reciprocating torque is generated by the reciprocating part of the connecting rod system. The schematic diagram of the movement of the connecting rod system is as follows in Figure 3.

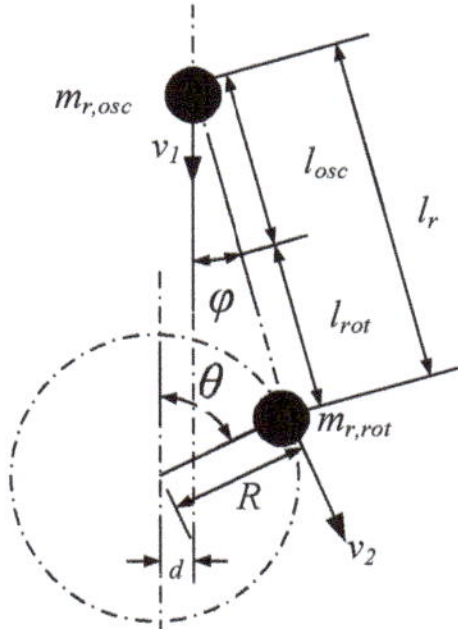

Figure 3. Crank and connecting rod mechanism.

$m_{r,osc}$ is the reciprocating mass of the system, including the piston group and the reciprocating part of the connecting rod. l_{osc} is the length of the reciprocating part of crank connecting rod mechanism. $m_{r,rot}$ is the rotation mass of the system, including the crankshaft and the rotation part of the connecting rod mechanism, and l_{rot} is the length of the rotating part of the connecting rod. l_r denotes the length of the connecting rod. R is radius of the crank. θ is the crankshaft rotation angle. φ denotes the connecting rod angle. v_1 is the velocity of the piston. v_2 is the linear velocity of the crank. d denotes the piston pin offset, which is neglected in this research.

According to the law of conservation of energy, the system's kinetic energy consists of the kinetic energy of the reciprocating mass and the rotation mass, which can be expressed as Equation (2) [23].

$$J\dot{\theta}^2 = m_{r,osc}v_1{}^2 + m_{r,rot}v_2{}^2 \tag{2}$$

The velocities of the reciprocating mass and the rotation mass can be expressed as:

$$\begin{cases} v_1 = R\dot{\theta}\left[sin\theta + \dfrac{\lambda_r sin2\theta}{2\sqrt{1-\lambda_r{}^2 sin^2\theta}}\right] \\ v_2 = R\dot{\theta} \end{cases} \tag{3}$$

where $\lambda_r = (R/l_r)$ denotes the crank radius to connecting rod length ratio [23].

So, according to Equations (2) and (3), the moment of inertia of the system can be expressed as:

$$J = m_{r,osc}R^2\left[sin\theta + \dfrac{\lambda_r sin2\theta}{2\sqrt{1-\lambda_r{}^2 sin^2\theta}}\right] + m_{r,rot}R^2 \tag{4}$$

T_r denotes the reciprocating torque [23].

$$T_r = m_{r,osc}R^2 f(\theta)\left[f(\theta)\ddot{\theta} + g(\theta)\dot{\theta}^2\right] \tag{5}$$

where:

$$f(\theta) = sin\theta + \dfrac{\lambda_r sin2\theta}{2\sqrt{1-\lambda_r{}^2 sin^2\theta}}$$

$$g(\theta) = cos\theta + \dfrac{\lambda_r sin2\theta}{\sqrt{1-\lambda_r{}^2 sin^2\theta}} + \dfrac{\lambda_r{}^3 sin^2(2\theta)}{4\sqrt{1-\lambda_r{}^2 sin^2\theta}}.$$

The total reciprocating torque can be written as:

$$T_r = \sum_{k=1}^{N} T_r^{(k)} = m_{r,osc}R^2 \sum_{k=1}^{N} f(\theta-\phi_k)[f(\theta-\phi_k)\ddot{\theta} + g(\theta-\phi_k)\dot{\theta}^2] \tag{6}$$

where:

$$\phi_k = \frac{4\pi}{N}(k-1).$$

N is the number of cylinders, in this case $N = 4$. ϕ_k ($k = 1, \ldots, N$) denotes the phase of the kth cylinder [13].

3.1.2. Indicated Torque Estimation

Indicated torque is generated at two process, the compression process and the combustion process. The indicated torque during combustion process is difficult to estimate. However, the compression process can be considered as a polytropic process, the in-cylinder pressure can be estimated using the manifold absolute pressure (MAP) sensor and the intake air temperature (IAT) sensor, assuming that the in-cylinder pressure can be approximated by MAP at the timing of intake valve closing and corrected by volumetric efficiency. The MAP sensor and IAT sensor are already standard sensors equipped in stock engines. The estimated pressure then can be used to calculate the indicated torque during compression process, shown in Figure 4.

$$P_{cyl}V^\kappa = const \tag{7}$$

where P_{cyl} denotes the in-cylinder pressure; V denotes the gas volume; and κ is the polytropic process factor, which is taken to be 1.3 in this research within the compression stoke [21].

$$V = V_c + (R + l_r - Rcos\theta - l_r\sqrt{1 - \lambda_r^2 sin^2\theta})\frac{\pi B^2}{4}. \tag{8}$$

where V_c is the combustion chamber volume, and B is the cylinder diameter [22].

The total indicated torque is:

$$T_{ind} = \sum\nolimits_{k=1}^{N} T_{ind}^{(k)} = \frac{\pi B^2}{4}R\sum\nolimits_{k=1}^{N}[P_{cyl}^{(k)}f(\theta - \phi_k)]. \tag{9}$$

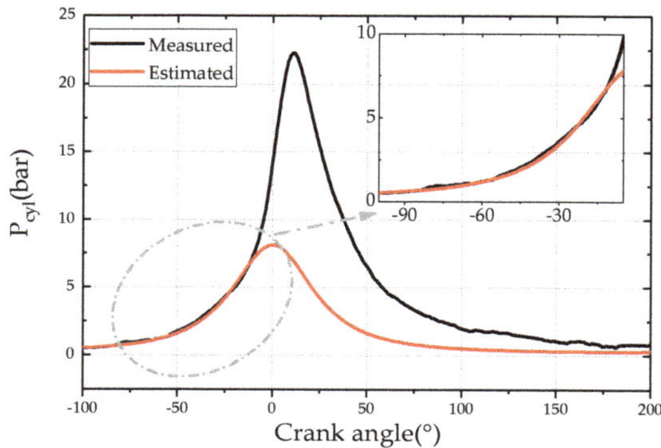

Figure 4. Comparison of the estimated and measured in-cylinder pressure during the compression stroke.

3.1.3. Load Torque Estimation

For an internal combustion engine, the load torque (T_{load}) involves two parts: The brake torque (T_e) and the friction torque (T_{fric}). T_{load} a slow-varying variable compared to the combustion process, so T_{load} can be considered approximately as a constant in one cycle. In the crankshaft dynamic model, both T_{load} and T_{ind} are unknown variables, which makes T_{load} difficult to estimate. In Section 3.1.2, T_{ind} during the compression process can be estimated, which can be defined as gas torque (T_{gas}). For four-cylinder engines and engines with less than four cylinders, there is a unique crankshaft angle window, where the sum of the other three cylinders can be neglected [21]. Within this particular crankshaft angle window, the total indicated torque can be calculated from the cylinder in the compression phase. A T_{load} estimator can be designed in this window, where there is no combustion, even for all cylinders. A bench test was done to locate this angle window, and experimental result shows that 50 crank angle before TDC to 20 crank angle before TDC is the unique angle window to estimate the total gas torque for all cylinders. Meanwhile, this angle window can be used to estimate T_{load} and $\Delta\xi(\theta)$.

In the unique crankshaft angle window from 50° before TDC to 20° before TDC, the only unknown variables are T_{load} and $\Delta\xi(\theta)$. So the T_{load} estimation issue can be regarded as a system parameter estimation issue. The least-squares method is a method for identifying system parameters. However, there are few sampling points in the unique window, and the least-squares method is challenging for online applications. So T_{load} estimation algorithm using the recursive least-squares method is proposed, and the algorithm can process the sampling data in multiple cycles. The estimated load torque ($\hat{T}_{load}$) at the end of last cycle will be set as the initial load torque into the engine load torque estimator. $\hat{T}_{load}$ and $\Delta\xi(\theta)$ are key variables for T_{ind} observer in the next section.

The specific algorithm is shown in Figure 5. The manifold absolute pressure at intake valve close (IVC) P_{MAP_IVC} is used to calculate T_{gas}.

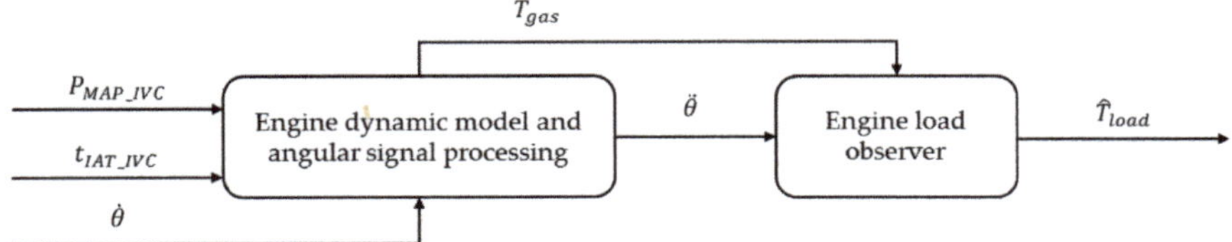

Figure 5. Schematic diagram of the load torque estimation method based on angular speed.

During the unique crankshaft angle window, Equation (1) can be transformed as:

$$T_u = J_e\ddot{\theta} + T_{load}. \tag{10}$$

where:

$$T_{load} = T_{fric} + T_e, T_u = T_{gas} - T_r, J_e = J + \Delta\xi(\theta)$$

For the K times of successive sampling, Equation (10) can be:

$$\begin{bmatrix} T_u(1) \\ \vdots \\ T_u(K) \end{bmatrix} = \begin{bmatrix} \ddot{\theta}(1) & 1 \\ \vdots & \vdots \\ \ddot{\theta}(K) & 1 \end{bmatrix} \begin{bmatrix} J_e \\ T_{load} \end{bmatrix} + \begin{bmatrix} \varsigma(1) \\ \vdots \\ \varsigma(K) \end{bmatrix}. \tag{11}$$

where ς is white noise with mean value 0.

So Equation (11) can be described as:

$$T_K = \Phi_K\Theta_K + \varsigma_K \tag{12}$$

$$T_K = \begin{bmatrix} T_u(1) \\ \vdots \\ T_u(K) \end{bmatrix}, \Phi_K = \begin{bmatrix} \ddot{\theta}(1) & 1 \\ \vdots & \vdots \\ \ddot{\theta}(K) & 1 \end{bmatrix}, \Theta_K = \begin{bmatrix} J_{eK} \\ T_{loadK} \end{bmatrix}, \varsigma_K = \begin{bmatrix} \varsigma(1) \\ \vdots \\ \varsigma(K) \end{bmatrix}$$

According to the recursive least-squares method:

$$\hat{\Theta}_K = P_K \Phi_K{}^T T_K \tag{13}$$

$$P_K = \left(\Phi_K{}^T \Phi_K\right)^{-1}$$

For K+1, when a new $\ddot{\theta}(K+1)$ and $T_u(K+1)$ are calculated, we define $\Psi_{K+1} = [\ \ddot{\theta}(K+1) \quad 1\ ]$, so:

$$P_{K+1} = \left(P_K{}^{-1} + \Psi_{K+1}\Psi_{K+1}{}^T\right)^{-1}. \tag{14}$$

Then the $\hat{\Theta}$ can be estimated using recursive least-squares method:

$$\begin{cases} Q_{K+1} = P_K\Psi_{K+1}\left(1 + \Psi_{K+1}{}^T P_K\Psi_{K+1}\right)^{-1} \\ P_{K+1} = P_K - Q_{K+1}\Psi_K{}^T P_K \\ \hat{\Theta}_{K+1} = \hat{\Theta}_K + Q_{K+1}\left(T_{K+1} - \Psi_{K+1}{}^T\hat{\Theta}_K\right) \end{cases} \tag{15}$$

The convergence of $\hat{\Theta}$ requires a certain amount of T_u and $\ddot{\theta}$. Therefore, the recursive process is expanded to multiple cycles, that is, the initial value of the estimated parameter $\hat{\Theta}$, including T_{load} and $\Delta\xi(\theta)$, is the result of the last estimated value from the previous cycle.

3.2. Indicated Torque Observer

3.2.1. Engine Management Model

This engine management model is a serious model to calculate the initial indicated torque $T_{ind\ ini}$, and the model is the base of the extended state observer (ESO) to estimate the indicated torque. The engine management model contains the thermal efficiency model, the crankshaft dynamic model, etc. [30].

The indicated work comes from the combustion process of the delivered fuel, and can be modeled as a function of fuel heating value, the delivered mass of the gasoline and the operation conditions [30]. The indicated work can be expressed as:

$$W_{ind} = m_f q_{LHV} E_{ff} \tag{16}$$

where W_{ind} is indicated work, m_f is the delivered gasoline mass, q_{LHV} is gasoline heating value. E_{ff} is thermal efficiency, which can be expressed as:

$$E_{ff} = \left(1 - \frac{1}{r_c^{\kappa-1}}\right)\cdot\min(1,\lambda)\cdot\eta_{ign}\left(\theta_{ign}\right)\cdot\eta_{ig,ch}(\dot{\theta}, V_d) \tag{17}$$

where r_c denotes the compression ratio, V_d is the engine displacement, θ_{ign} is the position for the ignition timing, η_{ign} is the ignition efficiency, $\eta_{ig,ch}(\dot{\theta}, V_d)$ is the heat transfer efficiency between the real and the ideal cycles, λ denotes the air/fuel ratio, and $\min(1,\lambda)$ describes that the fuel mass cannot fully utilized in the case of a rich mixture [30].

The engine dynamic model used in this section is already mentioned in Section 3.1

3.2.2. Indicated Torque Observer Design

Here, an indicate torque observation method is built using ESO. The indicated thermal efficiency E_{ff} is the key parameter to calculate the indicated torque. However, the E_{ff} is a state parameter of the combustion process, which is impossible to measure. So an ESO is built on the base of the engine management model, the uncertainty of the indicated thermal efficiency ΔE_{ff} is proposed to compensate the total error of the engine management model. The indicated torque observer based on ESO is shown in Figure 6.

The intake air quantity (Q_{mass}) used to calculate the initial indicated torque, along with the T_{load} estimated in Section 3.1.3, are processed by crankshaft dynamic model into engine rotation speed $\dot{\theta}$. The engine speed simulated by the crankshaft dynamic model is compared with the engine speed measured. The error between the simulated engine speed and the measured engine speed is obtained by ESO-based indicated thermal efficiency observer to calculate the ΔE_{ff}. Once the ΔE_{ff} is calculated, the indicated torque $\hat{T}_{ind}$ can be calculated using ΔE_{ff}, Q_{mass}, and λ. The detailed solution consists of system modeling in Equation (18), the model-based ESO design as Equation (20), and the ESO parameter tuning as Equation (23).

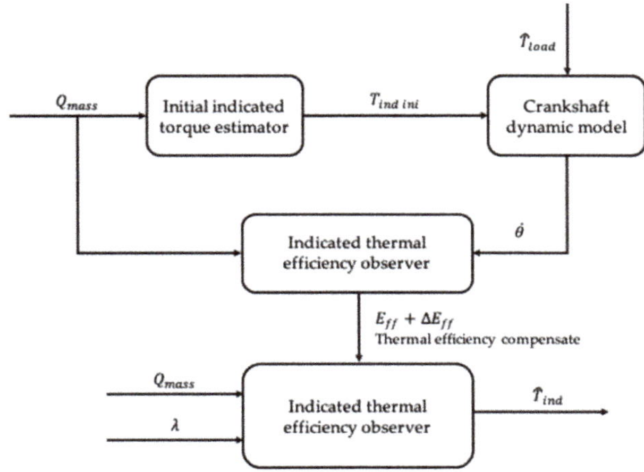

Figure 6. Schematic diagram of the basic structure of the indicated torque observer.

The crankshaft dynamic model can be described as a first order linear system, as in Equation (18):

$$\ddot{\theta} = \frac{4\varphi}{4\pi\lambda[J + \Delta\xi(\theta)]}\left(E_{ff} + \Delta E_{ff}\right)Q_{mass} - \frac{1}{J + \Delta\xi(\theta)}T_{fric} - \frac{1}{J + \Delta\xi(\theta)}T_e. \tag{18}$$

The ESO can be described as:

$$\begin{cases} \dot{z} = Ax + Bu + L(y - \hat{y}) \\ \hat{y} = Cz \end{cases} \tag{19}$$

where z is the estimated value of x, and $\hat{y}$ is the estimated value of y [31]. So, Equation (18) can be described as:

$$\begin{cases} \dot{z}_1 = a_1\left(E_{ff} + z_2\right)Q_{mass} - \frac{1}{J+\Delta\xi(\theta)}T_{fric} - \frac{1}{J+\Delta\xi(\theta)}T_e + \beta_1(y - \hat{y}) \\ \dot{z}_2 = \beta_2(y - \hat{y}) \\ \hat{y} = z_1 \end{cases} \tag{20}$$

where:

$$a_1 = \frac{4\varphi}{4\pi\lambda[J + \Delta\xi(\theta)]}.$$

State matrix A, B, C, and gain matrix L can be described as:

$$A = \begin{bmatrix} 0 & a_1 Q_{mass} \\ 0 & 0 \end{bmatrix}, B = \begin{bmatrix} a_1 E_{ff} \\ 0 \end{bmatrix}, C = \begin{bmatrix} 1 & 0 \end{bmatrix}, L = \begin{bmatrix} \beta_1 \\ \beta_2 \end{bmatrix} \tag{21}$$

To make the indicated thermal efficiency observer in Equation (19) converge, the characteristic roots of the polynomial $A - LC$ should be located in the left half of the complex plane.

$$\left| \lambda I - \overline{A} \right| = s^2 + \beta_1 * s + \beta_2 * a_1 * Q_{mass} \tag{22}$$

According to [31], the observer gains, β_1 and β_2, can be solved by the pole configuration method, and the concept of the bandwidth was introduced. The observer gain factors β_1 and β_2 represent the observation speed of the observer, which is the bandwidth. With larger bandwidth, the observer can acquire more information, but more system noise is also sampled by the observer. In Equation (23), ω_0 must be a positive value to ensure the convergence of the observer.

$$(s + \omega_0)^2 = s^2 + \beta_1 * s + \beta_2 * a_1 * Q_{mass} \tag{23}$$

$$\beta_1 = 2\omega_0, \beta_2 = \frac{\omega_0^2}{a_1 * Q_{mass}}.$$

By adjusting the bandwidth parameter ω_0 to adjust the system gains β_1 and β_2, the observer can get a better observation result. So the complicated parameter adjustment problem of the system gains β_1 and β_2 can be simplified into one parameter ω_0 problem, which can greatly reduce the workload of the indicated thermal efficiency observer parameter adjustment.

3.3. A Self-Learning Observer for Brake Torque Estimation

The T_{load} of a GDI engine (the total torque of the brake torque and friction torque) was estimated. In order to calculate the brake torque (T_e), the engine friction torque (T_{fric}) must be calculated first. An empirical engine friction model is established to change the look-up tables in traditional control algorithms. The engine friction model's parameters could be identified during certain operation points of the engine, such as in idle and fuel cut-off operating conditions, which do not only save a large amount of calibration work, but can also update the friction torque, along with the engine operations. Finally, both T_{fric} and T_e can be calculated from load torque (as in Section 3.1.3).

Engine friction torque includes the friction loss of accessories in mechanical systems, the friction loss of crankshaft bushings, piston liners and piston rings, the friction loss of valve trains, etc. A detailed engine friction torque model could better describe the relationship between T_{fric} and engine working characteristics, but this model would be too complicated and unsuitable for online applications. In traditional applications, a look-up table is used to describe friction torque. Thus, it requires a large amount of calibration work and is not easy to be corrected online. Therefore, a friction torque mean value model is proposed.

According to [32], T_{fric} increases with an increase in engine speed. This relationship can be described as in Equation (24):

$$T_{fric} = C_1 + C_2 \dot{\theta}_{avg} + C_3 \dot{\theta}_{avg}^2 \tag{24}$$

where $\dot{\theta}_{avg}$ is the engine's average speed; and C_1, C_2, and C_3 are the model's parameters.

T_{fric} is also related to oil temperature. The reference friction torque is calculated as T_{ref} at the oil viscosity of μ_{ref}; then, regardless of the initial oil viscosity, after an initial period of transient behavior, the T_{fric} can be expressed with an oil viscosity of μ [33], as follows:

$$\frac{T_{fric}}{T_{ref}} = \left(\frac{\mu}{\mu_{ref}} \right)^n. \tag{25}$$

where n is the model's parameter. n generally is taken within the range of 0.29 to 0.35.

In this study, the correction of the oil viscosity is simplified as follows:

$$T_{fric} = T_{ref} \cdot e^{C_4 \cdot t_{oil}} \tag{26}$$

where t_{oil} is the oil temperature of the engine, and the reference temperature is 0 °C.

Base on the reference model at a temperature of 0 °C, an engine friction model is proposed as:

$$T_{fric} = \left(C_1 + C_2\dot{\theta}_{avg} + C_3\dot{\theta}_{avg}{}^2\right)\cdot e^{C_4 \cdot t_{oil}}. \tag{27}$$

A friction torque look-up table requires a large amount of calibration work and cannot describe the aging issues as the engine operates. Thus, an engine friction model parameter self-learning algorithm is proposed (shown as Figure 7). The parameters are estimated based on engine idle and fuel cut-off operating conditions. Once the engine starts, it will retain idle speed operations. In this working condition, the engine's average speed will remain stable as the temperature of the cooling water and oil start to rise. In this process, the temperature parameter C_4 can be identified. During engine stop operations, the engine stops supplying fuel so the crankshaft rotation speed decreases over a few seconds. During this process, the temperature of the cooling water and oil barely change, so the engine speed parameters C_1, C_2, and C_3 can be identified.

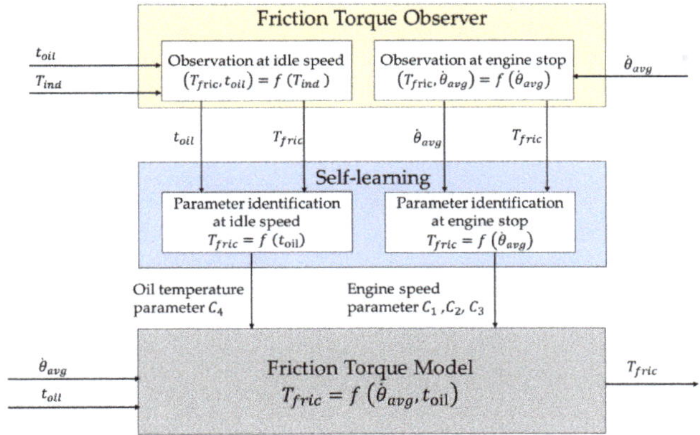

Figure 7. Basic structure of online learning algorithms of friction torque.

In engine idle speed working conditions, the crankshaft average speed remains stable and engine indicated torque is used to overcome friction loss. T_{ind} during idle speed operations is estimated in Section 3 using the engine management model and crankshaft instantaneous speed. Thus, the friction torque can be expressed as:

$$T_{fric} = T_{ind}. \tag{28}$$

In fuel cut-off operating conditions, the crankshaft dynamic system is only affected by friction. Thus, the process can be described as:

$$[J + \Delta\xi(\theta)]\ddot{\theta} = -T_{fric}. \tag{29}$$

During the idle and fuel cut-off operating conditions, the friction torque of the two working conditions can be estimated. However, this only covers a small part of the whole engine operation range. Once the parameters in the model are identified, the friction torque model can be expanded to cover a larger working range. While the friction torque model is a nonlinear model, it is very challenging to directly identify the speed parameter and oil temperature parameter at the same time in online applications. Therefore, it is necessary to combine the two working conditions—idle and fuel cut-off operating conditions—to identify the model parameters.

In idle speed working conditions, the relation between friction torque and oil temperature can be rendered as:

$$\frac{T_{fric1}}{T_{fric2}} = \frac{e^{C_4 t_{oil1}}}{e^{C_4 t_{oil2}}} = e^{C_4(t_{oil1} - t_{oil2})} \tag{30}$$

or

$$\Delta \ln T_{fric} = C_4 \cdot \Delta t_{oil}. \tag{31}$$

In fuel cut-off operating conditions, the oil temperature barely changes. The friction torque model can be described as:

$$0 = C_1\left(T_{fric1} - T_{fric2}\right) + C_2\left(T_{fric1}\dot{\theta}_{avg1} - T_{fric2}\dot{\theta}_{avg2}\right) + C_3\left(T_{fric1}\dot{\theta}_{avg1}^2 - T_{fric2}\dot{\theta}_{avg2}^2\right). \tag{32}$$

By combining the particularity of the specific operating conditions of the engine, the nonlinear coefficients in the model are temporarily eliminated, and the linearization of the model's coefficients is realized. During this process, the structure and parameters of the model do not change, and the relationship between the friction torque described in the model and the speed and oil temperature is not affected, so the accuracy of the model is not affected.

The parameters of the model are identified by the recursive least squares method. The description of the friction torque model can be abstracted into the following formula:

$$y = \varphi^t \cdot \hat{\theta} \tag{33}$$

where φ is the mode input, y is the model's output, and $\hat{\theta}$ stands for the parameters of the model.

The parameters' identification process using the recursive least squares method can be expressed as:

$$\begin{cases} \hat{\theta}(k) = \hat{\theta}(k-1) + K(k)\left[y(k) - \varphi^T(k)\hat{\theta}(k-1)\right] \\ K(k) = \dfrac{P(k-1)\varphi(k)}{1 + \varphi^T(k)P(k-1)\varphi(k)} \\ P(k) = \left[1 - K(k)\varphi^T(k)\right]P(k-1) \end{cases} \tag{34}$$

In engine idle speed working conditions, the system can be described as:

$$\begin{cases} y = \Delta \ln T_{fric} \\ \varphi = \Delta T_{oil} \\ \theta = C_4 \end{cases}, \tag{35}$$

and for fuel cut-off process as:

$$\begin{cases} y = 0 \\ \varphi = \left[T_{fric1} - T_{fric2}, T_{fric1}\dot{\theta}_{avg1} - T_{fric2}\dot{\theta}_{avg2}, T_{fric1}\dot{\theta}_{avg1}^2 - T_{fric2}\dot{\theta}_{avg2}^2\right] \\ \theta = [C_1, C_2, C_3] \end{cases}. \tag{36}$$

After the friction model's parameters are identified, a friction torque model can be acquired. On the basis of the identified friction torque, T_e can then be calculated from the aforementioned T_{load}.

4. Experimental Validation

4.1. Methodology

The validation experiment for engine torque estimation was conducted on the test bench mentioned in Section 2. Then the T_{load}, T_{ind}, T_{fric}, T_e estimation validation results were discussed one-by-one. Additionally, the real-time performance of the engine torque observer in a multi-core ECU was also discussed.

To validate the torque estimation algorithm, the torques were measured or calculated from the sensors equipped in test bench and the engine, as seen in Table 2.

Table 2. Validation data source for engine torque estimation.

Engine Torques	Validation Variable Source
Indicated torque (T_{ind})	Calculated based on P_{cyl} and engine geometry as shown in Equation (9)
Friction torque (T_{fric})	Obtained from the engine motoring experiment in the test bench
Brake torque (T_e)	Measured by in the dynamometer
Load torque (T_{load})	Calculated from T_{fric} and T_e (the sum of T_{fric} and T_e)

First, T_{load} was validated at steady state at the brake mean effective pressure (BMEP) of 6 bar at 1400 rpm and a BMEP of 5 bar at 1600 rpm. T_{load} was not straightforward to be measured accurately using the test bench, especially during transients, due to the existence of the rotational momentums of the engine and dynamometer. Additionally, the transient process was much slower for the load torque variation relative to the combustion process. Therefore, the load torque was validated only at steady state in this work. Then, to conduct the T_{load} estimation validation, an engine motoring experiment was carried out. Recall that T_{load} consists of two torques, T_{fric} and T_e, where T_{fric} is obtained from the engine motoring experiments and T_e is measured directly by the dynamometer. T_{load} is then obtained by summing up T_{fric} and T_e.

For the validation of the T_{fric}, an engine motoring experiment was conducted from 800 to 1800 rpm under multiple oil temperature. In the motoring test, the T_{fric} could be approximately measured by the dynamometer, and for the T_e validation, T_e was directly recorded by the dynamometer.

The validation of the indicated torque was carried out at both steady-state and transient conditions. This is because the dynamic of indicated torque during transients is complex, caused by the breathing and combustion process. As all four cylinders are equipped with in-cylinder pressure sensors, the T_{ind} for validation was calculated from the measured in-cylinder pressure P_{cyl} and engine geometry using Equation (9).

Finally, the real-time performance of the engine torque observer in a multi-core ECU, such as computational time and CPU loads, is discussed.

4.2. Load Torque Estimation Results

Figure 8a shows the load torque estimation results at the BMEP of 6 bar at 1400 rpm, the observation algorithm starts to approach the measured load torque in the first five cycles rapidly, and the load torque observer starts to converge within 40 cycles. After the observer converges, the estimated load torque still fluctuate within a small range of 3 Nm. The fluctuation is caused by the combustion cyclic variations. At a BMEP of 5 bar with the engine speed of 1600 rpm (Figure 8b), the observation algorithm starts to converge within 15 cycles.

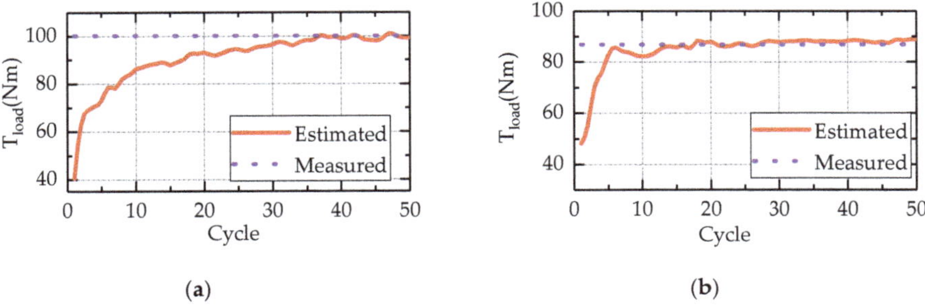

Figure 8. Estimation results of the engine load observer at 1400 rpm, 6 bar (**a**) and 1600 rpm, 5 bar (**b**).

Figure 9 is the estimation results of the disturbance factor ($\Delta\xi(\theta)$) of multiple operating points. It can be seen from the grey trend line that $\Delta\xi(\theta)$ increases with the increase of load at the same speed. This is because the larger load caused the elastic deformation of the crankshaft, which resulted in the change of $\Delta\xi(\theta)$. Under the same load, with the increase of engine speed, $\Delta\xi(\theta)$ also tends to increase. So the correlation analysis between $\Delta\xi(\theta)$ and engine operating variables was done. The correlation coefficient between $\Delta\xi(\theta)$ and the engine speed is 0.72. The correlation coefficient between $\Delta\xi(\theta)$ and throttle opening (θ_{THR}) is 0.85.

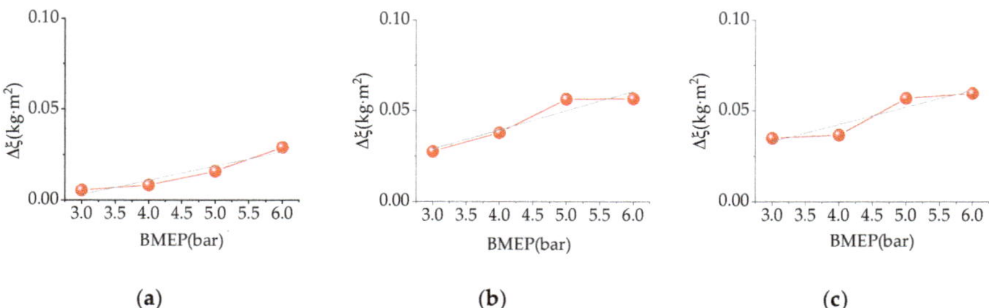

Figure 9. Estimation results of the disturbance factor at multiple operating points at 1200 rpm (**a**), 1400 rpm (**b**), and 1600 rpm (**c**).

$\Delta\xi(\theta)$ can compensate the error caused by the simplification of the crankshaft dynamic model. As $\Delta\xi(\theta)$ has a significant correlation with the engine speed ($\dot{\theta}_{avg}$) and throttle opening (θ_{THR}), a disturbance factor model was built. In the range of engine speed from 1000 rpm to 1600 rpm and BMEP from 3 bar to 6 bar, the trained disturbance factor model expression is:

$$\Delta\xi = -0.77 + \left(9.7 \times 10^{-4}\right)\dot{\theta}_{avg} + 0.01 \times \theta_{THR} - \left(3.3 \times 10^{-7}\right)\dot{\theta}_{avg}{}^2 - \left(7.4 \times 10^{-7}\right)\dot{\theta}_{avg} \times \theta_{THR} \tag{37}$$

4.3. Indicated Torque Estimation Result

4.3.1. Indicated Torque Estimation under Transient State

Figure 10a shows the indicated torque estimation results from a throttle opening of 7.2–10.8% at 1000 rpm. It can be seen that at the moment when the throttle starts to change, the estimated indicated torque fluctuates. This is because the ESO-based indicated torque observer not only used the engine speed and manifold pressure as the state feedback, but also used the derivative term of the engine speed and manifold pressure as feedback. After the throttle opening changes, there are several cycle delays between the estimated indicated torque and the measured indicated torque. However, when the indicated torque observation is stable, the estimation error between the estimated value and the measured value is within 3%.

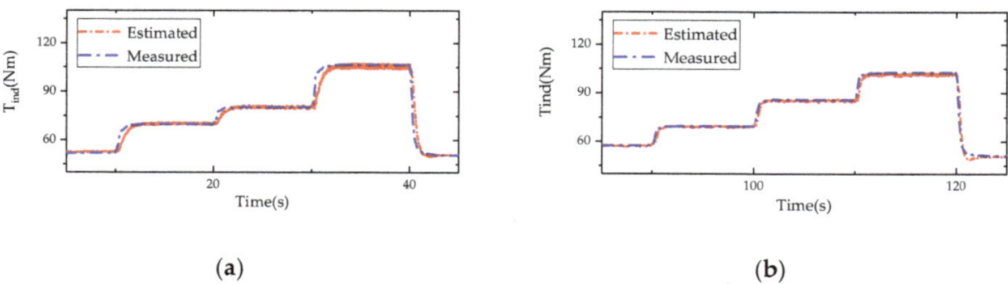

Figure 10. Comparison of the measured and estimated indicated torque at transient states. 1000 rpm with θ_{THR} of 7.2–10.8% (**a**); 1400 rpm with θ_{THR} of 9–11.7% (**b**).

Figure 10b shows the indicated torque estimation results from a throttle opening of 9–11.7% at 1400 rpm. Moreover, the estimation error between the estimated value and the measured value is also within 3%.

4.3.2. Indicated Torque Estimation under a Steady State

Figure 11a shows the indicated torque estimation result at 3 bar to 6 bar at 1000 rpm. In steady state operations, the estimated indicated torque has a certain deviation from the measure value. At 3 bar, the estimated indicated torque is about 1.83 Nm larger than the measured value, and the mean relative error is 3.7%. At 4 bar, 5 bar, and 6 bar, the mean relative error is 4.7%, 1.8%, and 3.6%, respectively. In summary, the mean relative error of the indicated torque estimation at the 3–6 bar operating point covered by 1000 rpm is within 5%. As the engine speed increases to 1400 rpm (Figure 11b), the mean relative error of the indicated torque estimation is within 5%, from 3–6 bar.

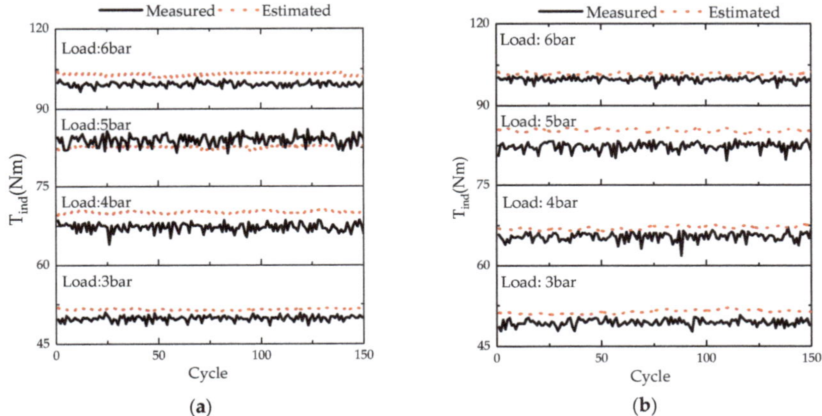

Figure 11. Comparison of the measured and estimated indicated torque at 1000 rpm (**a**) and 1400 rpm (**b**).

Figure 12 shows the indicated torque estimation result summary from 1000 to 1800 rpm; the average estimation accuracy under different loads could reach 96.1%. The estimation of the indicated torque not only provides a basis for the optimal control of the engine ignition's advance angle, but also lays a foundation for the friction torque model identification and brake torque estimation.

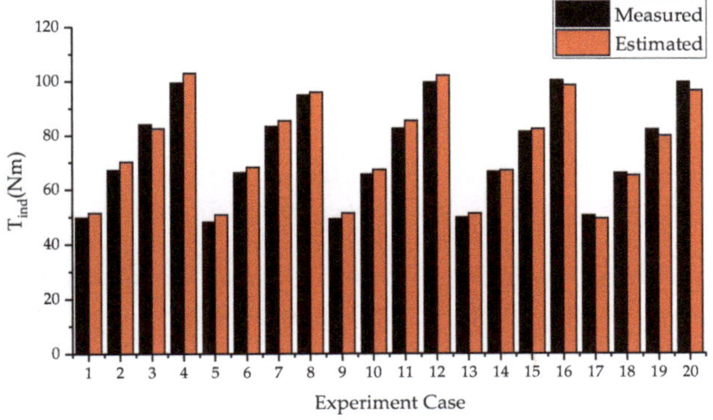

Figure 12. Comparison of the estimated indicated torque and measured indicated torque.

4.4. Friction Torque Estimation Results

4.4.1. Friction Model Parameter Identification Result

A test bench experiment was carried out to estimate the oil temperature parameter C_4 at an engine speed of 1000 rpm in an idle state. For the engine speed parameters C_1, C_2, C_3, an engine fuel cut-off experiment was carried out at an oil temperature of 55 °C. The trend of gradual convergence is presented, and convergence was achieved by 40 fittings during the engine stop process, as shown in Figure 13.

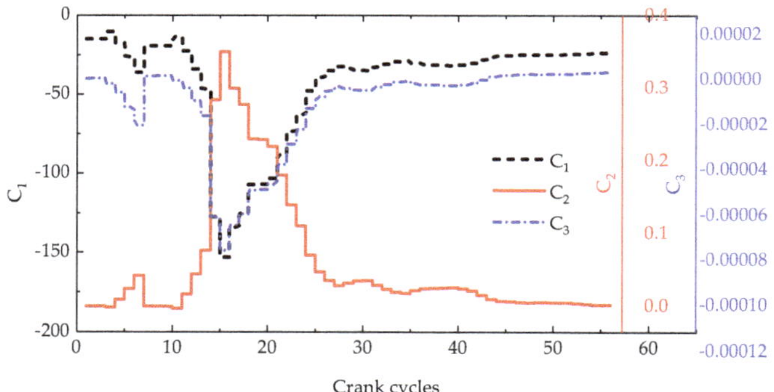

Figure 13. Friction model parameter identification from fuel cut-off operating conditions.

The final identified parameter values are shown in Table 3.

Table 3. Engine brake torque estimation task table.

Parameter	Fit from Observer	Fit from Motored Test
C_1	-23.86	-25.44
C_2,	2.45×10^{-3}	3.29×10^{-3}
C_3,	-5.20×10^{-6}	-5.45×10^{-6}
C_4	-4.67×10^{-3}	-5.10×10^{-3}

4.4.2. Friction Torque Estimation Validation

In order to verify the validity of the identification parameters and the accuracy of the friction torque mean model based on the identification parameters, an engine motoring experiment was carried out. The friction torque model obtained by the model parameter identification method can better describe the average value of the friction torque at different speeds and different oil temperatures.

As shown in Figure 14, by comparing the friction torque model identified by learning with the friction torque value obtained in the engine motoring experiment, the maximum deviation is 1.55 Nm and the average deviation is 1.27 Nm at an engine oil temperature of 80 °C. When the oil temperature is 70 °C, the maximum deviation is 1.39 Nm and the average deviation is 1.17 Nm. At an oil temperature of 60 °C, the maximum deviation of the friction torque is 0.75 Nm and the average deviation is 0.53 Nm. At an oil temperature of 50 °C, the maximum deviation of the friction torque output from the output of the friction torque model is 0.58 Nm and the average deviation is 0.27 Nm. As the oil temperature increases, the deviation between the friction torque model and the friction torque obtained by the reverse drag test tends to increase.

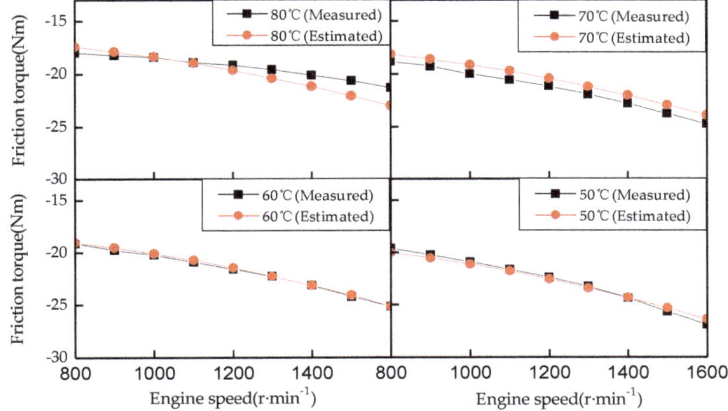

Figure 14. Comparison of the estimated friction torque and measured friction torque.

As shown in Figure 15, at different engine speeds, the friction torque model identified by learning was compared with the friction torque value obtained by the engine motoring experiment. At 800 rpm, the maximum deviation of the model compared with the reversed data is 1.48 Nm and the average deviation is 0.85 Nm. At 1000 rpm, the maximum deviation of the model is 1.41 Nm and the average deviation is 0.95 Nm. At 1200 rpm, the maximum deviation is 1.34 Nm and the average deviation is 0.91 Nm. At a speed of 1400 rpm, the maximum deviation of the model compared with the reversed data is 1.33 Nm and the average deviation is 0.92 Nm. At a speed of 1600 rpm, the maximum deviation of the lower model compared to the inverted data is 0.88 Nm and the average deviation is 0.48 Nm. With an increase of the rotational speed, the friction torque average model is relatively stable compared to the friction torque obtained by the reverse drag test, showing a decreasing trend.

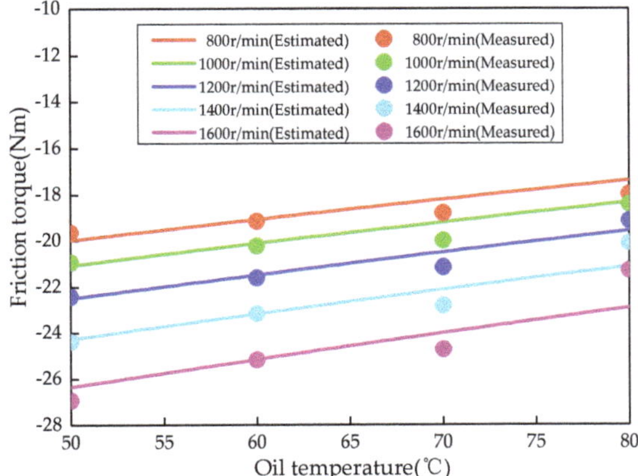

Figure 15. Comparison of the estimated friction torque and measured friction torque.

4.5. Brake Torque Estimation Result

On the basis of the identified friction torque, the engine brake torque (T_e) is decoupled from the load torque (T_{load}). A test bench experiment was done to validate the estimated brake torque compared to the reference brake torque measured by a dynamometer.

Figure 16a,b shows the results of an engine speed of 1000 rpm at 3 bar and 4 bar, respectively. Under a 3 bar load, $\hat{T}_e$ starts to approach the measured value rapidly in the first five cycles, and the brake torque estimates over 30 cycles tend to be stable. However, due to the combustion cyclic variations, $\hat{T}_e$ fluctuates. Between the 30th cycle and the 50th cycle, $\hat{T}_e$ fluctuates within 4 Nm and has good stability. At a load of 4 bar, the brake torque observation algorithm starts to approach the measured brake torque in the first 10 cycles and finally approaches stability after calculating 17 cycles. $\hat{T}_e$ fluctuates between the 17th and 50th cycles.

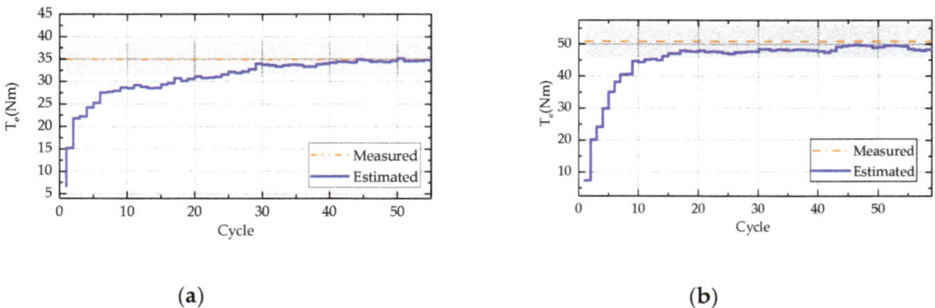

(**a**)   (**b**)

Figure 16. Estimation of the engine brake torque observer at 1000 rpm: 3 bar (**a**) and 4 bar (**b**).

Figure 17a,b shows the results of an engine speed of 1400 rpm and a load of 3 and 4 bar. At the load of 3 bar, $\hat{T}_e$ tends to converge after 13 cycles. The estimated value of the brake torque still fluctuates, and its fluctuation range is within 3 Nm. Under the 4 bar load, $\hat{T}_e$ convergence process is slower and approaches the measured value after 30 cycles. The amplitude of the fluctuation is about 4 Nm.

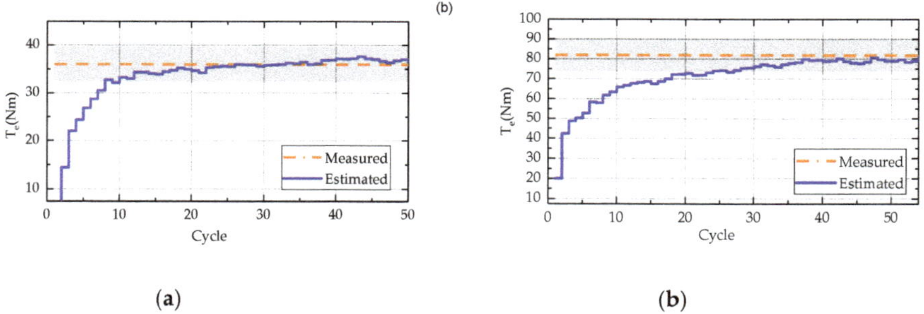

(**a**)   (**b**)

Figure 17. Estimation of the engine brake torque observer at 1400 rpm: 3 bar (**a**) and 4 bar (**b**).

Figure 18a,b shows the observations of the engine speed at 1600 rpm at 3 bar and 5 bar, respectively. Under a load of 3 bar, the brake torque converges to the reference value after 20 cycles, and the amplitude of the fluctuation after stabilization is about 5 Nm. Under a 5 bar load, the approach value is almost reached after five cycles, but after 15 cycles, the observer tends to be stable, and the amplitude of the brake torque after stabilization is small (concentrated around 3 N); moreover, the observation accuracy is promising.

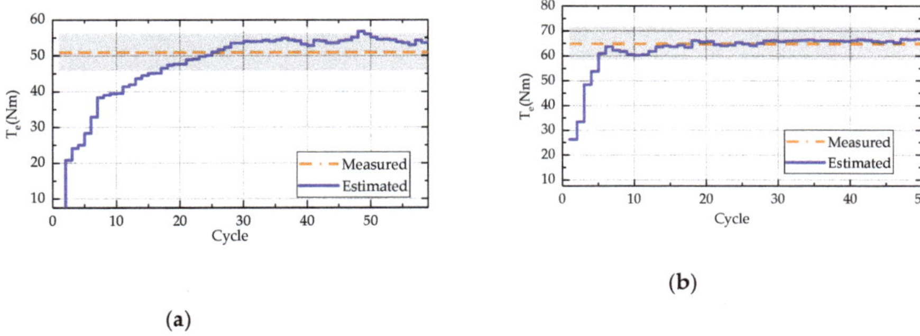

(a)   **(b)**

Figure 18. Estimation of the engine brake torque observer at 1600 rpm: 3 bar (**a**) and 4 bar (**b**).

Based on the above observations of steady-state conditions, the observation algorithm starts to converge rapidly during the first 10 cycles and basically stabilizes in the 20th to 30th cycles. The estimation accuracy of the steady-state test conditions increases to 95.8%. Even under steady-state conditions, there are still fluctuations in the observed brake torque because the brake torque is estimated based on the load torque estimation. Due to the cyclic variations of the engine combustion process, the cyclic fluctuation of the indicated torque further results in speed fluctuations, which in turn cause fluctuations in the estimations of the T_{load} and T_e.

4.6. Real-Time Performance of the Brake Torque Observer in a Multi-Core ECU

Considering the limited computational speed of the embedded platform, T_{load} observation, the T_{ind} observation, T_{fric} self-learning, and T_e calculation algorithm need to select methods suitable for operation on the embedded computing platform (specifications as Table 4). On the embedded computing platform, it is necessary to meet the characteristics of fast calculation speed, strong real-time performance, and the accuracy of observation.

Table 4. Embedded micro-controller specifications.

Variable	Value
Platform	Infineon AURIX TC275T-64F200W
CPU processor frequency	200 MHz
GTM processor frequency	100 MHz
RAM	472 KB
Flash ROM	4 MB

As shown in Table 5, the torque observation algorithm can be divided into four tasks in the embedded platform. The first task is a model-based Kalman filter for engine velocity and acceleration processing. The second task is the load torque estimation algorithm. The third task is the model-based indicated torque estimation algorithm. The fourth task is the friction torque self-learning and brake torque calculation task.

Table 5. Engine brake torque estimation task table.

Task Index	Task Content	Allocation
Task_0	Crankshaft speed signal processing	GTM processor
Task_1	T_{load} estimation	CPU
Task_2	T_{ind} estimation	CPU
Task_3	T_{fric} self-learning and T_e calculation	CPU

As shown in Figure 19, the velocity-related signal is processed utilizing a general timer module (GTM), which is isolated from the CPUs. The velocity processing results by GTM were shared in a public memory area that could be accessed by other CPUs, which could reduce the processing load of the CPU in the microcontrollers. The algorithm was converted to C code via Target-Link, compiled into executable files, and processed in an AURIX based ECU.

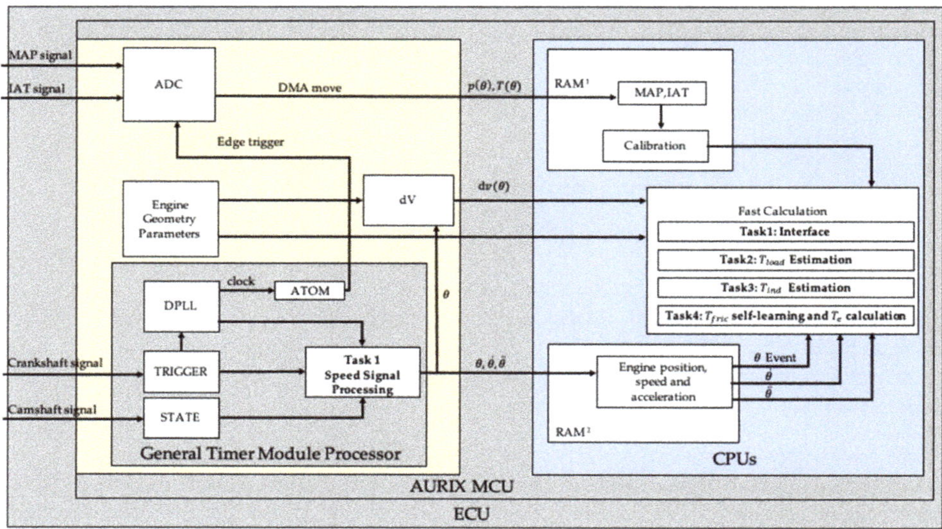

Figure 19. The software and hardware collaborative processing framework for brake torque estimation.

Third-party developing tools (shown as Figure 20) were used to measure the CPU load of the torque observation tasks. As shown in Table 6, task 0 has a CPU load of 0.55% because the main signal processing algorithms is implemented in the GTM processor. Task 1 has a CPU load of 16.04% and task 2's CPU load is 29.87%, which are the tasks of the load torque estimation and indicated torque estimation algorithms. Task 3 mainly processes the friction torque and brake torque calculation. This algorithm works at certain operating points, and a few calculations are needed, so the algorithm has a CPU load of 1.33%. Moreover, the whole torque estimation algorithm costs 3.2 ms per cycle, which is promising for real-time applications.

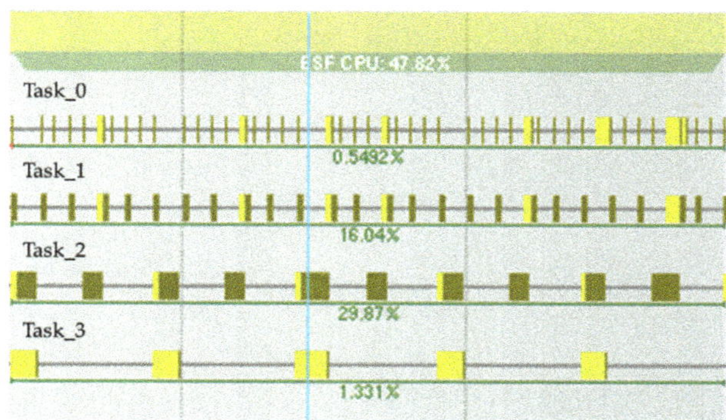

Figure 20. Task_0, task_1, task_2, and task_3 CPU load when executing the algorithm.

Table 6. Engine brake torque estimation task table.

Task Index	Task Content	CPU Load
Task_0	Crankshaft speed signal processing	0.55%
Task_1	T_{load} estimation	16.04%
Task_2	T_{ind} estimation	29.87%
Task_3	T_{fric} self-learning and T_e calculation	1.33%

5. Conclusions

In this paper, an approach for the online estimation of engine brake torque was proposed, utilizing the standard crankshaft instantaneous speed signal of stock engines. The main accomplishments are summarized below.

(1) An in-cycle crank angle-based crankshaft dynamic model was established, where a crank angle interval is chosen by experiments to estimate the load torque without influence from the combustion torque. A disturbance factor was designed to compensate for the deviation of the model from the real engine. Results show that the error of the estimated load torque is within 3 Nm.

(2) An indicated torque observer algorithm is proposed. The observer is an ESO, and the indicated torque observer was validated for both the steady state and transient state in experiments. Results show that the estimation error is less than 4%.

(3) A self-learning friction torque estimator was developed, which allows one to estimate the engine brake torque with the aforementioned sub-estimators. For this estimator, the parameters of the friction torque model were identified online in idle and fuel cut-off operating conditions. Experimental validation of the results show that the brake torque observation error is less than 5%.

(4) The proposed algorithm was implemented in a multi-core ECU with a cycle-triggered runnable. Results show that the corresponding computational time is 3.2 ms with a CPU computational load of 47.41%. This algorithm is suitable for real-time control applications, such as the optimal torque split control of HEVs.

Author Contributions: Q.T. and D.Z. carried out the engine test. K.S., Q.T. and D.Z. worked together on the experimental data analysis, model-based observer design, and paper writing. H.X. provided many valuable insights on the model-based observer design and validation.

Funding: This research was funded by the Great Wall Motors Company. This work was also supported by the National Natural Science Foundation of China under the grand number of 51906174.

Acknowledgments: The authors would like to thank Tao Chen of Tianjin University for several fruitful discussions and his useful insights during the project. The authors would like to thank Dongxian Song from Great Wall Motors, Kang Xu from UAES, and Patrick Leteinturier from Infineon Technologies for their insights. A special thank you to Zhiwei Yang from Tianjin University for assisting us in the management and maintenance of the test bench.

Conflicts of Interest: The authors declare no conflict of interest.

Nomenclature

Symbols

J	Moment of inertia
$\Delta\xi$	Disturbance factor
T_{ind}	Indicated torque
T_r	Reciprocating torque
T_{fric}	Friction torque
T_e	Brake torque
T_{gas}	Gas torque
θ	Crankshaft angle
$m_{r,osc}$	Mass of oscillating part of crank-rod mechanism and piston group
$m_{r,rot}$	Mass of rotating part of crank-rod mechanism
l_{osc}	Length of oscillating part of connecting rod
l_{rot}	Length of rotating part of connecting rod

P	Pressure
V	Gas volume
κ	Polytrophic process factor
Q_{mass}	Intake air mass
t_{oil}	Lubricating oil temperature
Subscripts	
MAP	Manifold absolute pressure
IAT	Intake air temperature
THR	Throttle
eng	Engine
avg	Average
ini	Initial
ff	Feed forward
cyl	Cylinder
k	Cylinder index

References

1. Tong, Y. *Study on the Coordinated Control Issue in Parallel Hybrid Electric System*; Beijing Tsinghua University: Beijing, China, 2004.
2. Yan, F.; Wang, J.; Huang, K. Hybrid Electric Vehicle Model Predictive Control Torque-Split Strategy Incorporating Engine Transient Characteristics. *IEEE Trans. Veh. Technol.* **2012**, *61*, 2458–2467. [CrossRef]
3. Kim, H.; Kim, J.; Lee, H. Mode Transition Control Using Disturbance Compensation for a Parallel Hybrid Electric Vehicle. *Proc. Inst. Mech. Eng. Part D J. Automob. Eng.* **2011**, *225*, 150–166. [CrossRef]
4. Yang, C.; Du, S.; Li, L.; You, S.; Yang, Y.; Zhao, Y. Adaptive real-time optimal energy management strategy based on equivalent factors optimization for plug-in hybrid electric vehicle. *Appl. Energy* **2017**, *203*, 883–896. [CrossRef]
5. Cavina, N.; De Ponti, F.; Rizzoni, G. Fast Algorithm for On-Board Torque Estimation. *SAE Tech. Pap.* **1999**, 10. [CrossRef]
6. Sellnau, M.C.; Matekunas, F.A.; Battiston, P.A.; Chang, C.F.; Lancaster, D.R. Cylinder-Pressure-Based Engine Control Using Pressure-Ratio-Management and Low-Cost Non-Intrusive Cylinder Pressure Sensors. *SAE Trans.* **2000**, *109*, 899–918.
7. Thor, M.; Egardt, B.; McKelvey, T.; Andersson, I. Using combustion net torque for estimation of combustion properties from measurements of crankshaft torque. *Control Eng. Pract.* **2014**, *26*, 233–244. [CrossRef]
8. Thor, M.; Egardt, B.; McKelvey, T.; Andersson, I. Parameterized Diesel Engine Combustion Modeling for Torque Based Combustion Property Estimation. *SAE Tech. Pap.* **2012**, 13. [CrossRef]
9. Larsson, S.; Andersson, I. Self-optimising control of an SI-engine using a torque sensor. *Control Eng. Pract.* **2008**, *16*, 505–514. [CrossRef]
10. Azzoni, P.M.; Minelli, G.; Moro, D.; Flora, R.; Serra, G. Indicated and Load Torque Estimation using Crankshaft Angular Velocity Measurement. *SAE Tech. Pap.* **1999**, 9. [CrossRef]
11. Zweiri, Y.H.; Seneviratne, L.D. Diesel engine indicated and load torque estimation using a non-linear observer. *Proc. Inst. Mech. Eng. Part D J. Automob. Eng.* **2006**, *220*, 775–785. [CrossRef]
12. Hamedovic, H.; Raichle, F.; Bohme, J.F. In-cylinder pressure reconstruction for multicylinder SI-engine by combined processing of engine speed and one cylinder pressure. In Proceedings of the IEEE International Conference on Acoustics, Speech, and Signal Processing, Philadelphia, PA, USA, 23 March 2005.
13. Li, Z.; Guo, Z.; Hu, C.; Li, A. On-line indicated torque estimation for internal combustion engines using discrete observer. *Comput. Electr. Eng.* **2017**, *60*, 100–115.
14. Mocanu, F.; Taraza, D. Estimation of Main Combustion Parameters from the Measured Instantaneous Crankshaft Speed. *SAE Tech. Pap.* **2013**, 16. [CrossRef]
15. Desbazeille, M.; Randall, R.B.; Guillet, F.; Badaoui, M.E.; Hoisnard, C. Model-based diagnosis of large diesel engines based on angular speed variations of the crankshaft. *Mech. Syst. Signal Process.* **2010**, *24*, 1529–1541. [CrossRef]
16. Grondin, O.; Letellier, C.; Maquet, J.; Aguirre, L.A.; Dionnet, F. Direct Injection Diesel Engine Cylinder Pressure Modelling via NARMA Identification Technique. *SAE Tech. Pap.* **2005**, 11. [CrossRef]

17. Guezennec, Y.G.; Gyan, P. A Novel Approach to Real-Time Estimation of the Individual Cylinder Combustion Pressure for S.I. Engine Control. *Int. Congr. Expo.* **1999**, 12. [CrossRef]
18. Ge, Y.; Li, G.; Di, X. Diesel Engine Torque Estimation Based on ENN. *SAE Tech. Pap.* **2015**. [CrossRef]
19. Geveci, M. An investigation of crankshaft oscillations for cylinder health diagnostics. *Mech. Syst. Signal Process.* **2005**, *19*, 1107–1134. [CrossRef]
20. Liu, F.; Amaratunga, G.; Collings, N. A Fourier Analysis Based Synthetic Method for In-cylinder Pressure Estimation. *Powertrain Fluid Syst. Conf. Exhib.* **2006**. [CrossRef]
21. Liu, F.; Amaratunga, G.A.J.; Collings, N.; Soliman, A. An Experimental Study on Engine Dynamics Model Based In-Cylinder Pressure Estimation. *SAE Tech. Pap.* **2012**, 19. [CrossRef]
22. Weissenborn, E.; Bossmeyer, T.; Bertram, T. Adaptation of a zero-dimensional cylinder pressure model for diesel engines using the crankshaft rotational speed. *Mech. Syst. Signal Process.* **2011**, *25*, 1887–1910. [CrossRef]
23. Wang, J.; Yang, F.; Ouyang, M. Cylinder by cylinder indicated torque and combustion feature estimation based on engine instantaneous speed and one cylinder pressure through error similarity analysis. *SAE Tech. Pap.* **2015**, 10. [CrossRef]
24. Kallenberger, C.; Hamedovic, H.; Zoubir, A.M. Evaluation of torque estimation using gray-box and physical crankshaft modeling. In Proceedings of the 2008 IEEE International Conference on Acoustics, Speech and Signal Processing, Las Vegas, NV, USA, 31 March–4 April 2008; Volume 1–12, p. 1529.
25. Kallenberger, C.; Hamedovic, H.; Zoubir, A.M. Comparison of the Extended Kalman Filter and the unscented Kalman filter for parameter estimation in combustion engines. In Proceedings of the 2007 15th European Signal Processing Conference, Poznan, Poland, 3–7 September 2007.
26. Citron, S.J.; O'Higgins, J.E. Cylinder-by-cylinder engine pressure and pressure torque waveform determination utilizing crankshaft speed fluctuations. *SAE Tech. Pap.* **1989**, 18. [CrossRef]
27. Kulah, S.; Donkers, T.; Willems, F. Virtual Cylinder Pressure Sensor for Transient Operation in Heavy-Duty Engines. *Sae Int. J. Engines* **2015**, *8*, 1029–1040. [CrossRef]
28. Stotsky, A. Adaptive Estimation of the Engine Friction Torque. *Eur. J. Control* **2007**, *13*, 618–624. [CrossRef]
29. Tong, Q.; Xie, H.; Zou, D.; Ruan, D. Indicated Torque Estimation of Gasoline Direct Injection Engine via Optimized Model-Based Observer. *IFAC-Papers OnLine* **2018**, *51*, 827–832. [CrossRef]
30. Guzzella, L.; Onder, C. *Introduction to Modelling and Control of Internal Combustion Engines Systems*; Springer Science & Business Media: Berlin, Germany, 2009.
31. Gao, Z. Scaling and bandwidth-parameterization based controller tuning. In Proceedings of the 2003 American Control Conference, Denver, CO, USA, 4–6 June 2003.
32. Patton, K.J.; Nitschke, R.G.; Heywood, J.B. Development and Evaluation of a Friction Model for Spark-Ignition Engines. *Soc. Automot. Eng.* **1989**, 24. [CrossRef]
33. Paul, J.S.; Leong, D.; Murphy, M. Contributions to Engine Friction During Cold, Low Speed Running and the Dependence on Oil Viscosity. *SAE Trans.* **2005**, *114*, 1191–1201.

Article

Numerical Investigation of 48 V Electrification Potential in Terms of Fuel Economy and Vehicle Performance for a Lambda-1 Gasoline Passenger Car

Federico Millo *, Francesco Accurso *, Alessandro Zanelli and Luciano Rolando

Department of Energy, Politecnico di Torino, Corso Duca degli Abruzzi 24, 10129 Torino, Italy
* Correspondence: federico.millo@polito.it (F.M.); francesco.accurso@polito.it (F.A.);
  Tel.: +39-011-090-4517 (F.M. & F.A.)

Received: 17 July 2019; Accepted: 1 August 2019; Published: 3 August 2019

Abstract: Real Driving Emissions (RDE) regulations require the adoption of stoichiometric operation across the entire engine map for downsized turbocharged gasoline engines, which have been so far generally exploiting spark timing retard and mixture enrichment for knock mitigation. However, stoichiometric operation has a detrimental effect on engine and vehicle performances if no countermeasures are taken, such as alternative approaches for knock mitigation, as the exploitation of Miller cycle and/or powertrain electrification to improve vehicle acceleration performance. This research activity aims, therefore, to assess the potential of 48 V electrification and of the adoption of Miller cycle for a downsized and stoichiometric turbocharged gasoline engine. An integrated vehicle and powertrain model was developed for a reference passenger car, equipped with a EU5 gasoline turbocharged engine. Afterwards, two different 48 V electrified powertrain concepts, one featuring a Belt Starter Generator (BSG) mild-hybrid architecture, the other featuring, in addition to the BSG, a Miller cycle engine combined with an e-supercharger were developed and investigated. Vehicle performances were evaluated both in terms of elasticity maneuvers and of CO_2 emissions for type approval and RDE driving cycles. Numerical simulations highlighted potential improvements up to 16% CO_2 reduction on RDE driving cycle of a 48 V electrified vehicle featuring a high efficiency powertrain with respect to a EU5 engine and more than 10% of transient performance improvement.

Keywords: hybrid electric vehicle; real driving emissions; fuel consumption; vehicle performance; electric supercharger; Lambda-1 engine; 48 V Mild Hybrid

---

1. Introduction

The regulatory framework for light-duty vehicles concerning both pollutants emissions and greenhouse gasses has become increasingly demanding in the last decade.

As far as pollutant emissions are concerned, the introduction of the Worldwide Harmonized Light-Duty Vehicles Test Procedure (WLTP) and of Real Driving Emissions (RDE) tests may represent a major challenge for car manufacturers. The new European Regulation [1] requires indeed the declaration of Auxiliary Emissions Strategies (AES) which may be activated during real driving operation and may have an impact on tailpipe pollutant emissions. Among these techniques, the fuel enrichment is a widely used practice in downsized gasoline engine to prevent engine damage from hot exhaust gases. As a matter of fact, downsized and turbocharged engines demonstrated to be an effective solution for fuel consumption reduction on type-approval driving cycles [2], but at the same time they require the exploitation of techniques (as spark timing delay and mixture enrichment) to increase the specific output power without component lifetime deterioration. In this context, embracing the strategy of delaying the spark timing for knock mitigation brings the temperature of the exhaust gases at a higher level, eliciting a further enrichment of the mixture. What is more, the use of mixture enrichment (i.e., moving from a stoichiometric to rich combustion) while cooling the in-cylinder charge and the exhaust gases, increases the CO emissions during high engine load operation, as highlighted by Clairotte et al. [3]. Moreover, future Regulation may prescribe a conformity factor for CO [4] limiting the exploitation of the fuel enrichment and pushing the development of the gasoline engines towards stoichiometric operation [5,6]. This aim can be achieved with the adoption of different powertrain technologies, like cooled exhaust manifold, high temperature resistant turbine as well as water injection, Miller cycle and advanced turbocharging, as explained by Glahn et al. [6].

On the other hand, concerning the greenhouse emissions, the average fleet target value for CO_2 has been set to be 95 g/km from 2021 [7], measured according to the new Worldwide Harmonized Light-Duty Vehicles Test Procedure (WLTP) and correlated to the value of the New European Driving Cycle (NEDC). Moreover, recently the European Parliament and the Council adopted Regulation (EU) 2019/631 [8] setting CO_2 emission performance standards for new passenger cars in the EU for the period after 2020. This Regulation prescribes a 15% reduction of the CO_2 emissions in 2025 over the 2020 target according to the WLTP procedure and a 37.5% reduction of the CO_2 emissions in 2030. To achieve these unprecedented CO_2 emissions, besides a remarkable improvement of the internal combustion engine efficiency, an increasing level of electrification of the powertrain is mandatory [9]. In this context, the employment of low voltage electrification solutions (i.e., 48 V electrification) has been proven to be the enabler for lower fuel consumption (and thus lower CO_2 emissions) at a reduced cost of the hardware and of the integration in the vehicle, as discussed by Bao et al. [10].

In such a framework, the use of a virtual environment for the evaluation of the benefits which can be achieved by innovative powertrains can reduce the cost and the time required for the assessment of different technology options, leaving the experimental activity only for validation and homologation purposes. Indeed, the fuel economy assessment for conventional and electrified powertrain concepts through numerical simulation is a widespread topic of many research activities: Bozza et al. [11] evaluated the effects of various engine techniques both on the overall performance map and on the vehicle CO_2 emission; Lee et al. [12] presented a development and validation of a 48 V Mild-Hybrid Electric Vehicle (MHEV) model exploiting the advantage of the co-simulation approach; in the work presented by Mamikoglu et al. [13], fuel economy for different engine and powertrain technologies for a compact class car were assessed on different driving cycles. Additionally, in the work introduced by Benajes et al. [14], a vehicle model simulation has been devoted to the evaluation of the advantages from the combination of electrification and advanced combustion modes in a P0 48 V hybrid vehicle.

In these works, the engine was modeled as a map on the basis of the assumption of a quasi-static behaviour of the engine. However, as pointed out by Millo et al. [15], a map-based approach is suitable to predict fuel economy over NEDC cycle which is a moderately transient cycle, but it can show significant discrepancies on the fuel consumption evaluation on highly dynamic driving cycles, such as Worldwide Harmonized Light-Duty Vehicles Test Cycle (WLTC) and Real Driving Emissions test cycles (RDE). The introduction of WLTC and RDE driving cycles indeed requires the usage of a physically-based engine model, such as, for instance the Fast Running Model (FRM) [16], which is fully capable of capturing effects like the turbo-lag, and therefore better suited for the prediction of the fuel consumption and enables a complete assessment of vehicle performance. Only recently have FRM engine models been adopted as a fundamental tool for the powertrain and vehicle development. For example, in the work done by Andert et al. [17], the validation of a so-called "road-to-rig-to-desktop" methodology is presented and the potential in terms of accuracy of adoption of an integrated vehicle simulation featuring an FRM engine is shown. Additionally, in the research article by Dorsch et al. [18], the predictive capability of a fully physical FRM engine coupled with vehicle and control models is exploited for the evaluation of a novel predictive Spark Ignition (SI) combustion model. On top of that, following the path of the auxiliary electrification, the opportunity to employ and electrified boosting solution requires that, for vehicle performance evaluation, an FRM engine model is used in order to capture the fluid-dynamics (i.e., to accurately predict the turbolag effect). To this regard, in the work carried out by Griefnow et al. [19], an FRM engine is used, in a vehicle simulation, with the aim to develop a Model Predictive Control for an electric supercharger.

The research works presented so far are mainly focused on definition and validation of methodologies with the aim to improve the vehicle development process. Conversely, this work moves from this context using the so far discussed approach to suggest a possible development of a next generation gasoline powertrain concept. With this in mind, taking into account the novel introduction of WLTP and RDE regulations and availability of innovative powertrain technologies, the aim of this work is therefore to evaluate, through numerical simulation, the benefits on both vehicle performance and fuel consumption of various electrified powertrain concepts equipped with a fully stoichiometric engine. Three driving cycles were selected for fuel consumption evaluation: the NEDC, the Worldwide Harmonized Light-Duty Test Cycle (WLTC) and a more dynamic RDE driving cycle developed from the base of WLTC, defined as "standardized random test for an aggressive driving style" and known with the acronym RTS-95 [20]. Moreover, the transient performances of the vehicle were investigated both in term of 0 to 100 km/h acceleration and on constant gear elasticity maneuvers (80 to 120 km/h in VI gear and 60 to 80 km/h in VI gear). As far as the hybrid architecture is concerned, the authors selected for the analysis a P0 48 V Mild-Hybrid Electric Vehicle architecture. The proposed hybrid architecture features a Belt Starter Generator (BSG), a dual voltage electric network (12 V + 48 V), and a stoichiometric engine. Several technologies were investigated as countermeasure of the performance deterioration due to the stoichiometric combustion exploitation as, among others, the adoption of the Miller cycle and of the electric boosting.

2. Methodology

In this section a detailed overview of the developed virtual test rig is presented, starting from the description of the case study and then moving to the explanation of the engine and vehicle modelling and control for the proposed powertrain concepts.

2.1. Case Study

The test case is a compact Sport Utility Vehicle (SUV) with a gasoline engine equipped with a 6-speed manual transmission. The main vehicle specifications for each driving cycle (since type approval regulations prescribe different test conditions for the vehicle on NEDC and WLTP, different vehicle mass and vehicle power demand have to be considered for the same vehicle), are outlined in Table 1.

Table 1. Main vehicle characteristics for the New European Driving Cycle (NEDC), Worldwide Harmonized Light-Duty Vehicles Test Cycle (WLTC) and the standardized random test for an aggressive driving style (RTS-95).

Parameter	Unit	NEDC	WLTC	RTS-95
Vehicle Mass	kg	1470	1630	1630
Rolling Radius	m	0.333	0.333	0.333
Power demand @ 100 km/h	kW	18	20.5	20.5
Electrical load (engine on)	W	220	400	400
Electrical load (engine off)	W	120	120	120

For this analysis, the engine chosen is a 1.4 liter EU5 gasoline turbocharged engine, whose main features are given in Table 2. This engine reaches a maximum brake torque of 250 Nm at 2250 RPM and a rated power of 121 kW at 5500 RPM, featuring a Compression Ratio (CR) of 9.8 and a Variable Valve Actuation (VVA) technology.

Table 2. Main technical specifications for the reference engine.

Parameter	Unit	Value
Displacement	cm^3	1368
Compression Ratio	-	9.8
Rated Power	kW	121 at 5500 RPM
Rated Torque	Nm	250 at 2250 RPM
Fuel Metering System	-	Port Fuel Injection
Air Management System	-	VVA

2.2. Engine and Vehicle Model

A virtual test rig was developed by using the commercially available software GT-SUITE. The virtual test rig integrates a vehicle model, a 1D-CFD FRM engine, a 6-gear manual transmission and an electric network. The 1D-CFD engine FRM was developed and correlated with the experimental data of the reference engine in a previous work of the authors [15]. The vehicle driver is simulated by means of a proportional-integral controller that defines the required power, the brake pedal and the clutch position. Since the selected transmission is manual, the gear shift is imposed for the NEDC according to UNECE 83 [21], while for the WLTC and the RTS-95 the gear shift pattern was computed by means of the Heinz Steven's tool made available by the UNECE committee [22], according to the UE Commission Regulation [1].

The engine is controlled by means of throttle and wastegate controllers. They receive as targets the intake manifold pressure and the boost pressure defined in steady-state conditions. As far as the combustion process simulation is concerned, an imposed Wiebe profile has been adopted, that is function of MFB-50 and experimentally measured MFB-1090 (50% Mass Fraction Burned angle and the 10% to 90% Mass Fraction Burned angular duration, respectively). The target boost and intake manifold pressures, the MFB-50 and relative Air-to-Fuel ratio (Lambda) maps are previously defined in steady-state engine conditions considering the typical limitations for a turbocharged gasoline engine: compressor surge, knock, maximum turbine inlet temperature (T3) and maximum turbocharger speed. For the calibration of each part load operating point (about 550 engine steady state operating point per engine map) a virtual Engine Control Unit (vECU) was developed for the exploitation of the engine maximum performances, while taking into account all the hardware limitations. This vECU operates directly on combustion timing, mixture enrichment, throttle, valve actuation and turbocharger waste gate opening. As a matter of fact, the MFB-50 is defined in such a way that, starting from a value equal to 8 CA aTDC—assumed to be the optimum value corresponding to the Maximum Brake Torque (MBT) timing—the combustion is delayed until the computed knock index falls below a safety threshold. On the other hand, the experimental MFB-1090 of the reference engine was adopted for the whole analysis. The Air-to-Fuel ratio was controlled for the reference engine in order to maintain the turbine inlet temperature in line with the maximum experimental one. The engine model integrated in the virtual test rig is controlled using as setpoints the above-mentioned engine maps for each time step as a function of the actual engine speed and target brake torque defined by the driver controller. Since the engine features a VVA system for the intake valves actuation, the proper valve lift profile is chosen as function of the actual engine speed and target brake torque.

The integration of the FRM in the virtual test rig provides a high level of accuracy for the fuel consumption evaluation also in highly dynamic driving cycles, as proved by Millo et al. [15]. It is worth noting that the fuel flow rate in this way is not estimated by interpolating a performance map, but is the result of a fully-physical engine model, that reproduces transient phenomena as the turbolag effect and the mechanical inertia of the components, which are crucial for the air charging operation and consequently for the fuel consumption. Moreover, in this way, it is also possible to evaluate vehicle performance during tip-in maneuvers for which mechanical and fluid dynamics transient phenomena as turbolag need to be correctly captured.

2.3. Hybrid Control Strategy

Differing from a conventional powertrain in which the Internal Combustion Engine (ICE) satisfies the total driver power demand, in a Hybrid Electric Vehicle (HEV) there is the possibility to decouple the engine output from the requested power, taking advantage of the available on-board energy stored in a battery and exploitable by an electric machine [23]. Typically, a high-level controller (supervisory controller) defines the HEV mode considering all the state variables (ICE status, vehicle speed, engine speed, battery limitation and State of Charge (SoC), etc.) for a given time instant. Specifically, the hybrid functionalities exploited in this work for the P0 48 V architecture and depicted in Figure 1 are:

- Stop-Start: the engine is switched off at a vehicle speed close to zero reducing or eliminating the idle phases (and the corresponding fuel consumption) during vehicle stops;
- Regenerative Braking: partial recovery of the vehicle kinetic energy by the Electric Machine (EM) during deceleration phases. In line with the work done by Zanelli et al. [24], in this work the power request to the BSG follows a rule-based approach defined by the brake pedal request. As a matter of fact, the electric machine power request is null below a brake pedal request of 10% and grows linearly up to the full exploitation of the braking power of the BSG when the brake pedal request is equal to 60%.
- Parallel Mode: during traction phases the vehicle power demand is split between ICE and EM, with a powersplit optimization performed by a proper Energy Management Strategy (EMS). This functionality can be further classified in a Torque-Assist mode and in a Load Point Moving mode. On the one hand, the Torque-Assist decreases the power requested to the ICE, fulfilling the driver request by means of the further contribution of the EM. On the other hand, the Load Point Moving increases the ICE power demand with respect to the driver power demand recovering the surplus by means of the EM, used as a generator, and storing it into the battery.

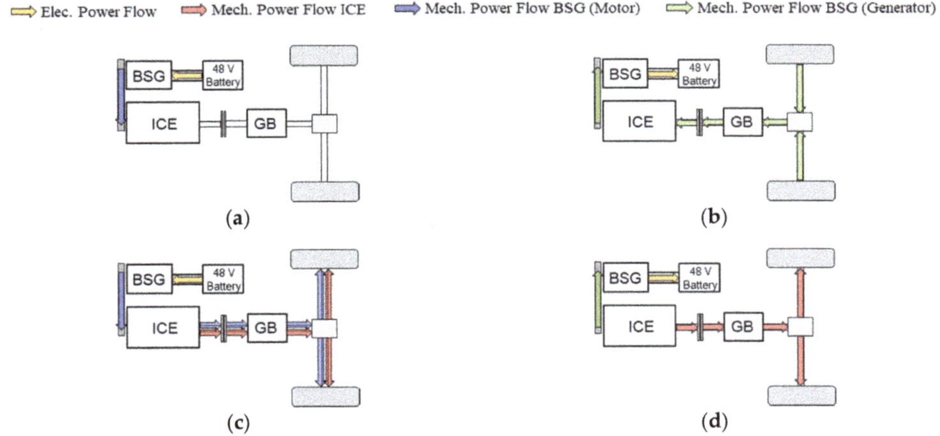

Figure 1. Schematic representation of the main Hybrid Electric Vehicle (HEV) modes for a P0 48 V architecture: (**a**) Stop-Start; (**b**) Regenerative Braking; (**c**) Torque-Assist; (**d**) Load Point Moving.

As far as the EMS is concerned, in this work the Equivalent Consumption Minimization Strategy (ECMS) [25] technique was adopted. This technique, as presented by Ali et al. [26], is a well-established method for the evaluation of a near optimal solution in real time. Additionally, the ECMS can be implemented in an Engine Control Unit (ECU) and requires a lower calibration effort with respect to a rule-based technique, as indicated by Mamun et al. [27]. The ECMS defines the optimum powersplit at each time step by minimizing an instantaneous cost function. As described by Millo et al. [28], an equivalent fuel consumption can be associated with the use of electrical energy: under the hypothesis of charge sustaining condition, this fuel consumption is equivalent to the fuel flow required by the ICE to re-establish the battery SoC at the previous level. The battery equivalent fuel consumption can be summed to the actual fuel consumption to obtain the instantaneous equivalent fuel consumption $\dot{m}_{f,eq}$ as shown in Equation (1):

$$\dot{m}_{f,eq} = \dot{m}_f + \dot{m}_{batt,eqv} = \dot{m}_f + \frac{s}{LHV} P_{batt} \tag{1}$$

where $\dot{m}_f$ is the engine instantaneous fuel consumption (fuel mass flow rate), LHV is the fuel Lower Heating Value, $\dot{m}_{batt,eqv}$ is the fuel consumption associated to the use of the electrical power, P_{batt} the power delivered by the battery. The term s is called equivalence factor and is used to convert electrical power into equivalent fuel consumption. For this work the equivalence factor was separately calibrated for each driving cycle in order to guarantee the charge sustaining condition. According to the EU Regulation (EC) 2017/1151 [1] for NOVC-HEV (Not Off-Vehicle Chargeable HEV) the value of CO_2 obtained over a type-approval driving cycle can be considered as is only if the charge sustaining condition over the cycle is fulfilled. Specifically, the charge sustaining condition is satisfied if the term $C_{criterion}$, defined in Equation (2) as the ratio between the depleted battery energy ΔE_{REES} and the fuel chemical energy E_{Fuel} required for the driving cycle is lower than 0.5% (i.e., if the battery energy is not depleted more than 0.5% of the energy of the fuel consumed over the driving cycle):

$$C_{criterion} = \frac{\Delta E_{REES}}{E_{Fuel}} < 0.005 \, i.e. \ < 0.5\% \tag{2}$$

The hybrid control strategy was implemented in MATLAB-Simulink environment featuring an on-line powersplit optimization: for each time-step the powersplit is defined as the combination of the P_{ICE} and P_{BSG} that minimizes the $\dot{m}_{f,eq}$ and at the meantime satisfies the driver power demand P_{Drv} defined by the driver controller. A schematic representation of the hybrid control strategy can be found in Figure 2.

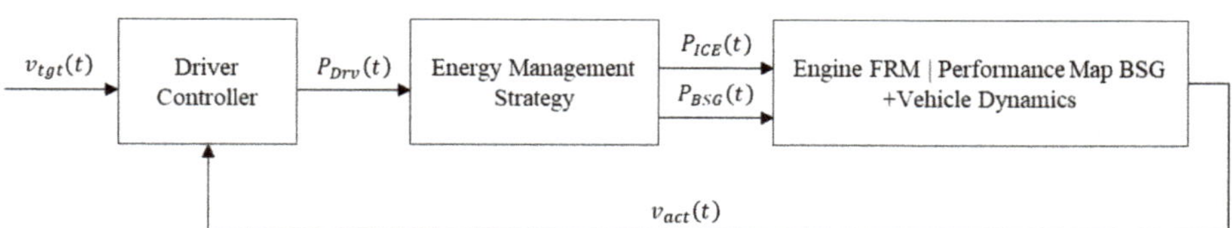

Figure 2. Schematic representation of the hybrid control strategy.

2.4. Analysed Powertrain Configurations and Simulation Matrix

Simulation results for the reference EU 5 engine show a wide operating region of the engine map in which the mixture enrichment technique is exploited. The adoption of a stoichiometric combustion on the entire engine map was therefore firstly investigated. As it is possible to observe from Figure 3, a dramatic reduction of the engine performance results from the stoichiometric combustion adoption. In Figure 3, the T3 temperature, the MFB-50 and the Lambda values are reported as function of the Brake Mean Effective Pressure (BMEP), both at 2000 RPM (Figure 3a) and 5500 RPM (Figure 3b).

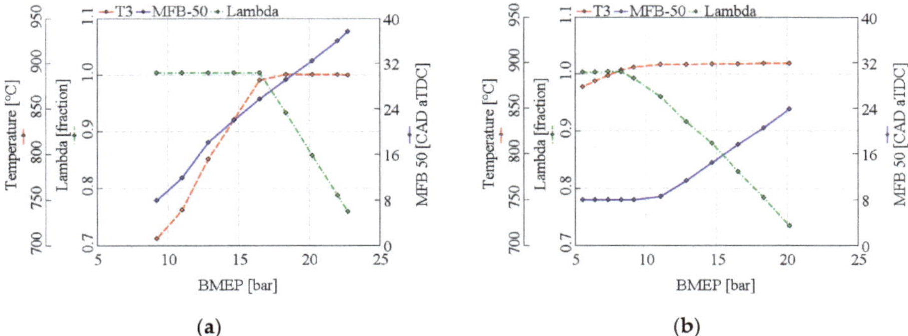

Figure 3. Load sweeps for the reference engine. Turbine inlet temperature in red dashed line, MFB50 in blue solid line and Lambda in green dash-dotted line as a function of Brake Mean Effective Pressure (BMEP): (**a**) engine speed 2000 RPM; (**b**) engine speed 5500 RPM.

As BMEP is increased, the spark timing has to be delayed to prevent engine knock, thus resulting in a delayed MFB-50, which in turns leads to an increase of exhaust gases temperature T3; to keep T3 below the turbine limit temperature, mixture fuel enrichment is necessary, operating the engine with an Air-To-Fuel ratio lower than the stoichiometric one. Therefore, if the engine operation has to be limited to the Lambda-1 region, the maximum BMEP level which can be achieved is limited to 17 bar at 2000 rpm and to 8 bar at 5500 rpm, with a dramatic reduction of the maximum engine torque at 2000 RPM and of the engine rated power at 5500 RPM.

The study carried out aims therefore to compare the reference powertrain, equipped with the EU5 turbocharged engine operating with fuel enrichment with three different powertrain configurations that feature a stoichiometric combustion on the whole engine map and that exploit several engine, hardware and control modifications to improve performance and fuel economy. The main characteristics of the powertrains are highlighted in Table 3 and will be described in the following paragraph.

Table 3. Analyzed Powertrains Configurations. BSG: Belt Starter Generator; eSC: electric SuperCharger.

Case	Label	Engine	Electric Network
1	Base	1.4 L EU5	12 V
2	ST-Conv.	1.4 L Stoich.	12 V
3	ST-MHEV	1.4 L Stoich.	12 V + 48 V (BSG)
4	HE-MHEV	1.4 L High Eff.	12 V + 48 V (BSG & eSC)

1. **Base**: conventional powertrain, equipped with the reference 1.4 L EU5 engine.
2. **ST-Conv**: conventional powertrain concept featuring a Lambda-1 engine (referred to as 1.4 L Stoich.). The maximum T3 was increased from 930 °C to 980 °C, assuming the adoption of an enhanced turbocharger system capable to withstand to exhaust gas temperatures up to 980 °C. Furthermore, a refined knock control, capable of fully exploiting the engine torque potential until the knock limited spark advance was also adopted, as reported by Millo et al. [28].
3. **ST-MHEV**: an electrified powertrain concept (P0 architecture), featuring the same Lambda-1 engine presented in 2 (1.4 L Stoich.) and integrating a 48 V BSG. The main technical data of the 48 V BSG and of the Lithium battery are reported in Table 4. The vehicle test mass was increased by 20 kg, representing the additional weight of the BSG and the Lithium battery employed in the dual voltage electric network.

Table 4. Specifications of Belt Starter Generator and 48 V Battery.

Belt Start Generator Parameter	Unit	Value
Nominal Voltage	V	48
Peak Power (Generation, Braking)	kW	18
Continuous Power (Generation)	kW	8
Peak Power (Motoring)	kW	12
Battery Parameter	**Unit**	**Value**
Voltage Range	V	$32.5 \div 54$
Current Range	A	$-250 \div 250$
Capacity	Ah	10 Ah

4. **HE-MHEV**: a high efficiency electrified powertrain was investigated. The CR of the engine was increased from the base value of 9.8 up to 12. For knock mitigation purpose a Miller cycle was introduced in the high load engine map, performing a Late Intake Valve Closure (LIVC) strategy, as experimentally investigated by Luisi et al. [29]. The valve closure delay of the intake valve was actuated keeping the valve open at the full lift for a specified angle duration and it was defined to maximize the brake torque at full load without exceeding the knock limit and the maximum T3 in the high load region. Moreover, a 48 V eSC was integrated (upstream of the turbocharger) in order to satisfy the high level of boost pressure required in the Low End Torque (LET) region (up to 3000 RPM). The eSC main characteristics are given in Table 5. The engine concept is referred to as 1.4 L High Eff. As far as the eSC control is concerned, a rule-based control developed by Zanelli et al. [24] was adopted, that defines the eSC activation and control based on the difference between target and actual boost pressure. Since the aim of this controller is to reduce the turbolag effect or satisfy the required boost level of the LIVC strategy, the controller converts the boost pressure gap in eSC target speed through the eSC compressor map. At this point a proportional integral controller defines an electric power request for the eSC motor in order to achieve the target eSC speed. Further information regarding the eSC control can be found in the work done by Zanelli et al. [24].

Table 5. Specifications of the 48 V eSupercharger (eSC).

Parameter	Unit	Value
Compressor Max Speed	RPM	75,000
Compressor Max Pressure Ratio	-	1.5
Electric Motor Max. Torque	Nm	0.6
Electric Motor Max. Power	kW	5.3

The vehicle performance and the fuel economy of the investigated powertrain configurations were assessed on several transient maneuvers and driving cycles.

As far as the vehicle performance is concerned, three typical maneuvers were selected for the analysis:

- 0–100 km/h
- 80–120 km/h in VI gear
- 60–80 km/h in VI gear.

In Figure 4, it is possible to appreciate the engine speed region related to each maneuver: the 0–100 km/h covers mainly the high-speed region, while the tip-in elasticity maneuvers 80–120 km/h and 60–80 km/h are in the LET region.

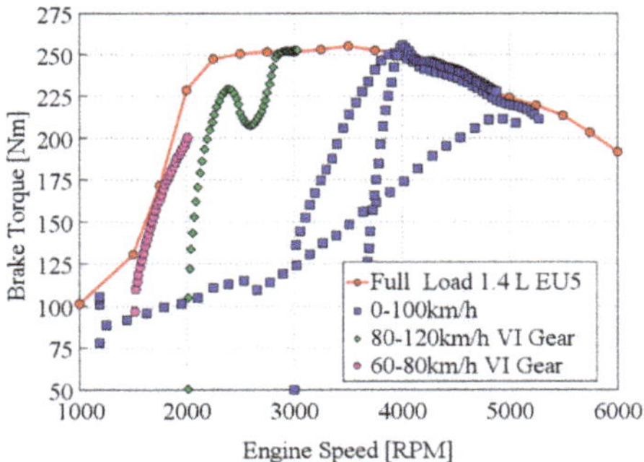

Figure 4. Transient maneuvers reported in the engine map for the reference engine: 0–100 km/h in blue, 80–120 km/h in sixth gear in green and 60–80 km/h in sixth gear in violet.

Moreover, in order to compare this a synthetic way, the results in terms of transient performance, a Performance Index (PI) were introduced, as defined in Equation (3).

$$PI = \frac{3600}{v_{max}} + t_{0-100} + t_{60-80} + t_{80-120} \tag{3}$$

The performance index includes, in addition to the above discussed transient maneuvers, the contribution of the maximum vehicle speed with the term $\frac{3600}{v_{max}}$ (where v_{max} is expressed in km/h), that represents the time taken by the vehicle to run 1 km at maximum speed and is therefore related to engine peak power.

Moving to the fuel economy investigation, the fuel consumption for the investigated powertrain configurations was evaluated on three driving cycles: NEDC, WLTC and RTS-95. In Table 6 the main characteristics of each driving cycle are reported.

Table 6. Driving cycles data [20].

Parameter	Unit	NEDC	WLTC	RTS-95
Duration	s	1180	1800	886
Distance	km	11.03	23.27	12.93
Average Velocity	km/h	43.1	46.5	52.5
Max Velocity	km/h	120.0	131.3	134.4
Average Acceleration	m/s^2	0.51	0.41	0.73
Max Acceleration	m/s^2	1.04	1.58	2.62

All the analysis has been performed excluding catalyst heating and cold start operation.

3. Results

In this section, the results in terms of both steady-state engine and vehicle transient performance are presented and discussed.

3.1. Steady-State Analysis

The performance of the developed engine concepts have been analyzed both at full load and part load engine operation. As far as the full load performance is concerned, in Figure 5, the brake torque and brake power of the three engines are reported.

The 1.4 L Stoichiometric engine, featuring an increased maximum turbine temperature and operating at knock limit, features a LET performance almost comparable to the reference engine

(114 Nm for the Lambda-1 concept against 110 Nm for the reference engine at 1000 RPM and 232 Nm at 2000 RPM for both engine concepts). However, the performance derating caused by the stoichiometric combustion adoption is not completely recovered in the high-speed region: a reduction of 11 kW in terms of peak power can be highlighted in Figure 5b (rated power of 110 kW for the 1.4 L Stoichiometric concept at 5500 RPM).

As far as the high-efficiency engine concept is concerned, a considerable improvement of the brake torque in the LET region is achieved. This is obtained thanks to the combination of the Miller cycle and the eSC exploitation, since the increased boost pressure level required by the LIVC strategy is obtained thanks to the eSupercharger. The two stages supercharging in the low engine speed region (< 3000 RPM) achieves a maximum available boost pressure up to 2.75 bar at 2000 RPM. The Miller cycle on the other hand reduces the knock likelihood enabling a significant improvement of the brake torque up to 255 Nm at 2500 RPM (almost equal to the reference non-stoichiometric engine) and improving the LET performance with respect to the reference EU5 engine of more than 40% at 1500 RPM (brake torque equal to 200 Nm). In addition, the LIVC strategy at high engine speed allows to almost completely recover the rated power of the engine (117 kW at 5500 RPM, only 4 kW lower than the reference EU5 engine).

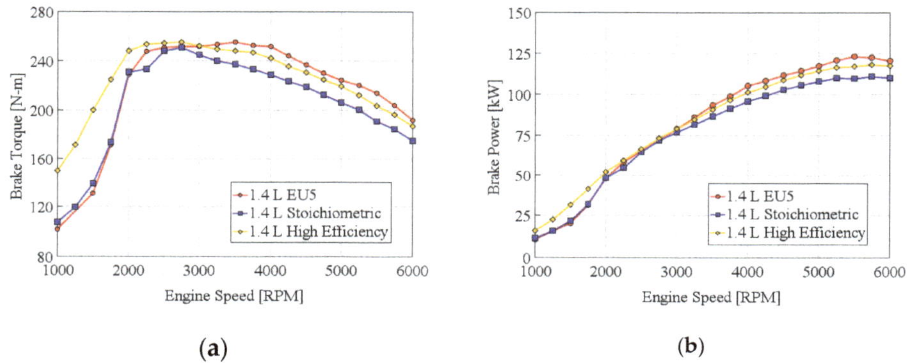

(a)   (b)

Figure 5. Full load performance for the proposed engine concepts: (**a**) brake torque; (**b**) brake power.

As far as the part load operation is concerned, an efficiency analysis was carried out. Focusing on the 1.4 L Stoichiometric engine concept, the refinement of the knock controller determines an advance of the MFB-50 with respect to the reference engine in the high load region, while at low load the MFB-50 is 8 CA aTDC in both cases. The more advanced combustion leads to a lower T3 temperature. In addition, for a given engine load, working with the advanced and more efficient combustion requires a lower amount of trapped air (i.e., a lower boost pressure). As an example, in Table 7, a comparison between the 1.4 L EU5 and the 1.4 L stoichiometric engine concepts is reported, in terms of MFB-50, T3, Lambda, Boost pressure and Brake Specific Fuel Consumption (BSFC). At 13 bar BMEP and 2000 RPM, the MFB-50 is reduced by 4.5 CA and the related BSFC advantage is 3.6 g/kWh (1.5%). This engine efficiency improvement is consistent with the reduction of the boost pressure.

At the higher load and engine speed operating point (18.5 bar BMEP @ 3000 RPM) the reference engine features a Lambda value equal to 0.87, in order to keep the exhaust gas temperature below the material limit. The knock controller of the Stoichiometric engine is able to operate the engine at a MFB-50 value lower than the reference case by 7.4 CA; with Lambda-1 operation the temperature reaches a value of 915 °C, higher than for EU5 engine due to the stoichiometric operation but still lower than the maximum admissible T3 (980 °C). A reduction of the boost pressure of 0.09 bar is achieved (5.4%). Considering the significant saving of fuel resulting from the stoichiometric operation, the overall BSFC advantage achieved is an impressive 18.3%.

Table 7. Part Load comparison between 1.4 L EU5 and 1.4 L stoichiometric engine concepts. aTDC: after Top Dead Center; BMEP: Brake Mean Effective Pressure; BSFC: Brake Specific Fuel Consumption; CA: Crank Angle; T3: Turbine Inlet Temperature.

Parameter	Unit	1.4 L EU5	1.4 L Stoich.
140 Nm (i.e., 13 bar BMEP—60 % of max load) at 2000 RPM			
MFB-50	CA aTDC	18.1	13.6
T3	°C	794	767
Lambda	-	1	1
Boost Pressure	bar	1.22	1.20
BSFC	g/kWh	240.0	236.4
200 Nm (i.e., 18.5 bar BMEP—80 % of max load) at 3000 RPM			
MFB-50	CA aTDC	25.8	18.4
T3	°C	892	915
Lambda	-	0.87	1
Boost Pressure	bar	1.65	1.56
BSFC	g/kWh	287.2	234.7

Considering the 1.4 L high efficiency engine concept, which features an increased CR from the base value of 9.8 to 12 and exploits a LIVC strategy for knock mitigation in the high load region, the BSFC difference with respect to the 1.4 L Stoichiometric engine is given in Figure 6. However, it has to be pointed out that in this comparison the electric power absorbed by the eSC is not considered for the efficiency investigation.

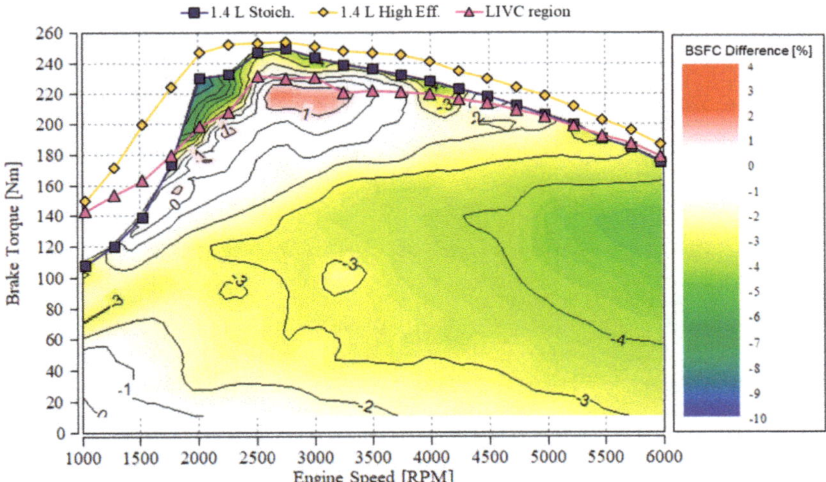

Figure 6. Percentage difference of BSFC between the 1.4L High Eff. Engine and the 1.4L Stoich. Engine.

As it is possible to appreciate from Figure 6, in the low to medium load region (up to 160 Nm), the BSFC improves, with more significant benefits at higher engine speeds: no advantages are obtained at 1000 RPM for 0 to 60 Nm of brake torque, while at 3000 RPM the benefit are in the order of 2.5% and at 5000 RPM it increases up to 3.5% on average in the same load range. At medium-high load, the increased knock tendency with the higher compression ratio results in delayed combustion phasing, and, as consequence, into a deterioration of the indicated engine efficiency. In the high load region, the BSFC advantages due to the LIVC strategy and CR 12 are in the order of 4% for engine speed higher than 3000 RPM and of more than 6% in the range 2000–3000 RPM, where the eSC activation reduces the engine backpressure, providing significant benefits both in terms of pumping losses and knock mitigation, thanks to the decrease of the residual gas fraction.

Finally, it is worth highlighting that, since the combustion duration was assumed to remain unchanged for all the investigated concepts, the benefits in terms of engine efficiency may be underestimated for the engine featuring an increased compression ratio at part load operation.

3.2. Transient Maneuvers Analysis

The results concerning the elasticity maneuvers are reported in Figure 7. Considering the acceleration from 0 to 100 km/h, the adoption of the stoichiometric combustion results in a performance gap despite the counteractions exploited for the 1.4 L stoichiometric engine (ST-Conv), as the high temperature turbine and the refined knock controller. The penalty in terms of brake torque in the high-speed region leads to an increase of about 0.5 s of the acceleration time. The electrification is able to reduce the performance gap to only 0.2 s for the stoichiometric engine with the BSG Torque-Assist (ST-MHEV) and to 0.1 s for the high-efficiency concept (HE-MHEV).

As far as the elasticity maneuvers are concerned, the ST-Conv powertrain is able to reduce the time required for the maneuvers by 0.8 s on the 80 to 120 km/h acceleration and to keep the same performance on the 60 to 80 km/h in VI gear with respect to the Base powertrain. The 48 V electrification (ST-MHEV) improves significantly the dynamic performance of the vehicle, decreasing the performance time of 1.4 s on the 80–120 km/h in VI gear and of 1.8 s on the 60–80 km/h with the same gear. The introduction of the eSC leads to a further reduction for the HE-MHEV concept with respect to the ST-MHEV of 0.4 s and 1.1 s, respectively on the 80–120 km/h and on the 60–80 km/h maneuver in VI gear.

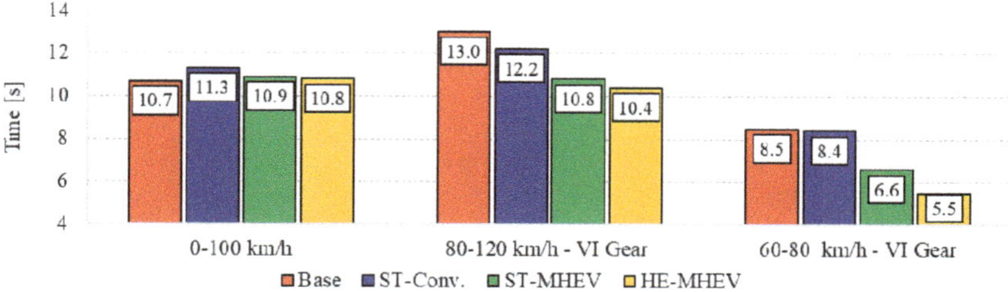

Figure 7. Performance time for the transient maneuvers investigated.

Even if the rated power of the Stoichiometric engine is considerably lower with respect to the reference EU5 engine, the higher engine efficiency leads to a reduction of the turbolag effect on the transient response of the vehicle, which achieves a higher level of brake torque with the same boost pressure level. In this regard, in Figure 8, the boost pressure and the ICE Torque are reported for all the investigated powertrain concepts in the first four seconds of the 80–120 km/h in VI gear maneuver. It is possible to observe that the boost pressure raises almost identically for the stoichiometric and the reference engines, while the ICE torque of the ST-Conv. powertrain is higher than the Base one. Differing from the ST-Conv and ST-MHEV powertrains, the HE- MHEV concept is able to achieve a remarkable improvement of the transient response of the vehicle: as it is possible to notice from Figure 8, the boost pressure increases almost immediately, thanks to the eSC action, achieving the 90% of the full load brake torque just two seconds after the tip-in start.

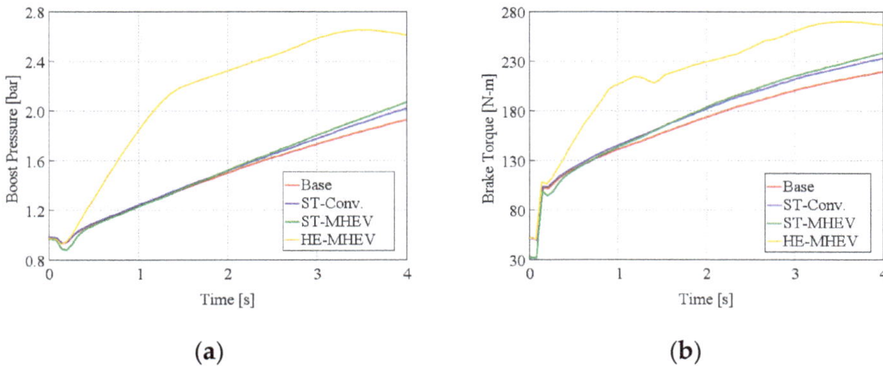

(a)   **(b)**

Figure 8. Detail of the 80–120 km/h in VI gear maneuver: (**a**) Boost Pressure; (**b**) Internal Combustion Engine (ICE) Brake Torque.

In addition, the 48 V electrification has a strong impact on the vehicle performance. As an example, in Figure 9 an analysis of the Torque-Assist functionality of the ST-MHEV and the HE-MHEV during the 60–80 km/h elasticity maneuver is presented. Thanks to the 48 V BSG, the engine shaft torque is increased by 40 Nm for the ST-MHEV concept and by 30 Nm for the HE-MHEV. The difference between the two configurations derives from the battery power limitation: the electrically assisted supercharger is considered as an electric auxiliary by the EMS, thus having the priority over the BSG power request. In both cases the maximum battery power is equal to 10 kW, but the simultaneous utilization of the eSC and the BSG in the HE-MHEV concept reduces the maximum electrical power for the BSG to 8 kW. Nevertheless, even if the BSG torque is lower in HE-MHEV if compared to the ST-MHEV case, the benefits of the eSC and the Miller cycle in terms of engine output torque in the LET region result in a significant improvement (more than 1 s) in the 60–80 km/h in VI gear maneuver.

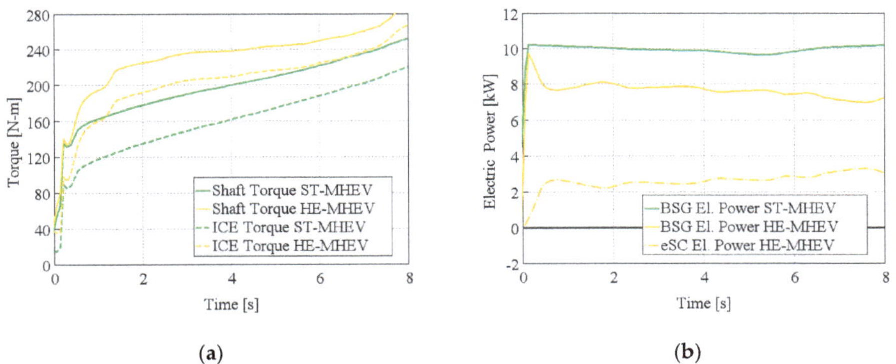

(a)   **(b)**

Figure 9. Details of the 60–80 km/h in VI gear maneuver: (**a**) shaft torque in solid lines, ICE torque in dashed lines; (**b**) BSG electric power in solid line, eSC electric power in dashed line.

3.3. Fuel Consumption Evaluation

In this section, the fuel consumption (although reported in terms of CO_2 emissions to allow an easier comparison with legislation targets) of the investigated concepts is reported and discussed. The CO_2 emissions values are reported in Figure 10. The reference vehicle emits 156 g/km on NEDC, 176 g/km on WLTC, 244 g/km on RTS95. These values are consistent with an extensive data acquisition campaign carried out and presented by the Joint Research Centre (JRC) of the European Union

Commission [30] in terms of absolute value of CO_2 emissions and in terms of fuel consumption ratio between NEDC and WLTC.

The ST-Conv. concept does not obtain any fuel consumption reduction on the NEDC. The reason is that the engine operating points of the vehicle performing the NEDC lie in the low load region of the engine map, in which the engine works with a stoichiometric combustion and with the same combustion timing both in the case of Base and ST-Conv powertrains. On the other hand, for the more dynamic WLTC and RTS-95 driving cycles, the ST-Conv concept achieves a considerable reduction of the CO_2 emissions: a reduction of 2.6% can be highlighted for the WLTC, while in the RTS-95 the benefit increases up to a remarkable value of 9.4%, reflecting the engine BSFC improvement in the high load region.

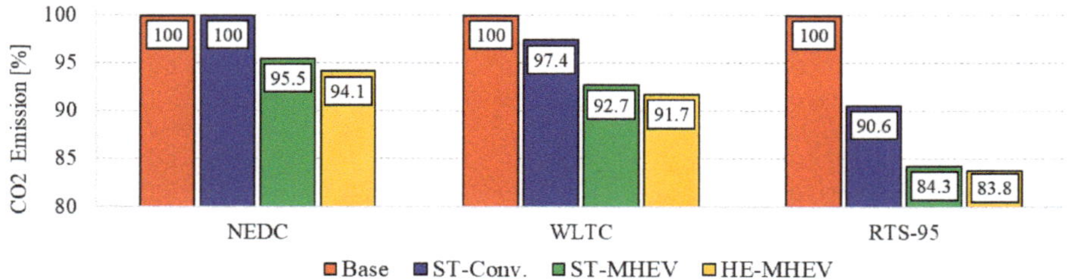

Figure 10. Normalized CO_2 emissions for the NEDC, WLTC and RTS-95.

Focusing on the impact of the 48 V electrification on the fuel economy, a significant fuel consumption reduction can be highlighted from Figure 10: the ST-MHEV powertrain concept features a reduction of about 4.5% on NEDC and WLTC, of 7.3% on RTS-95 with respect to the ST-Conv configuration and an overall improvement of about 8% on the WLTC and 15% on the RTS-95 with respect to the Base concept.

The benefits in terms of fuel economy for the ST-MHEV referred to the ST-Conv concept derive from the Torque-Assist hybrid functionality. A detailed analysis is proposed in Figure 11 for a section of the RTS-95 driving cycle (from 295 s to 375 s). On the top chart, the vehicle speed for the ST-Conv and the ST-MHEV is reported, as well as the gear number and the difference of fuel consumption between the two powertrains. On the bottom chart, the powersplit performed by the ECMS is presented. From the top chart is possible to highlight a slight deviation of the vehicle speed from the target for the ST-Conv concept during the acceleration phases (compliant with the regulation tolerance). The speed deviation is due to the turbolag effect, in particular after the gear-shifting operation. The high driver power demand resulting from the speed deviation is partially reduced by the BSG Torque-Assist functionality, resulting in a reduction of the fuel consumption together with a better replication of the target speed profile.

The load point moving functionality is exploited neither in the WLTC nor in the RT-95. Consequently, the overall energy employed by the BSG derives uniquely from the regenerative braking functionality. The recovered energy is reported for each driving cycle in Table 8.

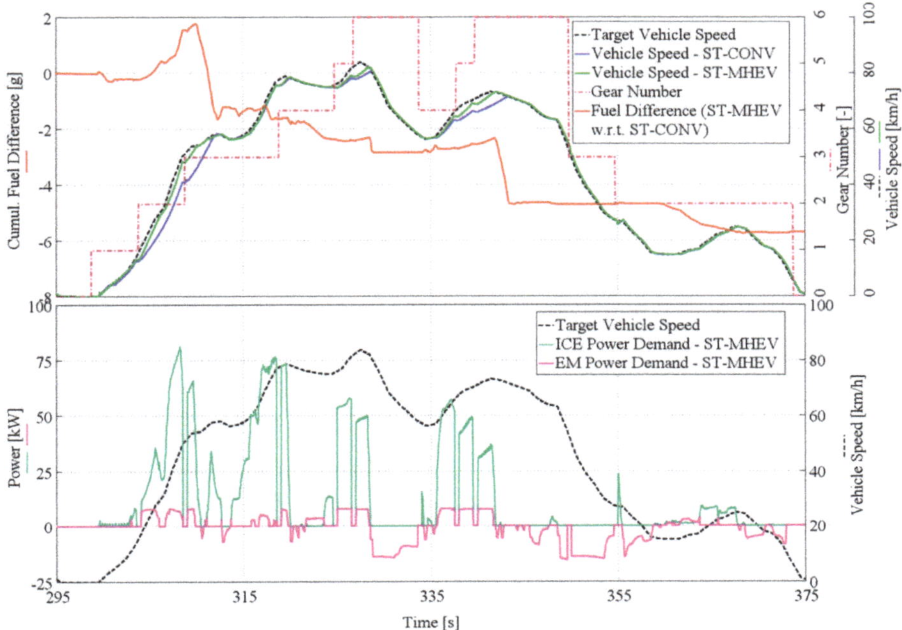

Figure 11. Detail of the hybrid control strategy operation in the RTS-95 driving cycle (from 295 s to 375 s): vehicle speed, gear number and cumulated fuel consumption difference on top chart; powersplit optimization in the bottom chart.

Table 8. Recovered energy on the NEDC, WLTC and RTS-95 for the ST-MHEV and the HE-MHEV powertrain concepts.

Parameter	Unit	NEDC	WLTC	RTS-95
Recovered Energy	Wh	102	288	330
Spec. Recovered Energy	Wh/km	9.7	12.5	25.7

For a deeper investigation of the 48 V electrification effectiveness on fuel consumption reduction, in Figure 12 a comparison of the energy delivered by the engine, grouped in load bins, for the ST-Conv and the ST-MHEV powertrain concepts is reported, both for the WLTC and for RTS-95 driving cycles. Firstly, a noteworthy difference of the distribution of the energy released by the engine between WLTC and RTS-95 can be observed from the charts. The WLTC requires most of the energy (52%) at an engine load between 30 and 60%. The energy contribution provided at a load higher than 70%,where there is a worsening of the engine efficiency, is more than the 18% of the total energy released by the engine for the ST-Conv. configuration. However, it is significantly reduced to 11% by the hybrid strategy in ST-MHEV. The powersplit defined by the ECMS aims therefore to reduce the energy released at high load by concentrating the ICE power in the maximum efficiency load region (i.e., between 60 and 70% of the engine full load), exploiting the electric energy coming from regenerative braking. A similar result can be pointed out for the RTS-95 driving cycle: the high load energy distribution is significantly reduced by the ECMS (from 40% to 26% of the overall energy released by the ICE). This hybrid strategy is more effective for the RTS-95 (7.3% of fuel consumption reduction with respect to the ST-Conv powertrain) than for the WLTC (4.5% of fuel consumption reduction) because of both the higher amount of available energy coming from regenerative braking (25.7 Wh/km against 12.5 Wh/km) and the difference of high load energy distribution for the two driving cycles.

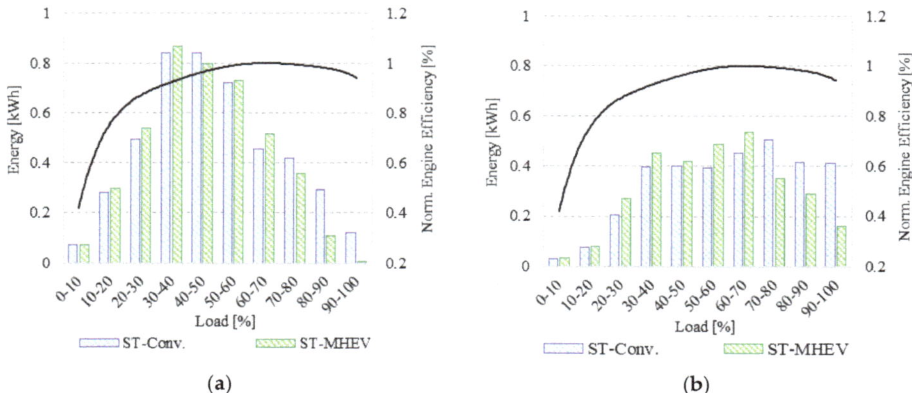

Figure 12. Energy delivered by the engine grouped in load bins; normalized engine efficiency at 2000 RPM on the right axis: (**a**) WLTC; (**b**) RTS-95.

As far as the HE-MHEV concept is concerned, the fuel consumption improvement is about 1.5% on NEDC and lower than 1% on the other driving cycles with respect to the ST-MHEV, as shown in Figure 10. The limited benefit on the NEDC is related to the reduced advantage in terms of engine efficiency coming from the adoption of a higher value of compression ratio in the low speed and low load engine map region (see Figure 6). On the other hand, considering the WLTC and the RTS-95, the exploitation of the eSC involves an absorption of electric power which results in a depletion of the 48 V battery. Consequently, the activation of the electric supercharger limits the power exploitable by the BSG for the powersplit operation. For explanation, the electric energy required by the BSG for Torque-Assist operation and by the eSC are given in Table 9.

Table 9. Electrical energy required by the BSG (Torque-Assist) and by the eSC on the NEDC, WLTC and RTS-95.

Parameter	Unit	NEDC	WLTC	RTS-95
BSG Electric Energy	Wh	51.0	126.8	216.9
eSC Electric Energy	Wh	7.2	46.1	49.7

Focusing on the difference in terms of energy management strategy between ST-MHEV and HE-MHEV, in Figure 13 the eSC activation, the BSG power and the fuel consumption difference between concepts is shown. As it is possible to appreciate from the bottom chart of Figure 13, the instantaneous mechanical power provided by the BSG in the case of HE-MHEV is lower compared to the ST-MHEV. In the high load region, the eSupercharger activation reduces the available battery power. It is therefore necessary to consider that, if on the one hand the engine operation in the high-load region is characterized by greater efficiency by exploiting the Miller cycle, on the other hand it is not possible to fully exploit the potential of the BSG Torque-Assist. Looking at the top chart of Figure 13, it is possible to evaluate the difference of fuel consumption between the two electrified powertrain concepts and the eSC activation. The electrical supercharging operation is mainly concentrated in the first part of the WLTC (up to 1150 s), that is characterized by low values of engine and vehicle speed and frequent accelerations. In this time frame, the engine efficiency advantage deriving from a more efficient combustion is completely overcome by the eSupercharger activation (that is defined by the rule-based strategy already discussed and not integrated in the EMS), whose energy demand limits the Torque-Assist functionality. The benefits deriving from the increased compression ratio with respect to ST-MHEV configuration are considerable in the time intervals between 1150 s and 1350 s and after

1550 s, where the engine speed is on average higher than 2500 RPM (higher efficiency benefit from the CR 12) and the eSC is not used.

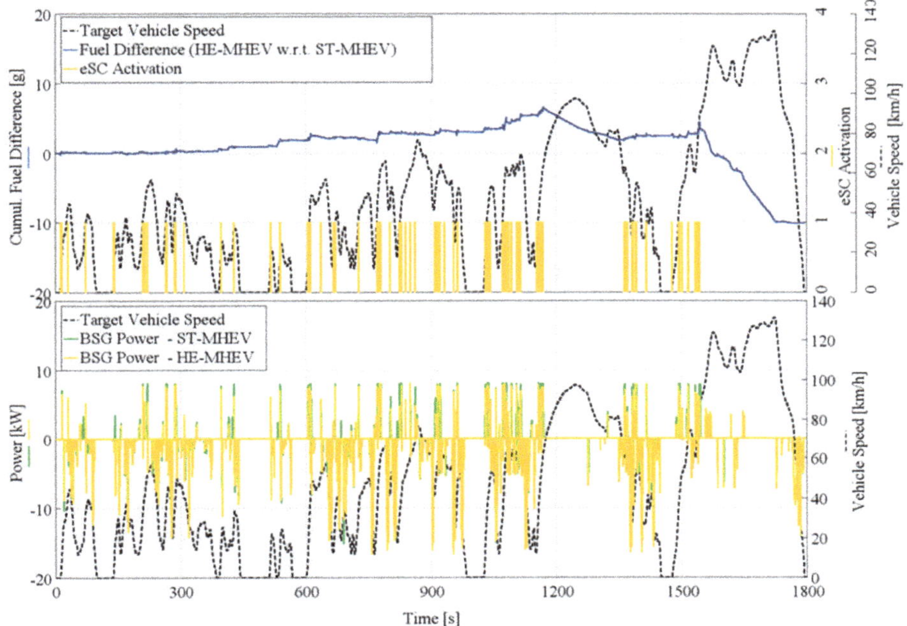

Figure 13. Focus on electric power management in the WLTC. Top chart: eSC activation and cumulated fuel difference (HE-MHEV w.r.t. ST-MHEV). Bottom chart: BSG mechanical power for ST-MHEV and for HE-MHEV.

3.4. Performance Summary

In order to compare, in a synthetic way, the results in terms of fuel consumption and transient performance, in Figure 14, a performance summary in terms of Performance Index versus Fuel Consumption on NEDC, WLTC and RTS-95 is reported. As previously explained, the defined PI also considers the maximum vehicle speed (reported for all the investigated powertrain concepts in Table 10). The maximum vehicle speed was evaluated in VI gear and exploiting the BSG Torque-Assist functionality for the electrified configurations.

Table 10. Maximum vehicle speed.

Parameter	Unit	Base	ST-Conv.	ST-MHEV	HE-MHEV
v_{max}	km/h	202	193	196	200

Looking at Figure 14, it is possible to appreciate that the performance worsening, caused by the adoption of the stoichiometric combustion on the entire engine map, was almost completely overridden (+1% PI) with the countermeasures investigated for the 1.4 L stoichiometric engine (increased turbine limit temperature and knock calibration update in the ST-Conv concept). The fuel consumption reduces by 2.6% on the WLTC and by 9.4% on the RTS-95 with respect to the reference case, while no difference is obtained in the NEDC. The 48 V electrification of the powertrain (ST-MHEV concept) leads to a significant improvement of the vehicle performance (PI reduction for the ST-MHEV up to 6.3% with respect the Base concept), while the fuel consumption benefits are about 4.6% on NEDC and WLTC and more than 6% on RTS-95 compared to the ST-Conv powertrain. Finally, the HE-MHEV concept, whose engine features an increased CR and exploits the potentiality of the eSC and of the Miller Cycle, dramatically reduces the turbolag effect due to the prompt supercharging response and is able to reduce the peak power gap of the engine taking advantage of the Miller cycle introduction. The result is an additional improvement of the PI by 4% with respect to the electrified concept ST-MHEV (achieving an impressive 10% of PI improvement compared to the reference Base vehicle); furthermore, the HE-MHEV concept is capable of reducing the fuel consumption by 1% compared to the ST-MHEV.

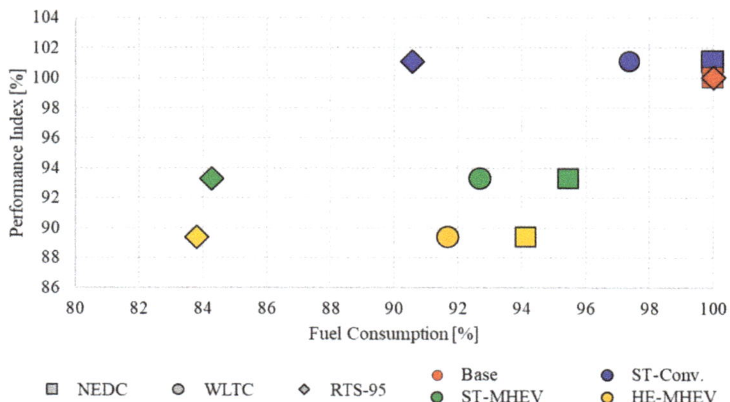

Figure 14. Performance Summary: Normalized Performance Index vs Normalized Fuel Consumption evaluated on NEDC, WLTC and RTS-95.

4. Conclusions

In this work the impact of RDE regulation on the development of Lambda-1 engine concepts and the 48 V electrification benefits in terms of fuel economy and vehicle performance for a gasoline passenger car were investigated through numerical simulation. Vehicle performance and fuel consumption were evaluated according to several transient maneuvers and three driving cycles, respectively. The reference engine (Base) was a turbocharged gasoline engine compliant with the Euro 5 emissions standards; three additional powertrains were developed in order to achieve the compliance with the RDE regulation. To do so, while in the meantime maintaining the engine performance at a comparable level with respect to the reference engine, some hardware and control modifications were investigated: the turbine inlet limit temperature was increased (+ 50 °C compared to Base engine); additionally, the knock calibration was refined targeting the engine operation at knock limited spark advance, resulting in a moderate advance of the combustion phasing in the high load region (ST-Conv). Afterwards, an increased compression ratio (from 9.8 to 12) was adopted in an electrified powertrain concept, together with a Miller cycle for knock mitigation and an electric supercharger.

Simulation results demonstrated that the stoichiometric engine concept integrated in a P0 48 V electrified powertrain (ST-MHEV) achieves a performance improvement up to 6% with respect to a conventional gasoline car in terms of vehicle acceleration. A significant reduction of the fuel consumption was achieved with the Lambda-1 electrified powertrain (7.3% on the WLTC and more than 15% on the RTS-95). The fuel economy benefits derived both from the increased engine efficiency and from the powersplit optimization performed by the energy management system; in particular, the hybrid control strategy aims to concentrate the engine operation in the maximum efficiency load region.

The possibility to exploit an electrified boosting system as the eSC (in the HE-MHEV powertrain) considerably improved the transient response of the engine, dramatically reducing the performance time in elasticity maneuvers (up to −11% on average). As far as the fuel consumption is concerned, the engine concept featuring the eSC coupled with other engine technologies like the Miller cycle and the increased CR, reduced the CO_2 emission of the vehicle of about 1% averaging the results over the NEDC, WLTC and RTS-95.

An evolution of this work will be the integration of the eSC activation strategy in the EMS going beyond the conventional rule-based approach. This will require a modification of the ECMS so that the overall electric power needed by the vehicle would be considered in the powersplit operation. Moreover, this approach could be expanded to take into account additional auxiliaries power load as, for example, an electric catalyst (requiring the updating of the vehicle virtual test rig to include also cold start condition).

Author Contributions: F.A. built the virtual Engine Control Unit for the engine steady state map computation and developed the stoichiometric engine concepts. Moreover, he performed the simulation and he wrote the first draft of the manuscript. A.Z. developed the vehicle virtual test rig and the rule based activation strategy for the electric Supercharger (eSC). L.R. developed the hybrid vehicle ECMS control strategy. F.M. was the project supervisor and coordinator, and he was involved in identifying the most meaningful comparisons. A.Z., L.R. and F.M.were involved in the internal review process of the manuscript.

Funding: The project was financially supported by FCA through the "Optimization of a 48 V hybrid architecture for a passenger car" research project (2018).

Acknowledgments: The valuable support provided to the research activity by FCA is gratefully acknowledged. In particular, the authors would like to thank Caterina Venezia, Davide Peci, Giovanni Guenna and Constantinos Vafidis for their precious and constant support as well as for their invaluable suggestions during the simulation activities.

Conflicts of Interest: The authors declare no conflict of interest.

Definitions/Abbreviations

aTDC	After Top Dead Center
BMEP	Brake Mean Effective Pressure
BSFC	Brake Specific Fuel Consumption
BSG	Belt Starter Generator
CA	Crank Angle
CF	Conformity Factor
CFD	Computational Fluid Dynamics
CR	Compression Ratio
ECMS	Equivalent Consumption Minimization Strategy
ECU	Engine Control Unit
EM	Electric Machine
EMS	Energy Management Strategy
eSC	Electric Supercharger
FRM	Fast Running Model
GB	Gearbox
HEV	Hybrid Electric Vehicle
ICE	Internal Combustion Engine
JRC	Joint Research Centre (European Union Commission)
Lambda	Relative Air-to-Fuel Ratio
LET	Low End Torque
LIVC	Late Intake Valve Closure
MBT	Maximum Brake Torque
MFB-1090	10–90% Mass Fraction Burned Angular Duration
MFB-50	50% Mass Fraction Burned Angle
MHEV	Mild-Hybrid Electric Vehicle
NEDC	New European Driving Cycle
NOVC-HEV	Not Off-Vehicle Chargeable Hybrid Electric Vehicle
PI	Performance Index
RDE	Read Driving Emissions
RTS-95	standardized random test for an aggressive driving style

SI	Spark ignition
SoC	State of Charge
SUV	Sport Utility Vehicle
T3	Turbine Inlet Temperature
vECU	Virtual Engine Control Unit
VVA	Variable Valve Actuation
WLTC	Worldwide Harmonized Light-Duty Test Cycle
WLTP	Worldwide Harmonized Light-Duty Test Procedure

References

1. COMMISSION REGULATION (EU) 2017/1151, OJ L 175, 7.7.2017, pp. 1–643. Available online: https://eur-lex.europa.eu/legal-content/EN/TXT/?uri=OJ:L:2017:175:TOC (accessed on 2 August 2019).
2. Leduc, P.; Dubar, B.; Ranini, A.; Monnier, G. Downsizing of Gasoline Engine: An Efficient Way to Reduce CO2 Emissions. *Oil Gas Sci. Tech.* **2003**, *58*, 115–127. [CrossRef]
3. Clairotte, M.; Valverde, V.; Bonnel, P.; Giechaskiel, B.; Carriero, M.; Otura, M.; Fontaras, G.; Pavlovic, J.; Martini, G.; Krasenbrink, A. *Joint Research Centre 2017 Light-Duty Vehicles Emissions Testing*; Publications Office of the European Union: Luxembourg, 2018; ISBN 978-92-79-90601-5. [CrossRef]
4. Scharf, J.; Thewes, M.; Balazs, A.; Fh, D.; Lückenbach, S.; Speckens, F. All Clean Gasoline Hybrid Powertrains—Real Driving Emissions, Lambda = 1 & Euro 7. In Proceedings of the Aachen Colloquium Automobile and Engine Technology, Aachen, Germany, 8–10 October 2018; pp. 935–956.
5. Görgen, M.; Fandakov, A.; Hann, S.; Keskin, M.T.; Urban, L.; Bargende, M. All lambda 1 gasoline powertrains. In *Internationaler Motorenkongress 2018*; Liebl, J., Beidl, C., Maus, W., Eds.; Springer: Wiesbaden, Germany, 2018; pp. 93–111. [CrossRef]
6. Glahn, C.; Koenigstein, D.A.; Hermann, I. Future for All?—Lambda-1-Combustion Systems of Small Powertrains for the High Volume Market. In Proceedings of the Aachen Colloquium Automobile and Engine Technology, Aachen, Germany, 8–10 October 2018; pp. 913–934.
7. COMMISSION REGULATION (EU) 2009/443, OJ L140, 5.6.2009, pp. 1–15. Available online: https://eur-lex.euro pa.eu/LexUriServ/LexUriServ.do?uri=OJ:L:2009:140:0001:0015:EN:PDF (accessed on 2 August 2019).
8. COMMISSION REGULATION (EU) 2019/631, OJ L111, 25.4.2019, pp. 13–53. Available online: https://eur-lex.europa.eu/legal-content/EN/TXT/PDF/?uri=CELEX:32019R0631&from=EN (accessed on 2 August 2019).
9. FEV—The Future Drives Electric? Available online: http://magazine.fev.com/en/fev-study-examines-drivetr ain-topologies-in-2030-2/ (accessed on 17 July 2019).
10. Bao, R.; Avila, V.; Baxter, J. Effect of 48 V Mild Hybrid System Layout on Powertrain System Efficiency and Its Potential of Fuel Economy Improvement. *SAE Tech. Paper* **2017**. [CrossRef]
11. Bozza, F.; De Bellis, V.; Teodosio, L.; Tufano, D.; Malfi, E. Techniques for CO2 Emission Reduction over a WLTC. A Numerical Comparison of Increased Compression Ratio, Cooled EGR and Water Injection. *SAE Tech. Paper* **2018**. [CrossRef]
12. Lee, S.; Cherry, J.; Safoutin, M.; Neam, A.; McDonald, J.; Newman, K. Modeling and Controls Development of 48 V Mild Hybrid Electric Vehicles. *SAE Inter. J. Altern. Powertrains* **2018**. [CrossRef]
13. Mamikoglu, S.; Andric, J.; Dahlander, P. Impact of Conventional and Electrified Powertrains on Fuel Economy in Various Driving Cycles. *SAE Tech. Paper* **2017**. [CrossRef]
14. Benajes, J.; García, A.; Monsalve-Serrano, J.; Martínez-Boggio, S. Optimization of the parallel and mild hybrid vehicle platforms operating under conventional and advanced combustion modes. *Energy Conver. Manag.* **2019**, *190*, 73–90. [CrossRef]
15. Millo, F.; Di Lorenzo, G.; Servetto, E.; Capra, A.; Pettiti, M. Analysis of the Performance of a Turbocharged S.I. Engine under Transient Operating Conditions by Means of Fast Running Models. *SAE Int. J. Engines* **2013**, *6*, 968–978. [CrossRef]
16. Gamma Technologies, Inc. *GT-SUITE Flow Theory Manual*; Gamma Technologies, Inc.: Westmont, IL, USA, 2019.

17. Andert, J.; Xia, F.; Klein, S.; Guse, D.; Savelsberg, R.; Tharmakulasingam, R.; Scharf, J. Road-to-rig-to-desktop: Virtual development using real-time engine modelling and powertrain co-simulation. *Inter. J. Engine. Res.* **2018**. [CrossRef]

18. Dorsch, M.; Neumann, J.; Hasse, C. Fully coupled control of a spark-ignited engine in driving cycle simulations. *Automot. Engine. Technol.* **2019**. [CrossRef]

19. Griefnow, P.; Andert, J.; Xia, F.; Klein, S. Real-Time Modeling of a 48V P0 Mild Hybrid Vehicle with Electric Compressor for Model Predictive Control. *SAE Tech. Paper* **2019**. [CrossRef]

20. DieselNet—Emission Test Cycles. Available online: https://www.dieselnet.com/standards/cycles/index.php#eu-ld (accessed on 17 July 2019).

21. DIRECTIVE 2007/46/EC, OJ L 263, 9.10.2007, p. 1–160. Available online: https://eur-lex.europa.eu/legal-content/EN/TXT/PDF/?uri=CELEX:02007L0046-20160101&from=IT (accessed on 2 August 2019).

22. Gearshift Calculation Tool. Available online: https://wiki.unece.org/display/trans/Gearshift+calculation+tool (accessed on 24 April 2019).

23. Onori, S.; Serrao, L.; Rizzoni, G. *Hybrid Electric Vehicles: Energy Management Strategies*; Springer: Berlin/Heidelberg, Germany, 2016. [CrossRef]

24. Zanelli, A.; Millo, F.; Barbolini, M.; Neri, L. Assessment through Numerical Simulation of the Impact of a 48 V Electric Supercharger on Performance and CO_2 Emissions of a Gasoline Passenger Car. *SAE Tech. Paper* **2019**. [CrossRef]

25. Paganelli, G.; Guerra, T.M.; Delprat, S.; Santin, J.J.; Delhom, M.; Combes, E. Simulation and assessment of power control strategies for a parallel hybrid car. *Proc. Inst. Mech. Eng. Part D J. Automobil. Eng.* **2000**, *214*, 705–717. [CrossRef]

26. Ali, A.; Söffker, D. Towards Optimal Power Management of Hybrid Electric Vehicles in Real-Time: A Review on Methods, Challenges, and State-Of-The-Art Solutions. *Energies* **2018**, *11*, 476. [CrossRef]

27. Mamun, A.A.; Liu, Z.; Rizzo, D.M.; Onori, S. An Integrated Design and Control Optimization Framework for Hybrid Military Vehicle Using Lithium-Ion Battery and Supercapacitor as Energy Storage Devices. *IEEE Trans. Transp. Electr.* **2018**, *5*, 239–251. [CrossRef]

28. Millo, F.; Rolando, L.; Pautasso, E.; Servetto, E. A Methodology to Mimic Cycle to Cycle Variations and to Predict Knock Occurrence through Numerical Simulation. *SAE Tech. Paper* **2014**. [CrossRef]

29. Luisi, S.; Doria, V.; Stroppiana, A.; Millo, F. Experimental Investigation on Early and Late Intake Valve Closures for Knock Mitigation through Miller Cycle in a Downsized Turbocharged Engine. *SAE Tech. Paper* **2015**. [CrossRef]

30. Pavlovic, J.; Marotta, A.; Ciuffo, B. CO_2 emissions and energy demands of vehicles tested under the NEDC and the new WLTP type approval test procedures. *Appl. Energy* **2016**, *177*, 661–670. [CrossRef]

Article

Impact of Electrically Assisted Turbocharger on the Intake Oxygen Concentration and Its Disturbance Rejection Control for a Heavy-duty Diesel Engine

Chao Wu, Kang Song *, Shaohua Li and Hui Xie *

State Key Laboratory of Engines, Tianjin University, Tianjin 300072, China
* Correspondence: songkangtju@tju.edu.cn (K.S.); xiehui@tju.edu.cn (H.X.); Tel.: +86-1752-695-8480

Received: 7 July 2019; Accepted: 1 August 2019; Published: 5 August 2019

Abstract: The electrically assisted turbocharger (EAT) shows promise in simultaneously improving the boost response and reducing the fuel consumption of engines with assist. In this paper, experimental results show that 7.8% fuel economy (FE) benefit and 52.1% improvement in transient boost response can be achieved with EAT assist. EAT also drives the need for a new feedback variable for the air system control, instead of the exhaust recirculation gas (EGR) rate that is widely used in conventional turbocharged engines (nominal system). Steady-state results show that EAT assist allows wider turbine vane open and reduces pre-turbine pressure, which in turn elevates the engine volumetric efficiency hence the engine air flow rate at fixed boost pressure. Increased engine air flow rate, together with the reduced fuel amount necessary to meet the torque demand with assist, leads to the increase of the oxygen concentration in the exhaust gas (EGR gas dilution). Additionally, transient results demonstrate that the enhanced air supply from the compressor and the diluted EGR gas result in a spike in the oxygen concentration in the intake manifold (X_{oim}) during tip-in, even though there is no spike in the EGR rate response profile. Consequently, there is Nitrogen Oxides (NOx) emission spike, although the response of boost pressure and EGR rate is smooth (no spike is seen). Therefore, in contrast to EGR rate, X_{oim} is found to be a better choice for the feedback variable. Additionally, a disturbance observer-based X_{oim} controller is developed to attenuate the disturbances from the turbine vane position variation. Simulation results on a high-fidelity GT-SUTIE model show over 43% improvement in disturbance rejection capability in terms of recovery time, relative to the conventional proportional-integral-differential (PID) controller. This X_{oim}-based disturbance rejection control solution is beneficial in the practical application of the EAT system.

Keywords: electrically assisted turbocharger; variable geometry turbocharger-exhaust gas recirculation; oxygen concentration; active disturbance rejection control

1. Introduction

Modern diesel engines are normally equipped with a variable-geometry turbocharger (VGT) or fixed geometry turbocharger (FGT) [1,2]. However, the FGT, and to a lesser extent VGT, suffer from an undesired response dynamic caused by the sluggish turbine power response, popularly known as turbo-lag. Turbo-lag is observed through the slow boost pressure (p_2) and torque responses, leading to undesired transient emissions spikes and fuel penalty. In addition, the poor transient boost response, together with smoke limited fueling, also leads to poor drivability performance [3]. Moreover, part of the exhaust energy is also wasted, either through the exhaust bypass valve or widely opened VGT vane, to avoid over-boost at high engine load conditions. This results in the thermodynamic efficiency penalty for engines equipped with conventional turbochargers.

The electrically assisted turbocharger (EAT) is promising in enhancing boost response and recovering exhaust energy [4–11]. By mounting an electrical motor/generator (referred to as TEMG in this paper) in the shaft of conventional turbocharger, EAT is capable of bi-directional energy transfer. The EAT can assist the compressor via the motor mode for enhanced boost response at the early stage of tip-in, and recover the exhaust energy via the generator mode to avoid over-boost. The schematic of the diesel engine equipped with EAT is shown in Figure 1. The benefits of EAT are summarized as below:

(1) Improved boost response allows enhanced engine torque response [12,13], and enables down-speeding and downsizing [14,15]. Improvements of fuel economy (FE) can also be achieved mainly by reduced pumping losses. In a drive cycle simulation by Zhao et al. [16], a 6.44% fuel-saving was reported for a Hybrid electric vehicle equipped with EAT assisted diesel engine.

(2) Better utilization of exhaust enthalpy during regen mode via energy recuperation from turbocharger (TC) braking.

(3) Lower pre-turbine pressure (p_3) and thereby lower pumping loss can be achieved through reduced VGT throttling during assist mode. This improves the thermal efficiency of the engine [17].

(4) Proper use of assist and regen modes allows the TC to operate in optimal efficiency region [14].

(5) Improved regulation of intake oxygen concentration (X_{oim}). Compared to the conventional variable geometry turbocharger-exhaust gas recirculation (VGT-EGR) system, EAT offers additional control degree of freedom and therefore brings a direct result of improved manipulations of p_3 and p_2. This helps to improve EGR inert quality and reduce transient soot/Nitrogen Oxides (NOx) emissions [18].

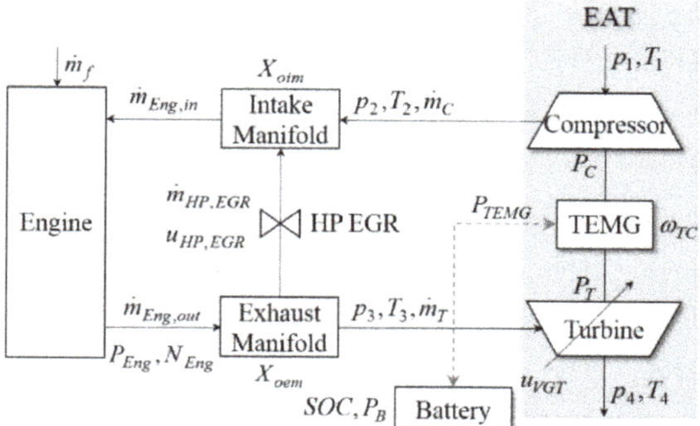

Figure 1. Schematic of the electrically assisted turbocharger (EAT)-assisted diesel engine. The nomenclature for all the symbols can be found in Nomenclature section.

For a tip-in maneuver with the EAT system, a fast boost response can be achieved in assist mode without the need for an equivalent increase in exhaust turbine power. Therefore, a wider open VGT vane position can be held for optimal turbine efficiency, relative to the unassisted VGT-EGR air-path system (referred to as nominal system in this paper henceforth) [19]. A wider open vane position leads to a decrease in the exhaust manifold pressure. The simultaneous decrease in p_3 and increase in p_2 reduce the ($p_3 - p_2$) margin which impacts both FE and the ability to flow high pressure EGR (HP-EGR).

Using the intake oxygen concentration as feedback variable was attempted by Nakayama et al. [20], Shutty et al. [21], and Millo et al. [22], for conventional VGT-EGR diesel engines. For engines equipped with EAT, however, fast boost response leads to increased fresh air flow relative to the nominal system, and wider open VGT vane position improves the engine volumetric efficiency for increased engine mass flow rate. Those two facts lead to increased dilution of the HP-EGR gas relative to the nominal system, which distinguished from conventional VGT-EGR engine. Therefore a higher HP-EGR flow rates may be of necessity to keep the intake oxygen concentration well matched with any hard constraints on engine out emissions. In other words, the EGR rate (X_{EGR}), as popularly used in the conventional VGT-EGR system, might not be an appropriate output for control for EAT [3]. So far no experimental assessment results are reported on the selection of control output variables for EAT assisted engines, according to the authors' best knowledge.

Regarding the control of X_{oim}, the traditional proportional–integral–derivative (PID) algorithm is fairly straightforward to use and consequently is of great popularity. For instance, Park [23] proposed a PID controller for X_{oim} control in a 2.2 L turbocharged passenger car diesel engine, where the proportional gain was determined according to the current NOx level. In another approach, Chen et al. [24] proposed an implicit PI-based X_{oim} controller by tracking the air-fuel ratio in the tail pipe.

However, the PID controller suffers from the time-consuming parameter tuning and gain-scheduling [25], and shows as poor robustness against operating condition variations and external disturbances. For EAT-assisted diesel engines, the transient response of the air system is much faster relative to the conventional VGT-EGR system. The frequent VGT vane position optimization also introduces a lot of disturbances to the X_{oim} controller. Those facts drive the need for an advanced control solution for X_{oim} with faster response and improved robustness against uncertainties.

In this paper, the benefits from using the X_{oim} as control output is assessed experimentally on a heavy duty Diesel Engine test bench. Then, an active disturbance rejection control [25,26] based X_{oim} controller (OADRC) is proposed, assisted by an existing X_{oim} observer [23]. The superiority of the proposed controller, against the conventional PID controller, is validated based on a previously calibrated high fidelity GT-SUITE simulation model [27].

The rest of the paper is organized as follows. The system layout and the experiment setup as well as the simulation platform used in this study are discussed briefly in Section 2. Experimental investigation on the fuel economy benefit and discussion on the selection of feedback variable for the HP-EGR system are provided in Section 3. The impact of EAT on X_{oim} control is analyzed in Section 4. The X_{oim} controller is developed and validated in Sections 5 and 6 respectively. Finally, the conclusions are summarized in Section 7.

2. Experiment Setup and Simulation Platform

A modern heavy-duty diesel engine equipped with common-rail injection system, HP-EGR and VGT, was used for the experimental evaluation. The engine specifications are tabulated in Table 1. Note that the original variable geometry turbocharger was replaced by the EAT and the specifications are listed in Table 2.

Table 1. Engine specifications.

Variable	Value
Displacement (liter)	6.7
Cylinders	V8
Compression ratio	16.2
Maximum injection pressure (MPa)	200
Bore (mm)	99
Stroke (mm)	108
Bore/Stroke Ratio	0.92
Maximum torque (Nm)/speed (rpm)	1166/2600
Rated power (kW)/speed (rpm)	328/2800

Table 2. EAT specifications.

Variable	Value
Operating voltage (V)	200~400
Maximum power (kW)	17 kW in continuous mode and 23 kW in intermittent mode
Shaft speed (krpm)	0~140
Type of cooling	water and oil cooling

The schematic diagram of the experimental platform is shown in Figure 2, where all the variables are explained in Nomenclature. The test bench is shown in Figure 3. The specifications of the test bench are listed in Table 3.

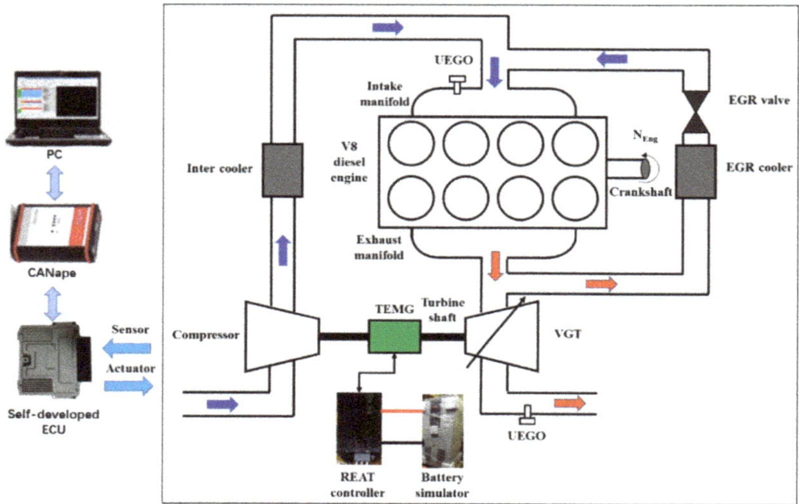

Figure 2. Schematic of experimental platform.

Figure 3. Test bench for diesel engine with EAT.

Table 3. Specifications of test bench.

Num	Device	Specifications	
1	AVL INDYS22 AC electric dynamometer	Max speed/power	8000 rpm/200 kW
2	AVL 735S/753 C fuel mass flow meter	Max measuring fuel consumption	125 kg/h
		Max measuring frequency	20 Hz
3	KWELL EVS-80-800 Battery Simulator	Max output voltage	800 V
		Rated power	80 kW
4	Bosch LSU 4.9 oxygen sensor	Measuring range	lambda 0.65–∞
5	Continental SNS14 NOx sensor	Measuring accuracy	±10% from 100 ppm to 1500 ppm

A high-fidelity simulation model was established in GT-SUITE [28]. The performance of the model was validated against experimental data from a FTP-75 test cycle. Validation results are shown in Figure 4. Model accuracy results are summarized in Table 4. The relative error (X_{error}) is defined as in Equation (1), where y_{meas} is the measured value in the experiment, y_{model} is the estimated value obtained from the GT-SUITE model.

$$X_{error} = \frac{y_{meas} - y_{model}}{y_{meas}} \times 100\% \tag{1}$$

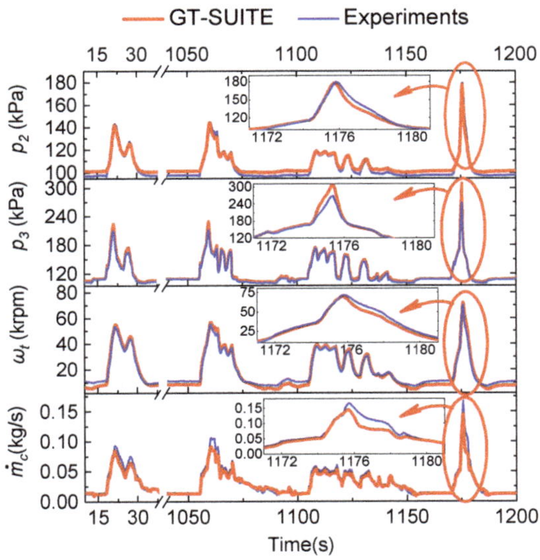

Figure 4. Validation of the GT-SUITE model over a segment of the hot start FTP-75 drive cycle in terms of boost pressure, pre-turbine pressure, and compressor mass flow rate [27].

Table 4. Estimation error of the GT-SUITE model.

Variable	GT-SUITE Model
p_2	1.8% points has error > 5 kPa
p_3	2.9% points has error > 10 kPa
$\dot{m}_c$	3.3% points has error > 0.01 kg/s
$\int P_{Eng} dt$	$X_{error} = 6.7\%$

3. Experimental Assessment on the Benefits from Using X_{oim} as Control Output Variable

3.1. Steady-State Investigation

In this section, steady-state experiment was carried out on the test bench to assess the superiority of regulating X_{oim} in terms of NOx emission reduction relative to the EGR rate. The impact of EAT on the HP-EGR flow and fuel efficiency were also investigated. The operating conditions were tabulated in Table 5. Note that in this paper 100% is fully open and 0% is fully closed for both the VGT vane and EGR valve.

Table 5. Test case set-up for steady-state investigation of assist power.

Test Case	N_{Eng}	T_{Eng}	p_2	X_{oim}	u_{VGT}	$u_{HP\text{-}EGR}$	P_{TEMG}
-	rpm	Nm	bar	%	%	%	kW
Case1	1400	265	1.24	15.5	Sweep from 15% to 55% open	Closed-loop control to maintain X_{oim}	Closed-loop control to maintain p_2

From the results shown in Figure 5, the following observations are clear:

1. When the VGT vane open is less than 22%, the electrical regeneration (E-regen) mode is turned on to brake the turbine in order to avoid over-boost. The electrical power in regeneration is about 0.46 kW at the cost of 45 *kPa* increase in $p_3 - p_2$, compared to the nominal system. Consequently, 12.1% increase in fuel consumption is seen with E-regen compared to the nominal system.

2. When the VGT vane is greater than 37% open, there is X_{oim} surplus even though the HP-EGR valve (the blue triangle curve) is already 100% open. This is a direct effect of reduced $p_3 - p_2$ caused by wide VGT vane open, which limits the HP-EGR flow capability.

3. For VGT vane ranging from 22% to 37%, with increasing VGT vane position, the assist power increases (from 0 kW to 0.78 kW) to compensate for the turbine power deficit in order to maintain the boost pressure. Consequently, the pumping loss (indicated by $p_3 - p_2$) decreases (from 4.22 kW to 0.86 kW) due to reduced p_3. Lower p_3 and unchanged p_2 elevates the thermodynamic efficiency (η_e) of the engine (8.5% improvement) through pumping loss reduction. This allows less fuel injection to maintain the desired torque, relative to the nominal system.

4. With VGT vane increasing from 22% to 37%, the incremental benefit in fuel economy decreases, since $p_3 - p_2$ gradually converges when the VGT open approaches 37%. The fuel saving with VGT vane 37% open is 7.8%, relative to the nominal system. Note that here the electrical energy is assumed to be free in the driveline electrical regeneration during vehicle braking. Therefore, the cost for the electrical power is neglected when evaluating the fuel economy.

5. It should be noted that equivalent brake-specific fuel consumption (Eq_BSFC), as defined in Equation (2), can also be used to evaluate the efficiency of the system when the cost of assist power has to be taken into consideration [3]. In Equation (2), t_0 and t_f are the time stamps of the start and end of the current test respectively, W_{Eng} is the engine work output, W_{TEMG}^{assist} and W_{TEMG}^{regen} are defined as $W_{TEMG}^{assist} = \int_{t_0}^{t_f} \max(0, P_{TEMG})dt$ and $W_{TEMG}^{regen} = \left| \int_{t_0}^{t_f} \min(0, P_{TEMG})dt \right|$ respectively. In terms of Eq_BSFC, the fuel economy benefit from using EAT is 5.9% relative to the conventional VGT-EGR engine.

$$Eq\_BSFC = \frac{\int_{t_0}^{t_f} \dot{m}_{fuel}dt}{W_{Eng} - W_{TEMG}^{assist} + W_{TEMG}^{regen}} \tag{2}$$

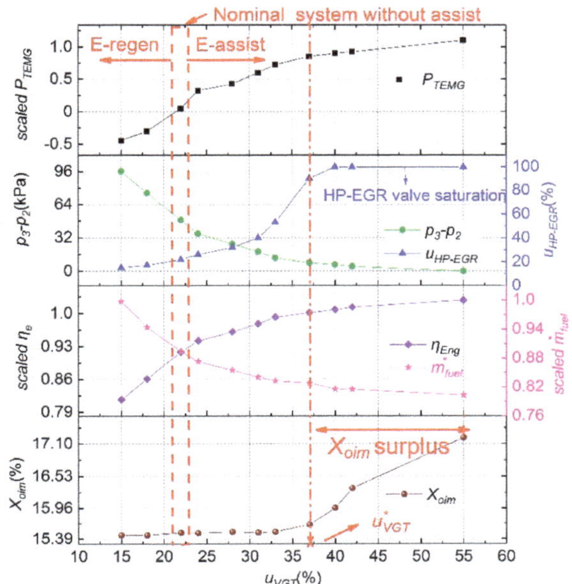

Figure 5. Engine fuel efficiency improvement and X_{oim} surplus.

In addition to FE improvement, another impact of reduced $p_3 - p_2$ is the improvement in the volumetric efficiency η_{vol} of the engine, as calculated according to Equation (3), where N_{Eng} is the engine speed, V_{Eng} is the engine displacement, M is the molar mass of gas in intake manifold, and p_2 and T_2 are intake manifold pressure and temperature, respectively. The engine mass flow rate $\dot{m}_{Eng}$ can be calculated based on Equation (4),where oxygen concentration in the recirculated exhaust gas (X_{oegr}) and X_{oim} are measured by the UEGO sensors, fuel injection rate $\dot{m}_{fuel}$ is obtained from the AVL 735 S fuel mass flow meter, X_{oair} is oxygen concentration in the air, AFR is air-fuel ratio. From Figure 6, it is seen that up to 13.6% improvement in η_{vol} is achieved with 1.01 kW E-assist power (P_{TEMG}) relative to the nominal system. This is primarily attributed to the decrease in $p_3 - p_2$, as clearly explained in the Equation (5) proposed in [29], where $C_{\eta_{vol}}(N_{Eng})$ is an engine speed dependent factor, r_c is compression ratio. Obviously, the decrease in $p_3 - p_2$ helps increasing η_{vol}.

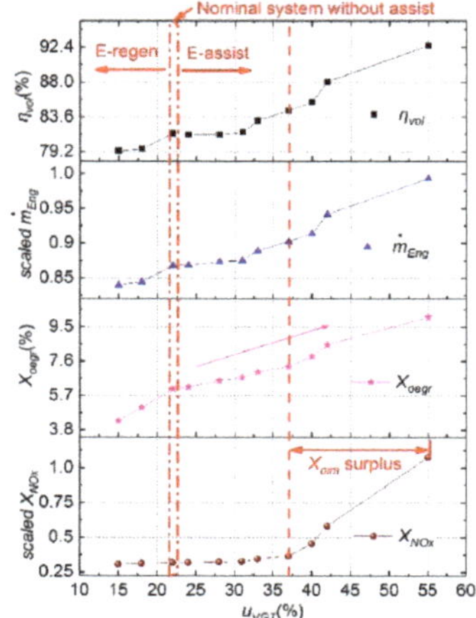

Figure 6. Volumetric efficiency improvement and in-cylinder composition change.

The increase in-cylinder charge amount caused by increased η_{vol}, together with reduced fuel amount (shown in Figure 5), elevates the oxygen concentration in the HP-EGR gas (EGR dilution), i.e., the increasing X_{oegr} as shown in Figure 6. Therefore, a 7.5% increase in EGR rate (from 36.30% in nominal system to 39.03% in EAT system) at 37% VGT open is needed to maintain the X_{oim} and NOx emission, relative to the nominal system (22% VGT vane open), as shown in Figure 7. This can be understood via the mathematical expression for X_{EGR} shown in Equation (6) [30], where $\dot{m}_{HP,EGR}$ is the HP-EGR mass flow rate. It is clearly seen from Equation (6) that the desired X_{EGR} increases with increasing X_{oegr} with X_{oim} fixed and X_{oair} being a constant.

$$\eta_{vol} = \dot{m}_{Eng} \cdot \frac{120}{N_{Eng}} \cdot \frac{RT_2}{p_2 V_{Eng} M} \tag{3}$$

$$\dot{m}_{Eng} = \frac{\dot{m}_{fuel}(AFR\, X_{oair} + X_{oegr})}{X_{oim} - X_{oegr}} \tag{4}$$

$$\eta_{vol}(p_2, p_3) = C_{\eta_{vol}}(N_{Eng}) \frac{r_c - \left(\frac{p_3}{p_2}\right)^{1/\gamma}}{r_c - 1} \tag{5}$$

$$X_{EGR} = \frac{\dot{m}_{HP-EGR}}{\dot{m}_{Eng}} = \frac{X_{oair} - X_{oim}}{X_{oair} - X_{oegr}} \tag{6}$$

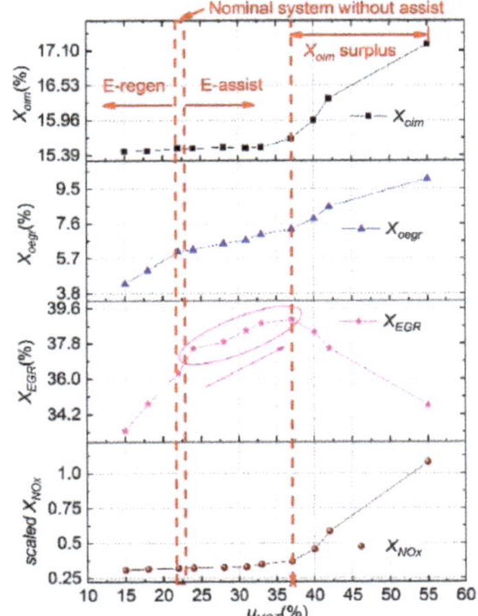

Figure 7. Comparison of three performance variables with NOx concentration.

The primary reason to regulate X_{oim} is that X_{oim} has much higher correlation with NOx emission than X_{EGR} and X_{oegr} in the EAT system, as shown in Figure 8. For EAT assisted air system, the HP-EGR is diluted due to elevated engine mass flow rate, which is not taken into account in the conventional EGR rate based control. In contrast, the X_{oim} based control is able to maintain X_{oim} using increased X_{EGR}, which is beneficial for NOx control for EAT assisted air system.

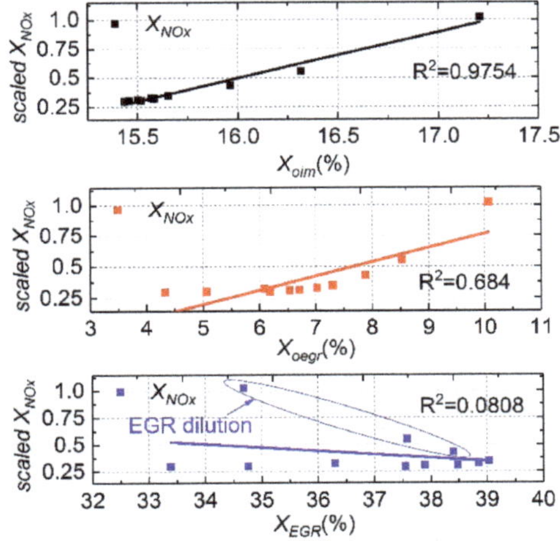

Figure 8. Correlation between NOx concentration and three performance variables.

Therefore, steady-state results show that X_{oim} is a better control output for the air system relative to the X_{EGR} in EAT assisted air system in diesel engines.

3.2. Transient Results Analysis

In addition to the steady-state experiment, transient tests were also carried out on the test bench under the operating conditions shown in Table 6. The three actuators were manipulated as shown in Figure 9.

Table 6. Test case set-up for transient investigation of assist power.

Test Case	N_{Eng}	T_{Eng}	u_{VGT}	$u_{HP\text{-}EGR}$	P_{TEMG}
-	rpm	Nm	%	%	kW
Case1	1200	120→260	56→36	24→14	0
Case2	1200	120→260	56→36	24→14	Closed-loop control to trace N_T

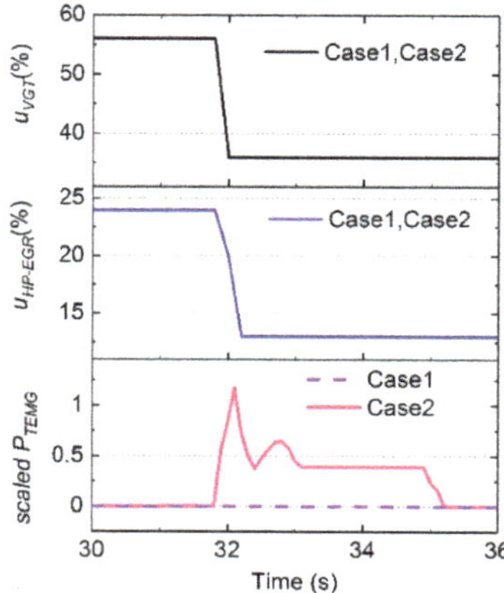

Figure 9. Regulations of three actuators during the transient process.

For conventional VGT-EGR system, as shown in Figure 10, the settling time for p_2 is about 4.6 s. Consequently, there is X_{oim} undershoot in the time period from 31.6 s to 36 s, where the X_{EGR} response is smooth. In the process of tip-in, there is a slight NOx spike.

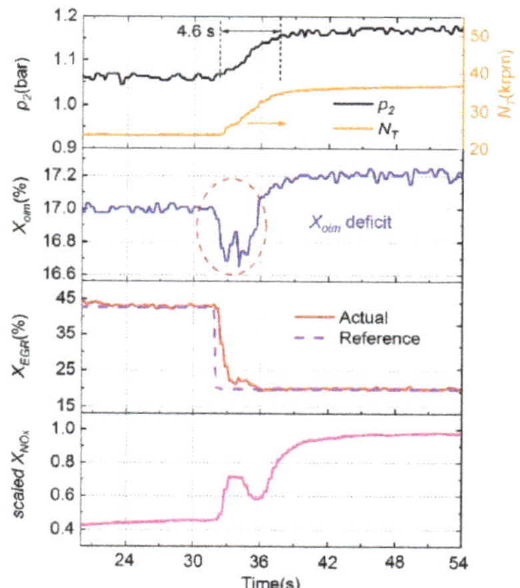

Figure 10. Tip-in test for the nominal system without assist.

For the EAT assisted engine, the p_2 response is significantly enhanced with the settling time shorten to 2.2 s (52.1% improvement relative to the nominal system without E-assist), as shown in Figure 11. This implies enhanced fresh air or oxygen response. Consequently, there is a X_{oim} spike shown in the time period from 32.8 s to 34.8 s. Therefore, it is not surprising to see a NOx spike in the same time interval. What is interesting is that the X_{EGR} response is still smooth without any overshoot or undershoot. This implies that there is no sign of oxygen surplus by observing the EGR rate for EAT assisted engines. This makes it difficult to suppress the NOx spike by controlling the EGR rate since the EGR rate based control is not able to take into account the EGR dilution caused by VGT vane position optimization with EAT. Therefore, transient results confirmed that X_{oim} is a better control output in the air system of EAT assisted engines, relative to EGR rate, since the NOx spike can be removed by simply avoiding the X_{oim} overshoot.

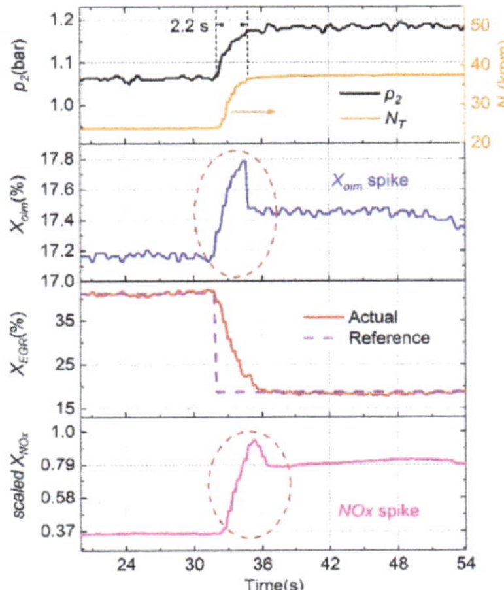

Figure 11. Tip-in test for the EAT system

4. Theoretical Analysis of the Impact of EAT on X_{oim} Control

From the aforementioned analysis, both steady-state and transient response confirmed the superiority of using X_{oim} as the control output for EAT assisted engines. However, interactions, between the electrical power of EAT and the states in the intake and exhaust manifolds, bring up complex disturbances to the X_{oim} control loop. This will be investigated based on the fundamentals of the boosting system in this section.

Based on the mass conservation of the intake manifold, the X_{oim} dynamics can be modeled as [27]:

$$\dot{X}_{oim} = \frac{RT_2}{p_2 V_{im}}(\dot{m}_C X_{oair} + \dot{m}_{HP-EGR} X_{oegr} - \dot{m}_{Eng} X_{oim}) \tag{7}$$

where $\dot{m}_C$ is the compressor mass flow rate, $\dot{m}_{HP-EGR}$ is the HP-EGR mass flow rate, $\dot{m}_{Eng}$ is the engine mass flow rate, V_{im} is the volume of intake manifold. At steady-state condition, $\dot{m}_{Eng}$ can be approximated by the sum of $\dot{m}_{HP-EGR}$ and $\dot{m}_C$. Detailed discussion about the unknown terms, namely $\dot{m}_C$, $\dot{m}_{HP-EGR}$ and X_{oegr}, are shown below. This purpose of the following analysis is to understand the impact of EAT assist on the X_{oim} control from the perspective of physics.

1. The compressor mass flow rate $\dot{m}_C$ can be obtained from the definition of the compressor efficiency (η_C) in [29], as shown in Equations (8) and (9).

$$P_{C,ideal} = \dot{m}_C \Delta h_{0s} = \dot{m}_C c_{p,c} T_1 \left\{ \left(\frac{p_2}{p_1}\right)^{\frac{\gamma-1}{\gamma}} - 1 \right\} \tag{8}$$

$$\eta_C = \frac{P_{C,ideal}}{P_C} = \frac{\left(\frac{p_2}{p_1}\right)^{\frac{\gamma-1}{\gamma}} - 1}{\frac{T_2}{T_1} - 1} \tag{9}$$

where $c_{p,c}$ is the specific heat capacity of air, p_1 and T_1 are pre-compressor pressure and temperature, respectively, $P_{C,ideal}$ is the isentropic power for compressing the air from p_1 to p_2, i.e., power requirement during an isentropic process, and P_C is the actual power that the compression process consumes, which can be approximated by the steady-state energy balance relation:

$$P_C = P_T + P_{TEMG} - P_{Loss} \tag{10}$$

where P_T is the turbine power, P_{Loss} is the mechanical loss of the EAT. Based on Equations (8)–(10), the compressor mass flow rate is modeled as function of P_{TEMG} as below:

$$\dot{m}_C = \frac{\eta_C(P_T + P_{TEMG} - P_{Loss})}{c_{p,c}T_1\left(\left(\frac{p_2}{p_1}\right)^{1-\frac{1}{k}} - 1\right)} \equiv g_{\dot{m}_C}(P_{TEMG}, P_T, p_2) \tag{11}$$

2. The HP-EGR flow rate $\dot{m}_{HP-EGR}$ can be estimated by the model proposed in [29] as:

$$\dot{m}_{HP-EGR} = \theta_{HP-EGR}\frac{A_{HP-EGR}C_D p_3}{\sqrt{RT_3}}\sqrt{1 - \left(\frac{p_2}{p_3}\right)^2} \tag{12}$$

where θ_{HP-EGR} is the HP-EGR valve position, C_D and A_{HP-EGR} are the flow coefficient and cross-sectional area of the HP-EGR valve, respectively, p_3 and T_3 are exhaust manifold pressure and temperature respectively.

The p_3 in Equation (12) can be obtained through the turbine efficiency definition and the Equation (10) as show in Equation (13) [29]:

$$P_T = P_C + P_{Loss} - P_{TEMG} = \eta_{tm}\dot{m}_T c_{p,e}T_3\left(1 - \left(\frac{p_4}{p_3}\right)^{1-\frac{1}{k}}\right) \tag{13}$$

where η_{tm} is the turbine efficiency, $\dot{m}_T$ is the turbine mass flow rate, $c_{p,e}$ is the specific heat capacity of exhaust gases, and p_4 is the post-turbine pressure. Specifically, the p_3 model shown in Equation (14) as:

$$p_3 = p_4\left(1 - \frac{P_C + P_{Loss} - P_{TEMG}}{\eta_{tm}\dot{m}_T c_{pe}T_3}\right)^{1-\frac{1}{k}} \equiv g_{p_3}(P_{TEMG}, P_C, T_3, \dot{m}_T) \tag{14}$$

Substituting Equation (14) into Equation (12) transforms the HP-EGR flow model into:

$$\begin{aligned}\dot{m}_{HP-EGR} &= \theta_{HP-EGR}\frac{A_{HP-EGR}C_D g_{p_3}(P_{TEMG},P_C,T_3,\dot{m}_T)}{\sqrt{RT_3}}\sqrt{1 - \left(\frac{p_2}{g_{p_3}(P_{TEMG},P_C,T_3,\dot{m}_T)}\right)^2} \\ &\equiv \theta_{HP-EGR}g_{\dot{m}_{HP-EGR}}(P_{TEMG}, p_2, \dot{m}_T, T_3)\end{aligned} \tag{15}$$

It can be inferred that increased assist power (P_{TEMG}) reduces the EGR flow rate through reduced pressure differential (or p_3) across the EGR value, as shown in Equation (14).

3. The X_{oegr} can be approximated as the engine outlet port oxygen concentration (X_{ocyl_out}), which is estimated based on [27]:

$$X_{oegr} \approx X_{ocyl\_out} = \frac{\dot{m}_{Eng}X_{oim} - \dot{m}_{fuel}AFRX_{oair}}{\dot{m}_{Eng} + \dot{m}_{fuel}} \tag{16}$$

where *AFR* is the air-to-fuel ratio, $\dot{m}_{Eng}$ and η_{vol} can be estimated by Equations (3) and (5) respectively. By substituting Equations (3), (5), and (14) into Equation (16), the X_{oegr} can be modeled as a function of P_{TEMG} as:

$$X_{oegr} = \frac{g_{\eta_{vol}}(N,p_2,g_{p_3}(P_{TEMG},P_C,T_3,\dot{m}_T))N_{Eng}p_2V_{Eng}}{120RT_2(\dot{m}_{Eng}+\dot{m}_{fuel})}X_{oim} - \frac{\dot{m}_{fuel}AFRX_{oair}}{\dot{m}_{Eng}+\dot{m}_{fuel}}$$
$$\equiv g_{X_{oegr}}(P_{TEMG},N,p_2,\dot{m}_{fuel},\dot{m}_{Eng}) \tag{17}$$

Finally, substituting Equations (11), (15), and (17) into Equation (7), the X_{oim} model is derived as below:

$$\dot{X}_{oim} = \frac{RT_2}{p_2V_{im}}(\dot{m}_C X_{oair} + \dot{m}_{HP-EGR}X_{oegr} - \dot{m}_{Eng}X_{oim})$$
$$= \frac{RT_2}{p_2V_{im}}[g_{\dot{m}_C}(P_{TEMG},P_T,p_2)X_{oair} + \theta_{HP-EGR}g_{\dot{m}_{HP-EGR}}(P_{TEMG},p_2,\dot{m}_T,T_3) \times g_{X_{oegr}}(P_{TEMG},N,p_2,\dot{m}_{fuel},\dot{m}_{Eng}) - \dot{m}_{Eng}X_{oim}] \tag{18}$$

It should be noted that the turbine and compressor efficiency are implicitly used in Equation (18), both of which vary with the operating point. For simplicity, they are read in turbine and compressor maps respectively, which are extrapolated from manufacturer's data.

By observing Equation (18), it is clear that the EAT assisting power affects the X_{oim} dynamic primarily from three different aspects:

- Enhanced air response through the term $\dot{m}_C = g_{\dot{m}_C}(P_{TEMG},P_T,p_2)$, since P_{TMEG} increased the compressor power.
- HP-EGR flow deficit when the pressure differential across the HP-EGR valve is insufficient as illustrated by $\dot{m}_{HP-EGR} = \theta_{HP-EGR}g_{\dot{m}_{HP-EGR}}(P_{TEMG},p_2,\dot{m}_T,T_3)$, as P_{TEMG} affects the upstream pressure of the EGR valve.
- EGR gas dilution due to increased engine mass flow rate from elevated volumetric efficiency as indicated by $X_{oegr} = g_{X_{oegr}}(P_{TEMG},N,p_2,\dot{m}_{fuel},\dot{m}_{Eng})$, since P_{TEMG} leads to increased fresh air flow and reduces the fuel consumption.

Clearly, in order to attenuate the disturbances caused by the additional control degree of freedom (P_{TEMG}), an advanced disturbance rejection solution becomes necessary, as will be discussed in Section 5.

5. X_{oim} Controller Development

Since the universal oxygen sensor is mounted only on the exhaust side (UEGO), a X_{oim} observer is necessary. In this paper, the X_{oim} observer based on the measured p_2, X_{oegr}, and estimated $\dot{m}_C$, proposed by Park in [23], is used. A brief introduction and experimental validation are provided in the Appendix A for completeness. Then, an active disturbance rejection controller of oxygen concentration in intake manifold (OADRC) was developed with the control structure shown in Figure 12.

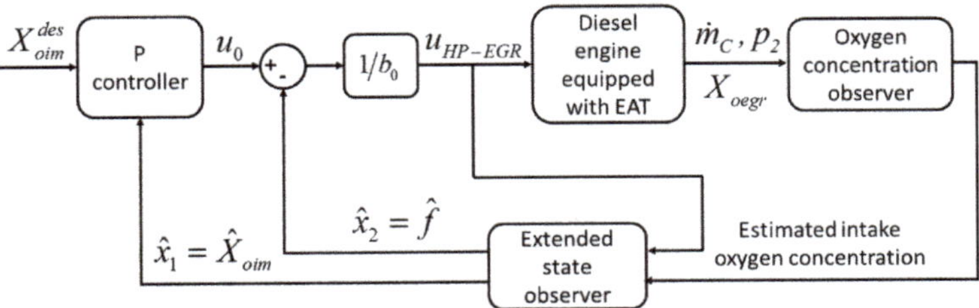

Figure 12. Control structure of the OADRC controller.

For OADRC controller derivation, Equation (18) is rewritten into Equation (19):

$$
\begin{aligned}
\dot{X}_{oim} &= \tfrac{RT_2}{p_2 V_{im}}[g_{\dot{m}_C}(P_{TEMG}, P_T, p_2) X_{oair} + \\
&\quad \theta_{HP-EGR} g_{\dot{m}_{HP-EGR}}(P_{TEMG}, p_2, \dot{m}_T, T_3) g_{X_{oegr}}(P_{TEMG}, N, p_2, \dot{m}_{fuel}, \dot{m}_{Eng}) \\
&\quad - \dot{m}_{Eng} X_{oim}] \\
&= \theta_{HP-EGR} \tfrac{RT_2}{p_2 V_{im}} g_{\dot{m}_{HP-EGR}}(P_{TEMG}, p_2, \dot{m}_T, T_3) g_{X_{oegr}}(P_{TEMG}, N, p_2, \dot{m}_{fuel}, \dot{m}_{Eng}) + \\
&\quad \tfrac{RT_2}{p_2 V_{im}}[g_{\dot{m}_C}(P_{TEMG}, P_T, p_2) X_{oair} - \dot{m}_{Eng} X_{oim}] \\
&= \theta_{HP-EGR} b_0 + f
\end{aligned}
\tag{19}
$$

where $b_0 = \frac{RT_2}{p_2 V_{im}} g_{\dot{m}_{HP-EGR}}(P_{TEMG}, p_2, \dot{m}_T, T_3) g_{X_{oegr}}(P_{TEMG}, N, p_2, \dot{m}_{fuel}, \dot{m}_{Eng})$ is the control input gain, and $f = \frac{RT_2}{p_2 V_{im}}[g_{\dot{m}_C}(P_{TEMG}, P_T, p_2) X_{oair} - \dot{m}_{Eng} X_{oim}]$ is denoted as the "total disturbance". It is clear that estimation errors in engine flow rate and HP-EGR flow rate as well as the external disturbance are all lumped as total disturbance.

Equation (19) is then rewritten in the state space form for ESO derivation, as shown in Equation (20):

$$
\begin{cases}
\dot{x} = Ax + Bu + E\dot{f} \\
y = Cx
\end{cases}
\tag{20}
$$

where $A = \begin{bmatrix} 0 & 1 \\ 0 & 0 \end{bmatrix}$, $B = \begin{bmatrix} b_0 \\ 0 \end{bmatrix}$, $C = \begin{bmatrix} 1 \\ 0 \end{bmatrix}^T$, $E = \begin{bmatrix} 0 \\ 1 \end{bmatrix}$, i.e., $x = [\, y \ \ f \,]^T$, $\begin{cases} x_1 = X_{oim} \\ x_2 = f \end{cases}$ are the states, $y = x_1$ is the output, u is the input ($u = \theta_{HP-EGR}$), namely the HP-EGR valve position.

Based on Equation (20), a linear first order extended state observer (ESO) is designed as in Equation (21), following the method shown in [31],

$$
\begin{cases}
\dot{\hat{x}} = A\hat{x} + Bu + L(y - \hat{y}) \\
\hat{y} = C\hat{x}
\end{cases}
\Rightarrow
\begin{cases}
\dot{\hat{x}} = (A - LC)\hat{x} + [\, B \ \ L \,] \begin{bmatrix} u \\ y \end{bmatrix} \\
\hat{y} = C\hat{x}
\end{cases}
\tag{21}
$$

where $L = \begin{bmatrix} \beta_1 \\ \beta_2 \end{bmatrix}$ is the tunable observer gain matrix, $A - LC = \begin{bmatrix} -\beta_1 & 1 \\ -\beta_2 & 0 \end{bmatrix}$, and $\begin{cases} \hat{x}_1 = \hat{X}_{oim} \\ \hat{x}_2 = \hat{f} \\ y = \hat{x}_1 \end{cases}$ are the observed states and output (measurement) estimate, respectively.

The tuning of L matrix is addressed using the bandwidth parameterized method proposed by Gao [31]. This is briefly introduced here for completeness. The characteristic polynomial for the ESO system, Equation (21), is $s^2 + \beta_1 s + \beta_2 = 0$. Since β_1 and β_2 are design variables, these can be chosen to make the ESO state space Hurwitz stable. If we recognize that the characteristic polynomial may be expressed as a damped second order system, that is: $s^2 + \beta_1 s + \beta_2 \equiv s^2 + 2\xi\omega_o s + \omega_o^2$ with ξ and ω_o being the damping coefficient and the natural frequency (bandwidth) of the ESO system respectively, the ESO observer gains, β_1 and β_2 are then easily determined by choosing a system bandwidth and damping as follows:

$$
\begin{cases}
\beta_1 = 2\xi\omega_o \\
\beta_2 = \omega_o^2
\end{cases}
\tag{22}
$$

Both ξ and ω_o can be time varying and a dynamic selection for ω_o is allowed. For simplicity, the damping ratio ξ is fixed at 1 in this paper. Hence, the eigenvalues for $A - LC$ in Equation (21) are determined by the system bandwidth ω_o. After convergence of ESO, $\hat{x}_1 \rightarrow x_1 \rightarrow X_{oim}$, $\hat{x}_2 \rightarrow x_2 \rightarrow f$, and then observed value of disturbance $\hat{f}$ will approach the actual disturbance f, which realizes the observation of the total disturbance of the system.

With the state observer properly designed, the plant is reduced to an integrator

$$
\dot{X}_{oim} = b_0 u + f = b_0 u + \hat{f} + (f - \hat{f}) \approx u_0
\tag{23}
$$

which is easily controlled by a P (proportional) controller

$$u_0 = k_p(r - \hat{x}_1) \tag{24}$$

where r is the set-point for the desired intake oxygen concentration, $\hat{x}_1$ is the observed value of oxygen concentration in intake manifold, k_p is the proportional gain, which is selected to place the pole of the closed-loop system at the controller bandwidth ω_c.

Finally, the control input, namely the HP-EGR valve position can be obtained as:

$$u = \frac{u_0 - \hat{x}_2}{b_0} = \frac{u_0 - \hat{f}}{b_0} \tag{25}$$

So far, the active disturbance rejection controller of oxygen concentration in intake manifold of the diesel engine equipped with EAT has been designed and the validation will be discussed in the next section.

6. Simulation Validation of the Proposed X_{oim} Controller

The OADRC controller was validated through the co-simulation between GT-SUITE and Matlab/Simulink. Compared with the traditional PID algorithm, the response speed and disturbance rejection ability of the OADRC controller were evaluated. Four evaluation metrics were proposed to quantify the control performance, including the settling time (defined as the time required for the response curve of intake oxygen concentration to reach and stay within a range of $\pm 2\%$ of the final intake oxygen concentration, denoted as T_S), steady-state error (defined as the difference between the desired intake oxygen concentration and the actual intake oxygen concentration when the response has reached steady state, denoted as E_{SS}), overshoot (defined as the maximum peak value of the actual intake oxygen concentration measured from the desired intake oxygen concentration, denoted as M_P) and recovery time (defined as the time it takes to recover to its prior intake oxygen concentration, denoted as T_R).

6.1. Intake Oxygen Concentration Step Response Test

To evaluate the tracking performance of the OADRC controller, a continuous step change of intake oxygen concentration were carried out at 1500 rpm engine speed and 18 mg/cycle fuel injection rate, with 80% VGT vane open. The desired X_{oim} (X_{oim}^{des}) was changed step by step as shown in Figure 13, with the demonstration of the control effect of OADRC controller and the total disturbance observed by the ESO. The estimated intake oxygen concentration in the first plot was the real-time output of the oxygen concentration observer mentioned in Appendix A.

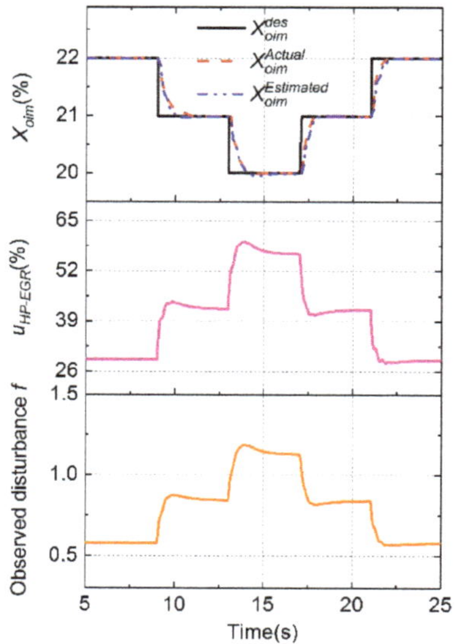

Figure 13. OADRC validation with step changes of desired intake oxygen concentration.

In the step response test of X_{oim} shown in Figure 14, it is seen that the proposed solution offers faster response compared to that of PID controller. The specific improvement is tabulated in Table 7.

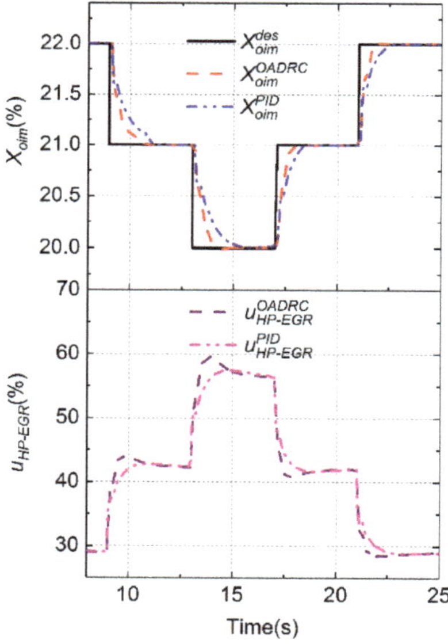

Figure 14. Comparison between OADRC and PID controller with step changes of intake oxygen concentration.

Table 7. Comparison between OADRC and proportional-integral-differential (PID) controller with step changes of intake oxygen concentration.

Item	OADRC	PID	Improvement with OADRC
Average T_S (s)	1.21	2.07	41%
Maximum T_S (s)	1.5	2.35	36%
Average E_{SS} (%)	0.05	0.05	0
M_P (%)	0	0	0

6.2. Disturbance Rejection Capability Test

In order to evaluate the disturbance rejection ability, the N_T steps from 20,000 rpm to 15,000 rpm and then to 10,000 rpm by manipulating the P_{TEMG}, as shown in Figure 15. It is observed that the maximum deviation of X_{oim} from X_{oim}^{des} is reduced by 15.6% using the proposed controller compared to the PID controller, in the time period from 8.1 s to 10.6 s. Detailed performance metrics are provided in Table 8.

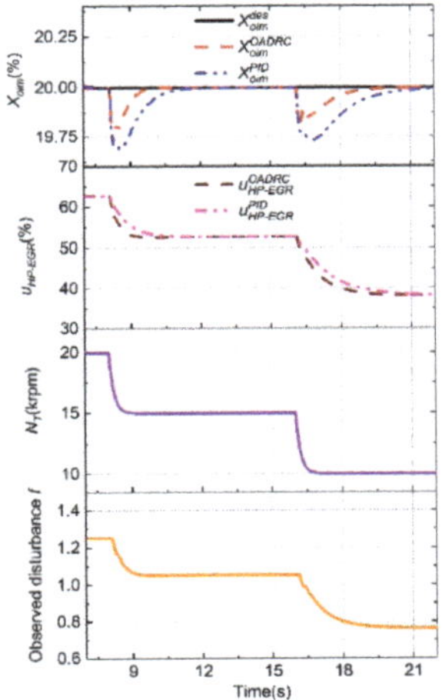

Figure 15. Disturbance rejection performance comparison with step change of N_T.

Table 8. Disturbance rejection performance comparison with step change of N_T.

Item	OADRC	PID	Improvement with OADRC
Average M_P (%)	0.24	0.31	22.6%
Maximum M_P (%)	0.27	0.32	15.6%
Average T_R(s)	1.96	3.47	43.5%
Maximum T_R(s)	2.48	4.5	44.9%

Another disturbance to X_{oim} comes from the VGT vane position, due to its influence on p_3, i.e., the upstream pressure of the HP-EGR valve. In Figure 16, the VGT vane position steps from 30% to 60% open, it is clear that the proposed solution shows reduced maximum X_{oim} deviation from the target and takes shorter time to converge to X_{oim}^{des}. Detailed performance assessment are available in Table 9.

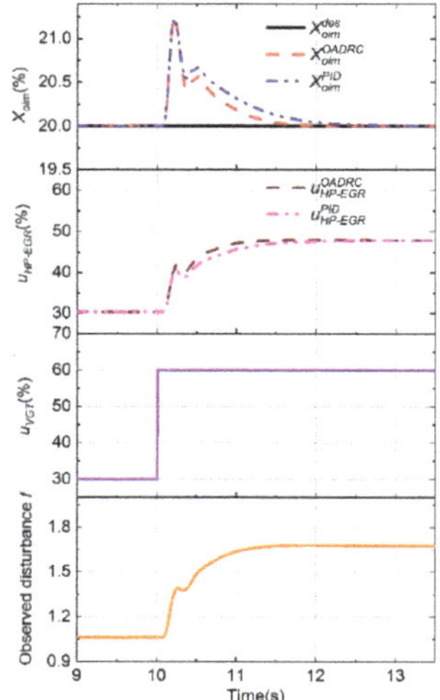

Figure 16. Disturbance rejection performance comparison with step change of variable-geometry turbocharger (VGT) vane.

Table 9. Disturbance rejection performance comparison with step change of VGT vane.

Item	OADRC	PID	Improvement with OADRC
M_P (%)	6.18	6.19	0.2%
T_R (s)	1.58	2.02	21.8%

7. Conclusions

In this paper, experimental assessment on the impact of EAT on the air system control was conducted, based on a 6.7 L diesel engine. A disturbance rejection controller for the EAT air system was proposed and validated in simulation.

(1) At 265 Nm engine load and 1400 rpm engine speed condition, up to 7.8% FE benefit can be achieved with the assist function of EAT(assuming the electrical energy is free in the driveline electrical regeneration during vehicle braking), through the reduction of pumping loss from wider VGT vane open, relative to the nominal system. Further FE benefit is limited by the HP-EGR deficit from over-reduced p_3 (low HP-EGR flow capability).

(2) Transient boost response with EAT assist is improved by 52.1% relative to the nominal system. The new dynamics of p_2 introduced by EAT must be considered to avoid intake oxygen concentration overshoot and NOx spike. The EGR requirements should be reconsidered because of the excess fresh air from a fast boost response.

(3) The EGR flow-based control objective should be transformed into the intake oxygen concentration-based control objective, which aligns well with the EAT system and the associated control design. EAT assist improved the volumetric efficiency of the engine (by up to 13.6% with 1.01 kW E-assist power at 265 Nm engine load and 1400 rpm engine speed in this paper), thereby increased the engine mass flow rate. This leads to HP-EGR gas dilution. EGR dilution and enhanced boost response makes the conventional EGR rate-based air system control not applicable to EAT system. Experimental results confirmed that X_{oim} works better as the control output for EAT system than EGR rate for NOx control.

(4) In order to attenuate the disturbances from EAT on X_{oim}, a disturbance rejection based controller was proposed and validated in a high-fidelity GT-SUITE model. Results showed over 36%

improvement in settling time and over 43% improvement in recovery time, relative to the conventional PID controller.

Author Contributions: C.W. and S.L. carried out the engine test, and set up the simulation platform. K.S. and C.W. worked together on the experimental data analysis, X_{oim} controller design and paper writing. H.X. provided a lot of valuable insights on the data analysis and controller validation.

Funding: This research was funded by Ford Motor Company.

Acknowledgments: The authors would like to thank Leon Hu, of Ford Motor Company for several fruitful discussions and his useful insights during the bi-weekly meeting. The authors would like to thank Devesh Upadhyay, Miachiel Van Nieuwstadt, Eric Curtis, and James Yi for their insights. A special thank you to Tongjin Wang from Tianjin University for assisting us in the management and maintenance of the test bench.

Conflicts of Interest: The authors declare no conflict of interest.

Nomenclature

Symbols

J	Rotational inertia
$\dot{m}$	Mass flow rate
P	Power
R	Ideal gas constant
T	Temperature
V	Volume
X_{oair}	Oxygen concentration in the air
X_{oim}	Oxygen concentration at the outlet of intake manifold
X_{ocyl}	Oxygen concentration in the cylinder after combustion
X_{ocyl_out}	Oxygen concentration at the exhaust port
X_{oegr}	Oxygen concentration in the recirculated exhaust gas
η	Efficiency
X_{EGR}	EGR rate
A	Cross-sectional area
$c_{p,c}$	Specific heat capacity of air
$c_{p,e}$	Specific heat capacity of exhaust gases

Subscripts

C	Compressor
Eng	Engine
$HP-EGR$	High pressure exhaust gas recirculation
T	Turbine
$TEMG$	Turbocharger shaft mounted electrical motor/generator
TC	Turbocharger
vol	Volumetric
1	Pre-compressor
2	Post-compressor
3	Pre-turbine
4	Post-turbine
max	Maximum value
im	Intake manifold
em	Exhaust manifold
cyl	Cylinder
des	Desired value

Appendix A

For X_{oim} observation, the existed observer, as developed by Park [23], was adopted. For completeness, it is explained briefly here.

Appendix A.1 Lyapunov-Based Observer Design

The state estimation observer is designed by rearranging the mean-value models of the oxygen concentration in the intake manifold and the exhaust manifold into the state-space form, as shown by:

$$\dot{X} = \begin{bmatrix} a_{11} & a_{12} \\ a_{21} & a_{22} \end{bmatrix} \hat{X} + \begin{bmatrix} w_1 \\ w_2 \end{bmatrix} \tag{A1}$$

$$\hat{y} = \begin{bmatrix} 0 & 1 \end{bmatrix} \hat{X} \tag{A2}$$

where $\hat{X} = \begin{bmatrix} \hat{X}_{oim} & \hat{X}_{oegr} \end{bmatrix}^T$, $a_{11} = \frac{R\hat{T}_2}{p_2 V_{im}}(-\hat{\dot{m}}_{Eng})$, $a_{12} = \frac{R\hat{T}_2}{p_2 V_{im}}\hat{\dot{m}}_{HP-EGR}$, $a_{21} = \frac{R\hat{T}_3}{\hat{p}_3 V_{em}}(\hat{\dot{m}}_{Eng} - \dot{m}_{fuel}AFR)$, $a_{22} = \frac{R\hat{T}_3}{\hat{p}_3 V_{em}}(-\hat{\dot{m}}_{Eng} - \dot{m}_{fuel})$, $w_1 = \frac{R\hat{T}_2}{p_2 V_{im}}\dot{m}_C X_{oair}$, $w_2 = 0$.

The form of the oxygen concentration observer is designed as:

$$\begin{aligned} \dot{\hat{X}} &= \begin{bmatrix} a_{11} & a_{12} \\ a_{21} & a_{22} \end{bmatrix} \hat{X} + \begin{bmatrix} w_1 \\ w_2 \end{bmatrix} + \begin{bmatrix} l_1 \\ l_2 \end{bmatrix}(y - \hat{y}) \\ &= A\hat{X} + W + L(y - \hat{y}) \end{aligned} \tag{A3}$$

where l_1 and l_2 are the observer gains, which are designed on the basis of the Lyapunov stability theorem.

Appendix A.2 Xoim Observer Validation

The X_{oim} observer was firstly built in Matlab/Simulink, followed by the automatic code generation using the Target-link from dSPACE Inc. Then the validation on the test bench was carried out to evaluate the real-time performance. The engine speed was fixed at 1400 rpm, with a 30% VGT vane open. A step change of HP-EGR valve position was manipulated from 25% to 15%, as shown in Figure A1.

After preliminary calibration of the model corresponding to the test bench, the average relative error of the estimated oxygen concentration in the intake manifold and exhaust manifold are 0.23% and 0.86% respectively. The detailed performance assessment are available in Table A1.

Table A1. Case setup and results for the validation of X_{oim} observer on the test bench.

Test Case	Fuel Injection Rate	$u_{HP\text{-}EGR}$	u_{VGT}	X_{oim}		X_{oegr}	
				Avg. X_{error}	Max. X_{error}	Avg. X_{error}	Max. X_{error}
-	mg/cycle	%	%	%	%	%	%
Case 1	18	25→15	30	0.23	2.53	0.86	3.15

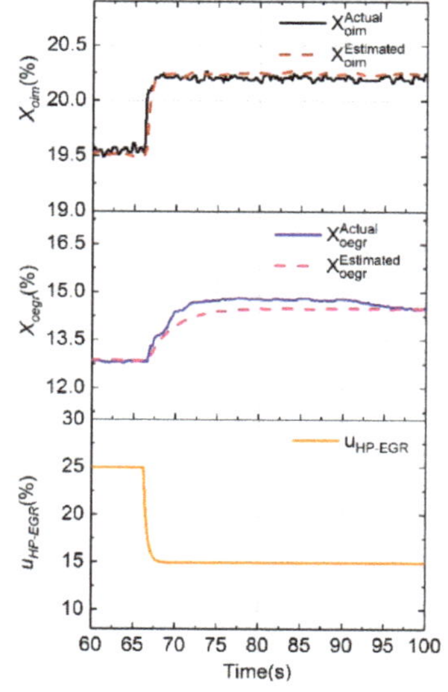

Figure A1. Experimental validation of oxygen concentration observer.

From the above experimental results, it can be seen that the oxygen concentration of intake manifold and exhaust manifold estimated by the observer can approximate the actual oxygen concentration accurately, which lays the foundation for the control of oxygen concentration in the intake manifold of diesel engine equipped with EAT.

References

1. Filipi, Z.; Wang, Y.; Assanis, D.N. Effect of variable geometry turbine (VGT) on diesel engine and vehicle system transient response. *SAE Tech. Pap.* **2001**. [CrossRef]
2. Dambrosio, L.; Pascazio, G.; Fortunato, B. Vgt turbocharger controlled by an adaptive technique. *IEEE/ASME Trans. Mechatron.* **2003**, *8*, 492–499. [CrossRef]
3. Song, K.; Upadhyay, D.; Xie, H. An assessment of performance trade-offs in diesel engines equipped with regenerative electrically assisted turbochargers. *Int. J. Engine Res.* **2018**, *20*, 510–526. [CrossRef]
4. Chiara, F.; Canova, M. A review of energy consumption, management, and recovery in automotive systems, with considerations of future trends. *Proc. Inst. Mech Eng. Part D J. Automob. Eng.* **2013**, *227*, 914–936. [CrossRef]
5. Aghaali, H.; Ångström, H. A review of turbocompounding as a waste heat recovery system for internal combustion engines. *Renew. Sustain. Energy Rev.* **2015**, *49*, 813–824. [CrossRef]
6. Dimitriou, P.; Burke, R.; Zhang, Q.; Copeland, C.; Stoffels, H. Electric turbocharging for energy regeneration and increased efficiency at real driving conditions. *Appl. Sci.* **2017**, *7*, 350. [CrossRef]
7. Newman, P.; Luard, N.; Jarvis, S.; Richardson, S.; Smith, T.; Jackson, R.; Rochette, C.; Lee, D.; Criddle, M. Electrical supercharging for future diesel powertrain applications. In Proceedings of the 11th International Conference on Turbochargers and Turbocharging, London, UK, 13–14 May 2014.
8. Katsanos, C.O.; Hountalas, D.T.; Zannis, T.C. Simulation of a heavy-duty diesel engine with electrical turbocompounding system using operating charts for turbocharger components and power turbine. *Energy Convers. Manag.* **2013**, *76*, 712–724. [CrossRef]
9. Millo, F.; Mallamo, F.; Pautasso, E.; Mego, G.G. The potential of electric exhaust gas turbocharging for HD diesel engines. *SAE Tech. Pap.* **2006**. [CrossRef]
10. Cipollone, R.; di Battista, D.; Gualtieri, A. Turbo compound systems to recover energy in ICE. *Int. J. Eng. Innov. Technol. (IJEIT)* **2013**, *3*, 249–257.
11. Briggs, I.; McCullough, G.; Spence, S.; Douglas, R. Whole-vehicle modelling of exhaust energy recovery on a diesel-electric hybrid bus. *Energy* **2014**, *65*, 172–181. [CrossRef]
12. Katrašnik, T.; Trenc, F.; Medica, V.; Markič, S. An analysis of turbocharged diesel engine dynamic response improvement by electric assisting systems. *J. Eng. Gas Turbines Power* **2005**, *127*, 918–926. [CrossRef]
13. Katrašnik, T.; Rodman, S.; Trenc, F.; Hribernik, A.; Medica, V. Improvement of the dynamic characteristic of an automotive engine by a turbocharger assisted by an electric motor. *J. Eng. Gas Turbines Power* **2003**, *125*, 590–595. [CrossRef]
14. Terdich, N.; Martinez-Botas, R. Experimental efficiency characterization of an electrically assisted turbocharger. *SAE Tech. Pap.* **2013**. [CrossRef]
15. Jain, A.; Nueesch, T.; Naegele, C.; Lassus, P.M.; Onder, C.H. Modeling and Control of a Hybrid Electric Vehicle with an Electrically Assisted Turbocharger. *IEEE Trans. Veh. Technol.* **2016**, *65*, 4344–4358. [CrossRef]
16. Zhao, D.; Stobart, R.; Dong, G.; Winward, E. Real-Time Energy Management for Diesel Heavy Duty Hybrid Electric Vehicles. *IEEE Trans. Control Syst. Technol.* **2015**, *23*, 829–841. [CrossRef]
17. Kolmanovsky, I.; Stefanopoulou, A.G. Evaluation of turbocharger power assist system using optimal control techniques. *SAE Trans.* **2000**, *109*, 518–528.
18. Glenn, B.C.; Upadhyay, D.; Washington, G.N. Control Design of Electrically Assisted Boosting Systems for Diesel Powertrain Applications. *IEEE Trans. Control Syst. Technol.* **2010**, *18*, 769–778. [CrossRef]
19. Jung, M.; Glover, K.; Christen, U. Comparison of uncertainty parameterisations for robust control of turbocharged diesel engines. *Control Eng. Pract.* **2005**, *13*, 15–25. [CrossRef]
20. Nakayama, S.; Fukuma, T.; Matsunaga, A.; Miyake, T.; Wakimoto, T. A New Dynamic Combustion Control Method Based on Charge Oxygen Concentration for Diesel Engines. *SAE Tech. Pap.* **2003**. [CrossRef]
21. Shutty, J.; Benali, H.; Daeubler, L.; Traver, M. Air System Control for Advanced Diesel Engines. *SAE Tech. Paper* **2007**. [CrossRef]

22. Millo, F.; Pautasso, E.; Pasero, P.; Barbero, S.; Vennettilli, N. An Experimental and Numerical Study of an Advanced EGR Control System for Automotive Diesel Engine. *SAE Int. J. Engines* **2009**, *1*, 188–197. [CrossRef]
23. Park, Y.; Min, K.; Chung, J.; Sunwoo, M. Control of the air system of a diesel engine using the intake oxygen concentration and the manifold absolute pressure with nitrogen oxide feedback. *Proc. Inst. Mech. Eng. Part D J. Automob. Eng.* **2015**, *230*, 240–257. [CrossRef]
24. Chen, S.K.; Yanakiev, O. Transient NOx emission reduction using exhaust oxygen concentration based control for a diesel engine. *SAE Tech. Paper* **2005**. [CrossRef]
25. Han, J. From PID to Active Disturbance Rejection Control. *IEEE Trans. Ind. Electron.* **2009**, *56*, 900–906. [CrossRef]
26. Han, J. *Active Disturbance Rejection Control Technique-the Technique for Estimating and Compensating the Uncertainties*; National Defense Industry Press: Beijing, China, 2008; pp. 197–270.
27. Song, K.; Upadhyay, D.; Xie, H. Control of diesel engines with electrically assisted turbocharging through an extended state observer based nonlinear MPC. *Proc. Inst. Mech. Eng. Part D J. Automob. Eng.* **2019**, *233*, 378–395. [CrossRef]
28. Yu, H.; Song, K.; Xie, H. Study on Optimal Control for the Air System of E-Turbo Assisted Diesel Engine Based on Model Prediction. *Chin. Intern. Combust. Engine Eng.* **2018**, *4*, 39–46.
29. Eriksson, L.; Nielsen, L. *Modeling and Control of Engines and Drivelines*; John Wiley & Sons: Hoboken, NJ, USA, 2014.
30. Song, K. *Active Disturbance Rejection Based High-Dilution Low-Temperature Combustion Control for P-DI Gasoline Engines*; Tianjin University: Tianjin, China, 2015.
31. Gao, Z. Scaling and bandwidth-parameterization based controller tuning. In Proceedings of the American Control Conference, Minneapolis, MN, USA, 14–16 June 2006; Volume 6, pp. 4989–4996.

Article

Design of an Electromagnetic Variable Valve Train with a Magnetorheological Buffer

He Guo, Liang Liu *, Xiangbin Zhu, Siqin Chang and Zhaoping Xu

School of Mechanical Engineering, Nanjing University of Science and Technology, Nanjing 210094, China; guoheyuyu@njust.edu.cn (H.G.); 117101021486@njust.edu.cn (X.Z.); changsq@njust.edu.cn (S.C.); xuzhaoping@njust.edu.cn (Z.X.)
* Correspondence: l.liu@njust.edu.cn; Tel.: +86-25-8430-3903

Received: 10 August 2019; Accepted: 17 October 2019; Published: 21 October 2019

Abstract: In this paper, an electromagnetic variable valve train with a magnetorheological buffer (EMVT with MR buffer) is proposed. This system is mainly composed of an electromagnetic linear actuator (EMLA) and a magnetorheological buffer (MR buffer). The valves of an internal combustion engine are driven by the EMLA directly to open and close, which can adjust the valve lift and phase angle of the engine. At the same time, MR buffer can reduce the seat velocity of the valve and realize the seat buffer of the electromagnetic variable valve. In this paper, the overall design scheme of the system is proposed and the structure design, finite element simulation of the EMLA, and the MR buffer are carried out. The electromagnetic force characteristics of the EMLA and buffer force of the MR buffer are measured, and the seat buffering performance is verified as well. Experiments and simulation results show that the electromagnetic force of the EMLA can reach 320.3 N when the maximum coil current is 40 A. When the current of the buffer coil is 2.5 A and the piston's motion frequency is 5 Hz, the buffering force can reach 35 N. At the same time, a soft landing can be realized when the valve is seated.

Keywords: camless; electromagnetic variable valve train; magnetorheological buffer; soft landing

---

1. Introduction

With global environmental issues being increasingly serious, the development of internal combustion engines is mainly focused on hybrid electric vehicles, electric vehicles, and alternative fuel vehicles. Hybrid electric vehicles combine a traditional internal combustion engine with a motor and hydraulic pump, which reduces harmful emissions of the internal combustion engine and effectively improves fuel economy. As one of the main components of hybrid electric vehicles, internal combustion engines still have an irreplaceable position [1–4]. Thus, the internal combustion engine will still comprise a large proportion of vehicles in the future.

As a variable valve mechanism, an electromagnetic variable valve train (EMVT) replaces the cam-driven valve mechanism in traditional engines and connects directly with the valve. Therefore, the limitation of the cam profile on valve timing is eliminated and the EMVT can adjust the valve lift and phase angle in real-time according to the engine operating conditions. The EMVT can also moderately increase the intake flow rate. Thus, the loss of pump gas and the power consumption of the valve train are reduced, and the fuel economy of the engine is improved. Xu et al. [5] have studied the influence of the EMVT on intake charge. The results show that the EMVT can improve the intake charging of the engine significantly. Fan et al. [6] realized the high compression ratio of the engine through EMVT. The results show that the thermal efficiency of the engine has an obvious improvement. Xinyu et al. [7] realized the optimization of exhaust valve motion by EMVT. The results show that the optimal exhaust valve opening motion can strengthen both the power performance and fuel economy at engine part loads. However, since the linear actuator is connected with the valve directly, the mechanical damping coefficient of EMVT is low. Thus, the control performance of the EMVT is easily affected by the external disturbances coming from the cylinder gas force, friction force, and spring force, etc. Thus, excessive valve seating velocity may be produced in practical application, and excessive valve seat velocity will not only cause valve seat rebound and vibration noise, but also cause damage of engine components. The life and control accuracy of the internal combustion engine will be affected as well.

The seat buffer for the variable valve can be realized by adding a buffer structure or applying a valve control strategy. Due to the low damping coefficient of the EMVT, it is easily disturbed by external disturbances in practical application. Thus, adding a buffer structure to reduce the valve seat velocity is an effective solution. In the study of electro-hydraulic variable valve systems, Li et al. [8] realized the soft landing of the valve through a "buffer" composed of a hydraulic valve, plunger, and plunger sleeve; Yang et al. [9] and Chen et al. [10] designed a cushion structure to reduce the force of the valve seating, and the valve was buffered by the principle of a variable throttle area; Tu et al. [11] adopted a one-way throttle valve for buffering, and a genetic algorithm is adopted to optimize the control parameters of the throttle valve. In the research of electromagnetic variable valves, MIT [12] added a mechanical buffer between the electromagnetic linear actuator and the valve to reduce the valve velocity and system energy consumption. A seat cushion can be realized by designing the valve seat trajectory rationally [13–15]. In order to design the landing trajectory of the valve reasonably, it is necessary to control the valve to follow the trajectory in the process of movement. A seat cushion can also be realized by the control the valve seat velocity. The valve velocity is taken as the control target [16–20], and a closed-loop control strategy is carried out.

In this paper, an EMVT with an MR buffer is proposed, which consists of an electromagnetic linear actuator (EMLA) and a magnetorheological buffer (MR buffer). The EMLA is a moving coil linear actuator [21]. The EMLA generates a linear driving force through an electrified coil in the permanent magnet field. The driving force is positively correlated with the current intensity in the coil, and the engine valves are driven by the driving force. The MR buffer is mainly used to reduce the velocity of the valve before seating. The buffer force can be adjusted by changing the coil current intensity of the MR buffer to meet different valve seating velocity requirements. Thus, the damping characteristic of the system is adjusted by the MR buffer as well. The main feature of the system is that a buffer is set between the EMLA and the valve, which can change the cushioning force according to the valve seat velocity, and the damping coefficient of the system is adjusted in real-time by the MR buffer in actual operation. Thus, the system stability is improved while the valve seat velocity is reduced, and the soft landing of the valve is realized. At the same time, the valve lift and phase angle can be adjusted by the EMLA as well.

According to the above description, the valve opening lift and phase angle of an internal combustion engine can be adjusted by using EMVT instead of the traditional cam mechanism. Therefore, the valve state can be changed in real time according to the operating conditions of the internal combustion engine, so that the internal combustion engine can achieve the best power output and fuel consumption under all operating conditions. This is obvious for the preservation of internal combustion energy.

2. System Overview

In order to replace the traditional cam mechanism and complete the engine ventilation process successfully, the EMVT with the MR buffer needs to meet the following requirements:

1. It has a large mechanical damping coefficient, which can overcome the influence of external interference and parameter changes.
2. The valve seating velocity is controlled within 0.1 m/s. The EMLA and MR buffer is controlled by the control unit, and the seat velocity of the valve is reduced by the MR buffer to ensure the valve seat velocity is effectively controlled.

Figure 1 shows the structure of the EMVT with the MR buffer. As can be seen from Figure 1, the system consists of three main parts: the EMLA, MR buffer, and valve components. The structure of the EMLA mainly includes the inner and outer magnetic yokes, permanent magnets, coil frame, and electromagnetic coil. The electromagnetic coil is wound on the coil frame in series, and the electromagnetic coil is in a magnetic field generated by the permanent magnets. The axial motion law of the EMLA is controlled by changing the current intensity applied on the electromagnetic coil, thereby the valve is controlled. Compared with the conventional linear actuator, this EMLA cannot only realize the adjustable valve phase angle, but also realize the adjustable valve lift and transition time. The main function of the MR buffer is to reduce the seat velocity and restrain the rebound of the valve. The MR buffer mainly consists of a piston, piston rod, buffer coil, cylinder, end cover, and MR fluid. A groove is dug into the piston to wrap the buffer coil, and MR fluid flows through the annular gap between the piston and the cylinder to generate the buffer force. After the piston coil is energized, the magnetic force lines are mainly concentrated at the piston. Then, magnetic force lines enter the cylinder through the annular gap, and return to the piston through the annular gap on the other side to form a closed loop. The area of the annular gap through which the magnetic field lines pass is called the active region of the MR buffer. The MR fluid near this region will produce the MR effect due to the passage of magnetic force lines and, thus, produce a certain shear yield strength, and the reciprocating motion of the piston drives the MR fluid flow through the annular gap to provide a controllable buffer force.

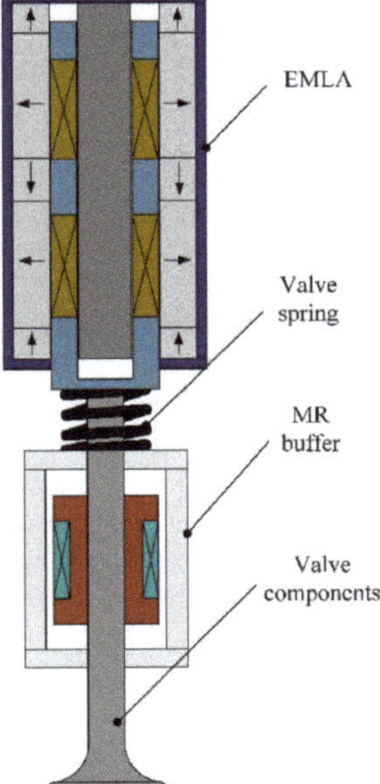

Figure 1. Structure of the EMVT with the MR buffer.

The EMLA and MR buffer are connected in series. A valve spring is installed between the EMLA and the MR buffer to provide a rebound force when the valve is closed. Thus, it can ensure the valve can be closed effectively.

The working principle of the EMVT with the MR buffer is described as follows: When the valve opens, the coil on the EMLA is energized. The coil is subjected to an axial electromagnetic force in the permanent magnetic field, which propels the valve to open. The opening time and lift of the valve controlled by the current intensity applied on the coil of the EMLA. When the EMLA drives the valve to close, the buffer coil is energized before seating. The MR buffer generates a buffer force of a specified size and the force direction is opposite to the valve velocity. Thus, the seat velocity of the valve can be reduced and valve seat buffer can be realized.

Considering that the EMLA and MR buffer are two complex multi-physical field coupling systems, the independent design of the two main components is considered in the system design. Then the design results are verified by experiment.

3. Structure Design

3.1. Structure Design of the Electromagnetic Linear Actuator

As the main driving component of the EMVT with the MR buffer, the EMLA shall meet the following requirements when being designed:

1. When the engine operates under high load, the pressure difference between the cylinder and the exhaust port can reach up to 8 bar when the exhaust valve is opened in the exhaust stroke [22]. Therefore, the maximum electromagnetic force of the EMLA needs to be over 300 N to ensure that the valve can be opened effectively.
2. In order to meet the requirements of the engine at high speed, the valve transition time (from 5% of the valve lift to 95% of the time experienced) can reach 4 ms, or even faster.
3. The EMLA is installed on the cylinder head of the engine, so the diameter and height should be limited to 40 mm and 70 mm, respectively, and can achieve a maximum power output under the volume limitation.
4. The change of the mover position should have less influence on the electromagnetic force of the EMLA.

In the finite element analysis software, the magnetic field of the air gap and electromagnetic force characteristics of the EMLA can be simulated by the finite element method, so that the structure of the EMLA can be reasonably designed. Figure 2a shows the magnetic finite element model of the EMLA established in MAXWELL provided by Ansoft in Pennsylvania, USA. The model mainly includes the inner and outer yokes, electromagnetic coil, permanent magnets, and air gap, where the coil is at the initial position and moves downward with a working stroke of 8 mm. The arrow at the permanent magnet region in Figure 2a represents the magnetization direction of the permanent magnets. Figure 2b shows the distribution of magnetic force lines and magnetic induction intensity obtained by simulation. As can be seen from Figure 2b, the magnetic force lines near the area of the electromagnetic coil are densely distributed and more uniform. The total magnetic force lines are symmetrically distributed. The color of different regions represents different magnetic induction intensities, and the magnetic induction intensity is mainly distributed in the inner yoke.

In the structure of the EMLA, different parameters such as the size of permanent magnets, the width of the air gap and the size of the inner and outer yokes will affect the electromagnetic force. The mass of the moving parts will affect the maximum acceleration of the moving parts, thus affecting the transition time of the EMLA. Under the condition of limited external size, the larger size of permanent magnets, and the smaller size of the air gap, a larger electromagnetic force can be generated. Excessive size of the permanent magnets will result in magnetic saturation of the yokes, and too small a size of the air gap will affect the installation space of the coil. Therefore, there are structural contradictions in the design of the EMLA. Thus, under the limited volume condition, the actuator should achieve as much power as possible, that is, to maximize the power density of the EMLA. In the finite element analysis software, the parametric analysis function is used to optimize the maximum acceleration that the moving parts can achieve. In the range of 40 mm diameter and 70 mm height, the air gap size, yokes size, and permanent magnet size are taken as optimization variables, and Table 1 shows the optimization results of the EMLA.

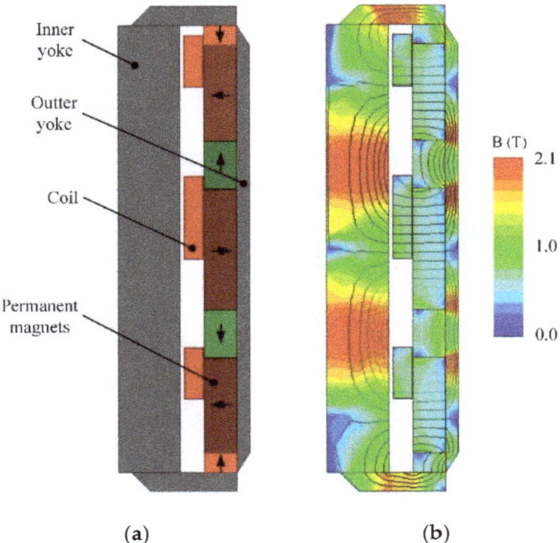

Figure 2. Magnetic field distribution of the EMLA: (**a**) Finite element model; (**b**) Magnetic field distribution.

Table 1. Main structure dimensions of the EMLA.

Parameter	Value
Stroke/mm	8
Diameter/mm	39
Height/mm	70
Volume/cm^3	84
Permanent magnet thickness/mm	8
Mass of moving parts/g	114

As the coil frame moves up and down in the magnetic field formed by the permanent magnets, it is necessary to make non-conductive epoxy resin materials to prevent large eddy current losses in the coil frame. The magnetic field of the EMLA is mainly produced by the permanent magnets. Therefore, the quality of the permanent magnet material is related to the performance of the EMLA. NdFeB has high residual flux density, coercive, and magnetic energy products. Thus, it can be used as the permanent magnet material of the EMLA. The magnetic force lines of the permanent magnet in the EMLA needs to form a loop, and the loop needs to pass through the inner and outer yokes. Therefore, soft magnetic materials with high permeability are needed. In this paper, a low carbon steel material with similar properties to pure iron is selected as the material of the inner and outer yokes.

Figure 3 shows the electromagnetic force characteristic curve obtained by simulation. Electromagnetic force characteristics are described as the relationship between the electromagnetic force and the coil current intensity and coil position. The magnitude of the electromagnetic force varies with the current intensity and position of the coil. As can be seen from Figure 3, when the current intensity in the coil is 40 A, the maximum electromagnetic force that the EMLA can achieve is 320.3 N. At the same time, the EMLA has a certain end effect. When the position of the coil changes, the magnitude of the electromagnetic force fluctuates about 23 N, about 7% of the electromagnetic force. Therefore, the change of the mover position have less influence on the electromagnetic force, and the structure design requirements of the EMLA are realized [22].

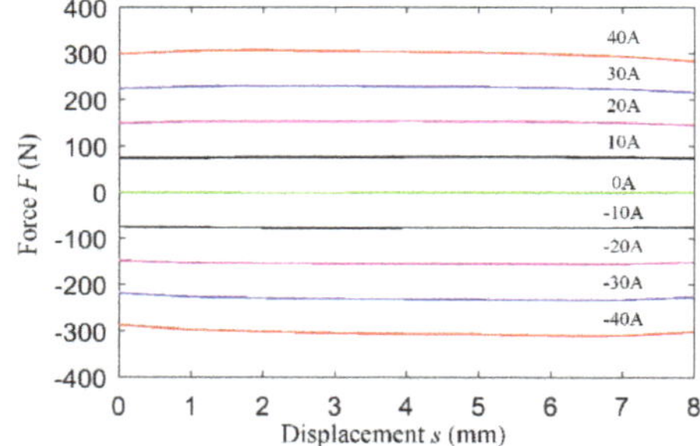

Figure 3. Electromagnetic force characteristic curve of the EMLA.

3.2. Structure Design of MR Buffer

The buffer force applied by the MR buffer mainly depends on the seat velocity of the valve and mass of the moving parts. Assuming a 5% displacement before valve closure as the seat buffer region, it is 0.4 mm before the valve seat, and according to the experimental data of the EMLA, the average seat velocity at this position is 0.5 m/s.

According to the law of dynamics, it can be obtained that:

$$\text{F} \times x = \frac{1}{2}m\left(v_1^2 - v_2^2\right) \tag{1}$$

where F is the buffering force when the valve is seated, x is the position when the valve is seated, m is the mass of the moving parts, v_1 is the velocity before the valve is seated, and v_2 is the velocity at the end of the valve. The total mass of the EMLA mover and valve components is 114 g. It is also assumed that the valve velocity at the end of the valve seating is 0 m/s. According to Equation (1), the required valve seat buffer force is 36.5 N. This data is calculated on the premise that 5% displacement before valve closure is used as seat buffer region. At this time, the EMLA drives the valve seat.

Figure 4 illustrates the structure and key parameters of the MR buffer. The main structure parameters affecting the buffer force include the piston diameter D, piston rod diameter d, piston effective length L, annular gap width h, and the performance parameters of the MR fluid, where the annular gap width h is the gap width between piston and cylinder, and the effective length L is the length of the other area on the buffer piston except the winding part. The structure design of the MR buffer is mainly based on the parameters mentioned above [23].

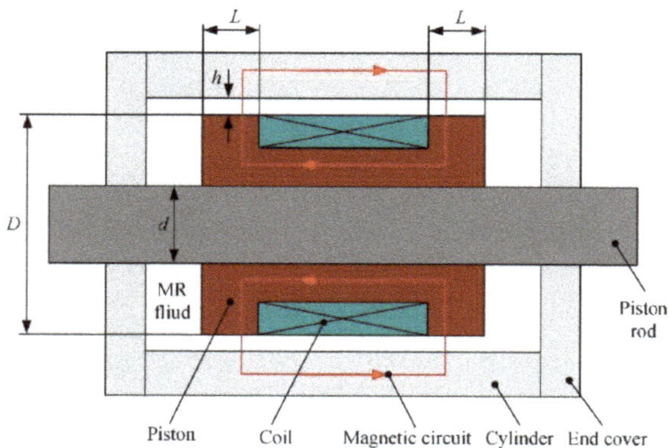

Figure 4. Structure of the MR buffer.

Considering the limited installation space of the buffer, the external size of the buffer should not be too large, and the diameter of the piston should be 20 mm. The piston rod is designed as a hollow rod so that the coil lead on the piston can be drawn from it. Therefore, the diameter of the piston rod should ensure that there is enough lead space. At the same time, the piston rod diameter should not be too large, otherwise it will affect the buffer force. Considering the above two factors, the diameter of the piston rod is 8 mm. The MR fluid was provided by Chongqing Materials Research Institute in Chongqing city, China. The magnetization curve is shown in Figure 5. Since the zero-field viscosity of the selected MR fluid is large (0.8 Pa*s), it is necessary to select a larger annular gap width to achieve a smaller zero-field buffer force. However, too large an annular gap width will have an impact on the buffer force. Therefore, considering the above two factors, the annular gap width is 2.5 mm. The effective length of the piston is one of the main factors affecting the buffer force. Excessive effective length will result in a large appearance size of the MR buffer. Too small an effective length will lead to unstable output of buffer force. Therefore, considering the above two factors, the effective length is 10 mm. The main structural parameters of the MR buffer are shown in Table 2.

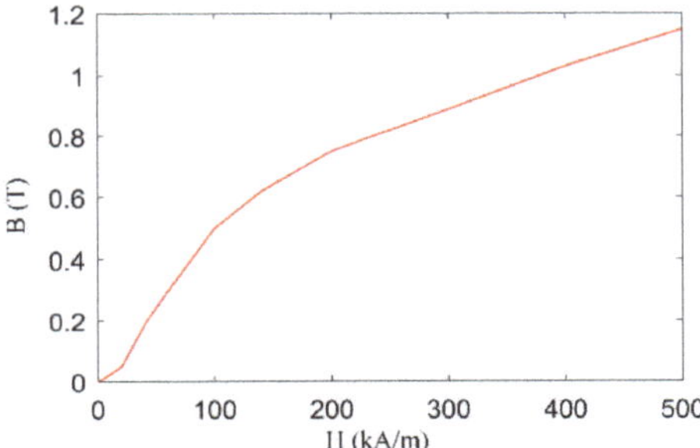

Figure 5. Magnetization characteristic curve of the MR fluid.

Table 2. Main structure dimensions of the MR buffer.

Parameter	Value
Cylinder inner diameter/mm	25
Piston diameter/mm	20
Piston rod diameter/mm	8
Effective length/mm	10
Annular gap width/mm	2.5
Cylinder outer diameter/mm	33
Stroke/mm	10

The finite element simulation of the magnetic field can be used to analyze the magnetic circuit of the MR buffer. Figure 6 shows the magnetic induction intensity and magnetic force line distribution of the MR buffer. Due to the axisymmetric structure of the MR buffer, only 1/2 of the finite element model of the MR buffer is established in the electromagnetic field simulation. As can be seen from Figure 6, the magnetic induction intensity at the annular gap is mainly concentrated near the effective length. The distribution of the magnetic induction intensity is uniform, and most of the magnetic force lines pass through the annular gap vertically.

The distribution curves of magnetic induction intensity at the annular gap under different coil current intensities are shown in Figure 7. As can be seen from Figure 7, the effective length of the buffer is the area where the magnetic field is mainly concentrated. The distribution of magnetic induction intensity in this region is more uniform, and the magnetic induction intensity outside the effective length is lower. With the increase of current intensity, the magnetic induction intensity at the effective length of the annular gap will also increase.

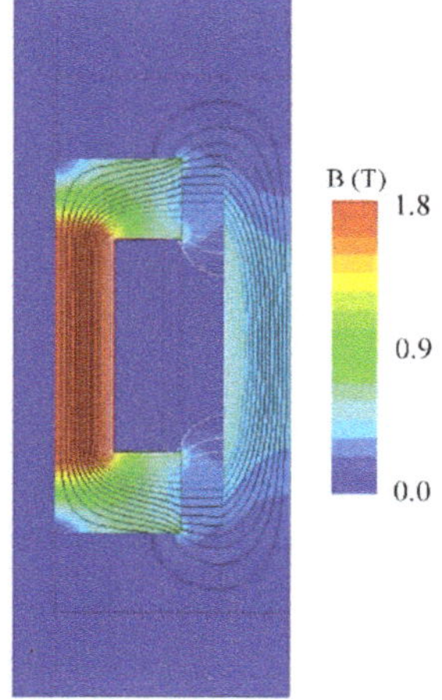

Figure 6. Magnetic field distribution of the MR buffer.

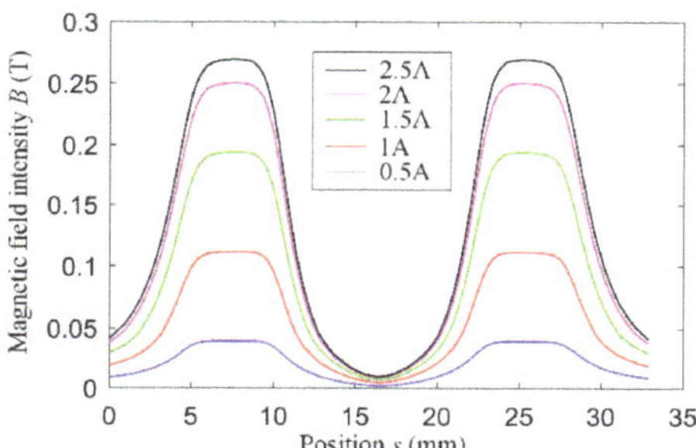

Figure 7. Magnetic induction intensity distribution curve in the annular gap.

4. Multi-Physics FE Modeling and Simulation of the MR Buffer

Figure 8 shows the multi-physical field model of the MR buffer established in COMSOL Multiphysics provided by COMSOL in Stockholm, Sweden. The model involves the coupling of the magnetic field and flow field. The moving grid module drives the piston up and down in the model to simulate the buffer force.

The Navier–Stokes equation is used in the fluid region and the Maxwell equation is used in the magnetic field region. The coupling relationship between magnetic field and flow field can be achieved by two variables, hydrodynamic viscosity η and magnetic inductance intensity B. The coupling mode is shown in Table 3. The multi-physical field simulation model takes the simulation results of the magnetic field as one of the input values of the flow field [24].

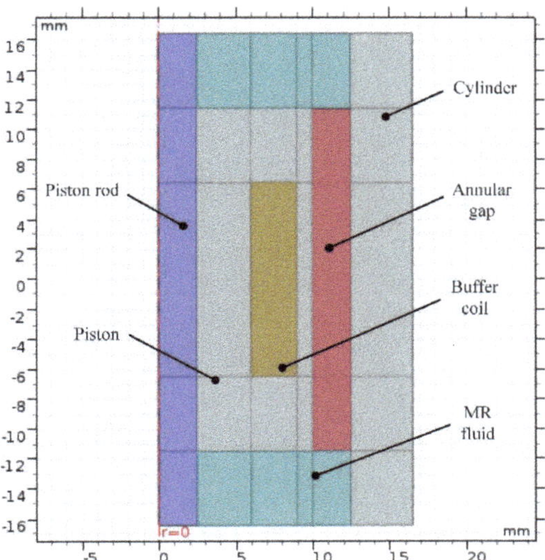

Figure 8. Multi-physical field model of the MR buffer.

Table 3. Multi-physical field equations.

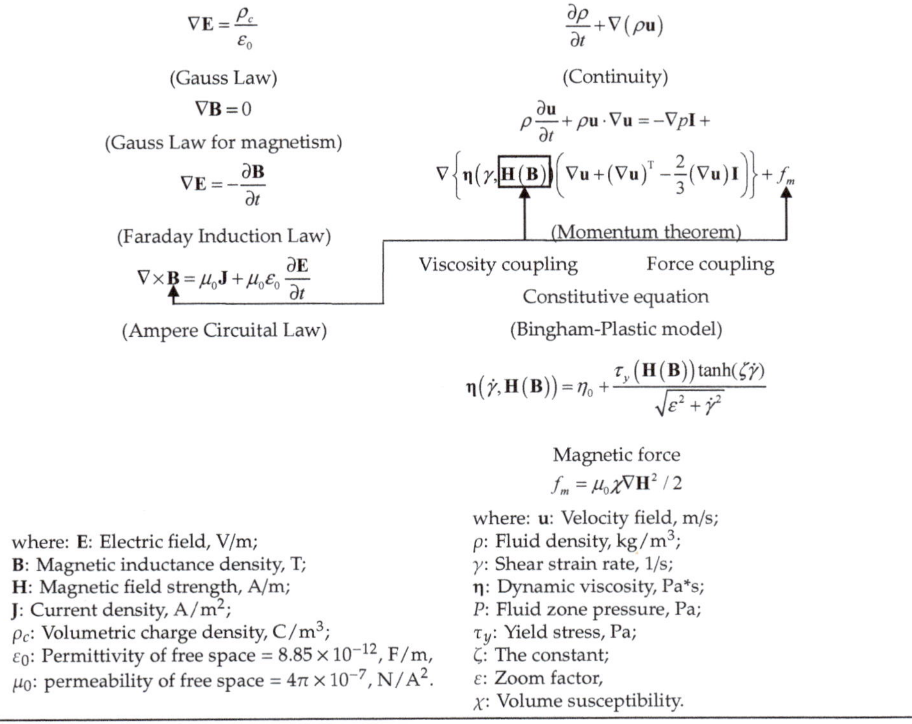

Electromagnetic Equations (Maxwell)	Fluid Dynamics Equations (Navier–Stokes)
$\nabla \mathbf{E} = \dfrac{\rho_c}{\varepsilon_0}$	$\dfrac{\partial \rho}{\partial t} + \nabla(\rho \mathbf{u})$
(Gauss Law)	(Continuity)
$\nabla \mathbf{B} = 0$	$\rho \dfrac{\partial \mathbf{u}}{\partial t} + \rho \mathbf{u} \cdot \nabla \mathbf{u} = -\nabla p \mathbf{I} +$
(Gauss Law for magnetism)	
$\nabla \mathbf{E} = -\dfrac{\partial \mathbf{B}}{\partial t}$	$\nabla \left\{ \eta\left(\gamma, \mathbf{H}(\mathbf{B})\right)\left(\nabla \mathbf{u} + (\nabla \mathbf{u})^{\mathrm{T}} - \dfrac{2}{3}(\nabla \mathbf{u})\mathbf{I} \right) \right\} + f_m$
(Faraday Induction Law)	(Momentum theorem)
	Viscosity coupling Force coupling
$\nabla \times \mathbf{B} = \mu_0 \mathbf{J} + \mu_0 \varepsilon_0 \dfrac{\partial \mathbf{E}}{\partial t}$	Constitutive equation
(Ampere Circuital Law)	(Bingham-Plastic model)

$$\eta\left(\dot{\gamma}, \mathbf{H}(\mathbf{B})\right) = \eta_0 + \frac{\tau_y\left(\mathbf{H}(\mathbf{B})\right)\tanh(\zeta\dot{\gamma})}{\sqrt{\varepsilon^2 + \dot{\gamma}^2}}$$

Magnetic force

$$f_m = \mu_0 \chi \nabla \mathbf{H}^2 / 2$$

where: **E**: Electric field, V/m;
B: Magnetic inductance density, T;
H: Magnetic field strength, A/m;
J: Current density, A/m^2;
ρ_c: Volumetric charge density, C/m^3;
ε_0: Permittivity of free space = 8.85×10^{-12}, F/m,
μ_0: permeability of free space = $4\pi \times 10^{-7}$, N/A^2.

where: **u**: Velocity field, m/s;
ρ: Fluid density, kg/m^3;
γ: Shear strain rate, 1/s;
η: Dynamic viscosity, Pa*s;
P: Fluid zone pressure, Pa;
τ_y: Yield stress, Pa;
ζ: The constant;
ε: Zoom factor,
χ: Volume susceptibility.

When the magnetic field is applied to the buffer, the MR fluid is characterized by Bingham fluid characteristics. The fluid flow and velocity distribution of a typical Bingham fluid between two parallel plates are shown in Figure 9. The MR fluid can be divided into three regions along the radial direction. The yield flow regions are on both sides and the plug flow region is in the middle. In the upper and lower regions near the plate, the MR fluid is yielding flow. At this region, the shear stress of the MR fluid is greater than the yield stress, and the MR fluid is in the flow state. The shear stress of the MR fluid in the middle plug flow region is less than its yield stress, and the MR fluid is in a relatively static state. In Figure 9, the active region is defined as the other region except the piston winding part at the annular gap. This region is the most obvious region where the MR fluid has MR effects. L is the length of the region. u and v are the velocity of the MR fluid and piston, respectively. h_1, h_2, and h are the height of yield flow region, the height of the plug flow region, and the width of the annular gap, respectively.

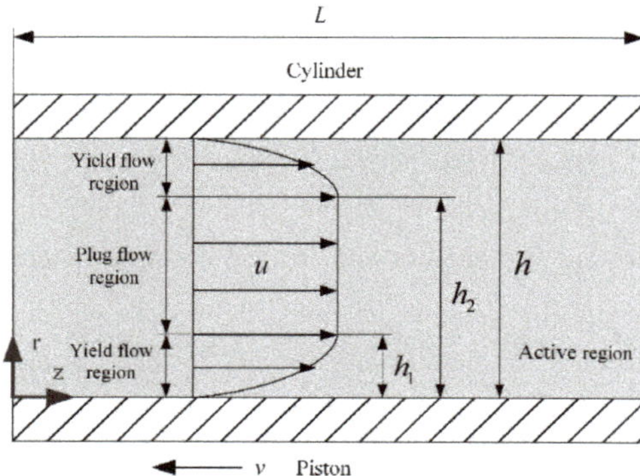

Figure 9. Fluid flow and velocity distribution of the MR fluid.

The Bingham model used to describe the properties of the MR fluid can be expressed as follows:

$$\begin{cases} \tau = \mu_0 \gamma + \tau_y, & \tau > \tau_y \\ \gamma = 0, & \tau \leq \tau_y \end{cases} \tag{2}$$

where μ_0 is the zero field viscosity of the MR fluid, τ is the actual yield strength of the MR fluid, γ is the shear rate of the MR fluid, and τ_y is the ultimate yield strength of the MR fluid.

However, due to the non-differentiability of the plug flow region in the model, it will lead to complications in the actual calculation process. One way to solve this problem is to first determine the boundary of the plug flow regions, and then describe the region as a region with a constant velocity vector. This approach, however, assumes a fully developed flow and become very difficult when dealing with complex annular gap geometries. Thus, it does not have universal applicability.

The method used in this paper is the Bingham plastic model proposed by Case et al. [25]. In this model, the continuous differentiability of yield stress is achieved by describing the plug flow region as a micro-flow region.

This model is described as:

$$\tau = \left(\frac{\tau_y \tanh(\xi\gamma)}{\sqrt{\varepsilon^2 + \gamma^2}} + \mu_0 \right) \cdot \gamma \tag{3}$$

where ξ, ε are constant coefficients.

By dividing the two sides of Equation (3) by the shear stress γ at the same time, the viscosity formula of the MR fluid can be obtained as follows:

$$\mu = \frac{\tau_y \tanh(\xi\gamma)}{\sqrt{\varepsilon^2 + \gamma^2}} + \mu_0 \tag{4}$$

In the multi-physical field simulation software, the correlation between the magnetic inductance intensity B and the fluid viscosity μ at each node of the annular gap can be set up by Equation (4), so as to realize the multi-physical field coupling calculation of the magnetic field and the flow field.

In the simulation, the mesh is divided into moving meshes. The motion equation of the piston part is set to a sine with a frequency of 5 Hz and an amplitude of 4 mm. At the same time, the current in the buffer coil is set to 0 A, 1 A, and 2.5 A, respectively. Constant coefficients ξ and ε are set to 0.03 and 0.1, respectively. In the model, the integral of the total stress along the axis on the boundary of piston and piston rod is the result of the buffer force obtained by simulation.

Figures 10–12 show the velocity distribution and dynamic viscosity distribution of the MR fluid in the MR buffer. As can be seen from the figure, the dynamic viscosity of the MR fluid increases significantly in the active region of the MR buffer, and compared with other fluid regions, the flow velocity of the active region decreases significantly. The results show that the MR fluid in the active region have an MR effect and the dynamic viscosity increases due to the influence of the magnetic field.

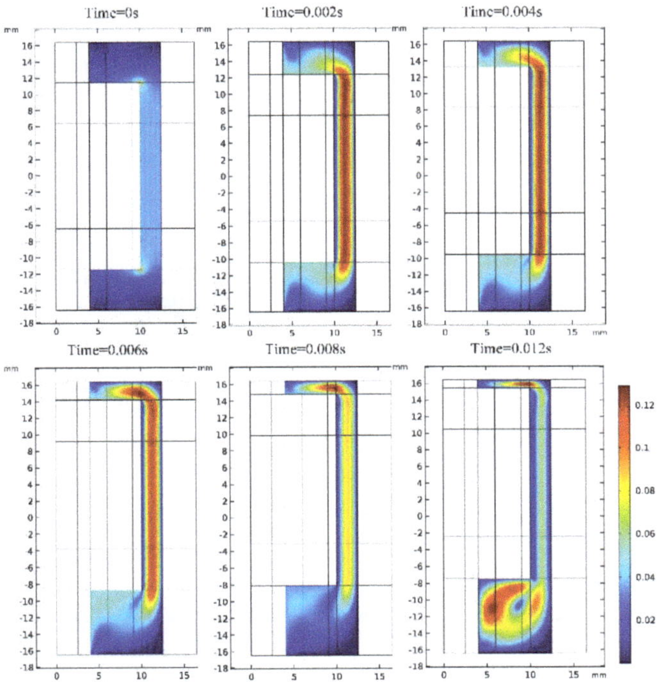

Figure 10. Velocity distribution map of the MR fluid when the current is 0 A.

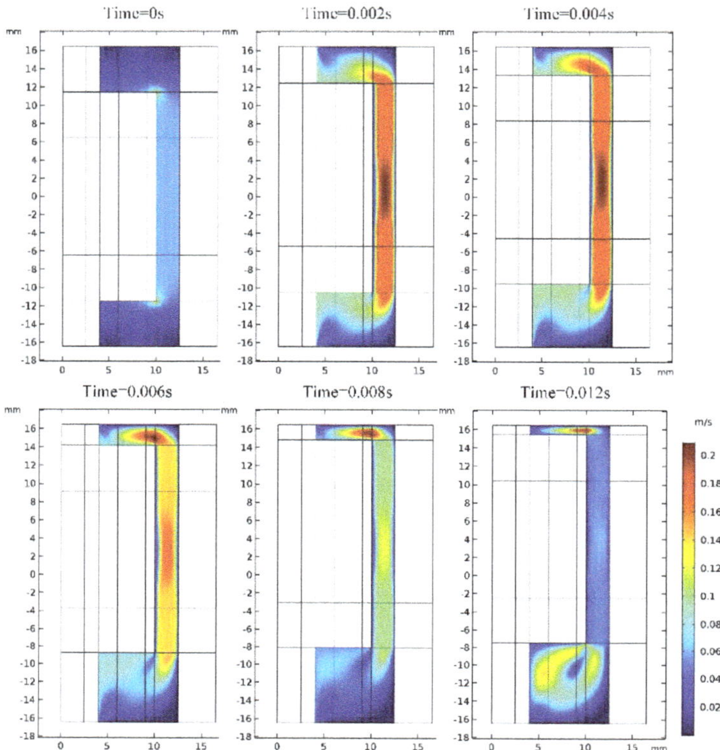

Figure 11. Velocity distribution map of the MR fluid when the current is 2.5 A.

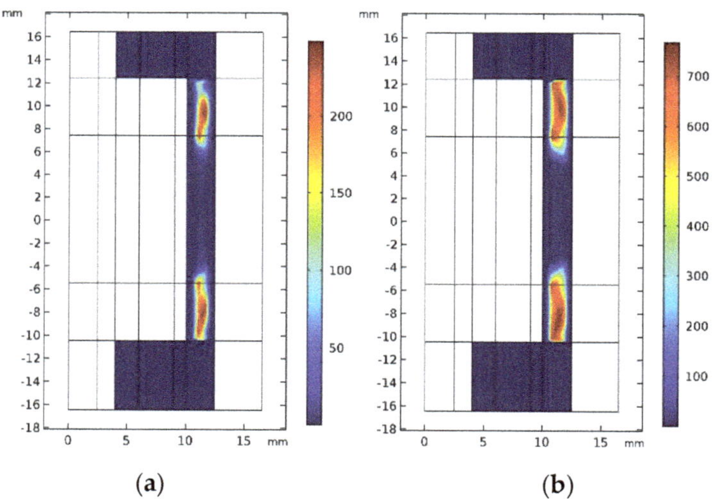

Figure 12. Dynamic viscosity distribution map of the MR fluid: (**a**) I = 1 A; (**b**) I = 2.5 A.

5. Validation with Experimental Results

5.1. Electromagnetic Force Characteristic Experiment of the EMLA

This experiment mainly measures the electromagnetic force characteristic curve of the EMLA, and compares with the simulation results to verify the accuracy of the simulation model. Figure 13 shows the experimental bench for measuring the electromagnetic force characteristic of the EMLA. The experimental bench includes the EMLA, displacement sensor (500LCIT series displacement sensors), force sensor (LSR type tension pressure sensor, with a maximum range of 200 N), adjusting screw rod, and baffle. The displacement sensor and the force sensor are used to measure the displacement and electromagnetic force of the mover, respectively. The measured data are fed back to the PC for data processing. The adjusting screw rod is mainly used to adjust the position of the mover. The position of the mover is changed by 1 mm for every turn of the screw. The EMLA is the control objective.

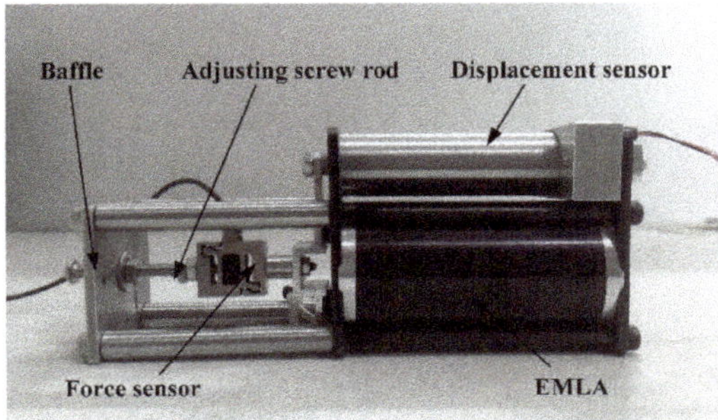

Figure 13. Experimental bench for the electromagnetic force characteristic.

In this experiment, the position of the mover is changed by adjusting the screw, and current intensities of 5 A, 10 A, and 15 A is applied to the electromagnetic coil to measure the electromagnetic force characteristic curve of the EMLA. Figure 14 shows a comparison between the measured electromagnetic force characteristic curve and the simulated curve. As can be seen from the figure, the electromagnetic force characteristic curves obtained from the experiment and simulation are in good agreement. The electromagnetic force is proportional to the current, and it changes very little with the change of the mover position. This shows that the magnitude of the electromagnetic force is only slightly affected by the change of the mover position, which meets the requirements of the structural design.

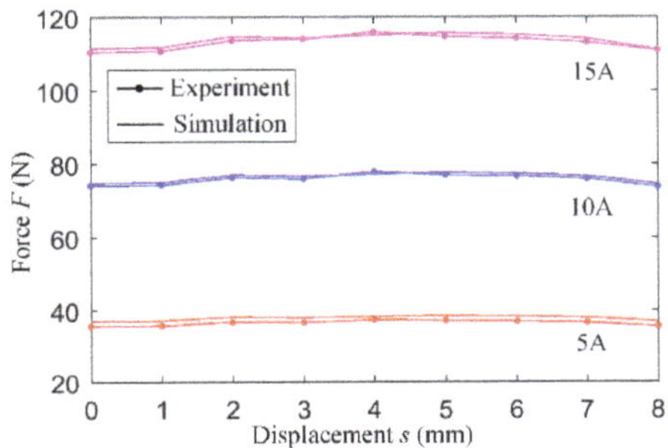

Figure 14. Electromagnetic force characteristic curve of the EMLA.

5.2. MR Buffer Force Measurement

This experiment mainly measures the buffer force of the MR buffer, where a constant current intensity is applied in the buffer coil, and the piston reciprocates in a sinusoidal curve. Then the experiment results and simulation results are compared and analyzed. An experimental bench for measurement is shown in Figure 15. The experimental bench includes the EMLA, displacement sensor, force sensor, and MR buffer. The displacement sensor and force sensor are used to collect the displacement data and buffer force data in the experiment and feed them back to the controller and PC for data processing. The EMLA and MR buffer are the control objectives. In the experiment, the EMLA drives the buffer piston to reciprocate at frequencies of 1 Hz, 3 Hz, and 5 Hz, respectively, and the sinusoidal wave amplitude is 4 mm. At the same time, currents of 1 A and 2.5 A are applied to the buffer coil, respectively.

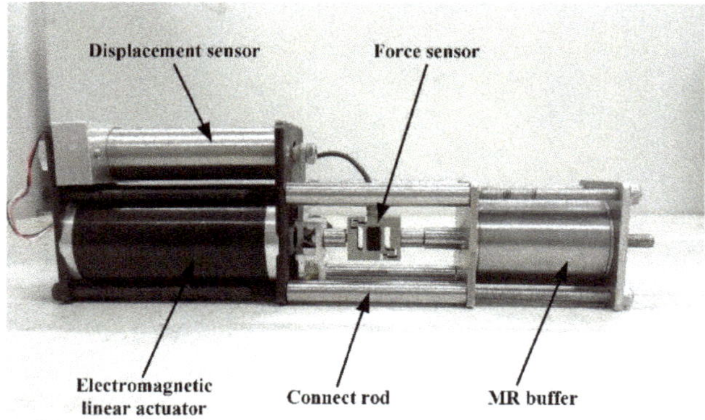

Figure 15. Experimental bench for the buffer force measurement.

Figure 16 shows a comparison between the experimental displacement–velocity curve and the target displacement–velocity curve of the MR buffer. As can be seen from Figure 16, a sudden change in the curve occurs during the start point and stroke end point. The reason for the sudden change is that the static friction will affect the displacement curve of the buffer piston at the beginning of the movement and reversal, which makes the experimental curve not follow the target curve.

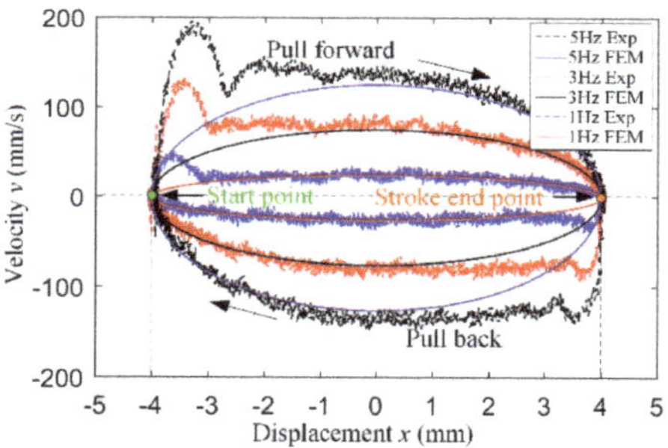

Figure 16. Displacement–velocity curve of the MR buffer.

Figures 17–19 show comparisons between the experimental buffer force curve and the simulation curve. As can be seen from the figures, when the piston motion frequency is 5 Hz and the buffer coil current is 1 A and 2.5 A, the maximum buffer force measured is 10 N and 35 N, respectively. At the same time, the buffer force measured by different coil current intensities is different, showing that the coil current intensity is the main factor affecting the buffer force. As can be seen from the figures, the simulation results are in good agreement with the experimental results. The main deviation occurs at the piston's starting and reversing position. Due to the sudden change of displacement at this location, the measured buffer force curve will produce a certain sudden change at this location, and the mutation becomes more obvious as the frequency increases.

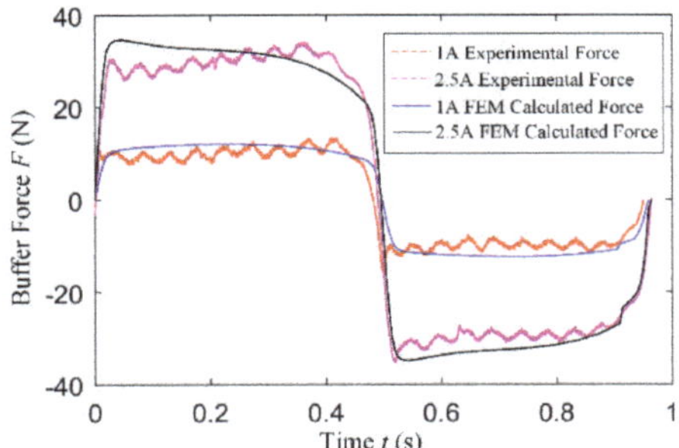

Figure 17. Buffer force curve at 1 Hz frequency.

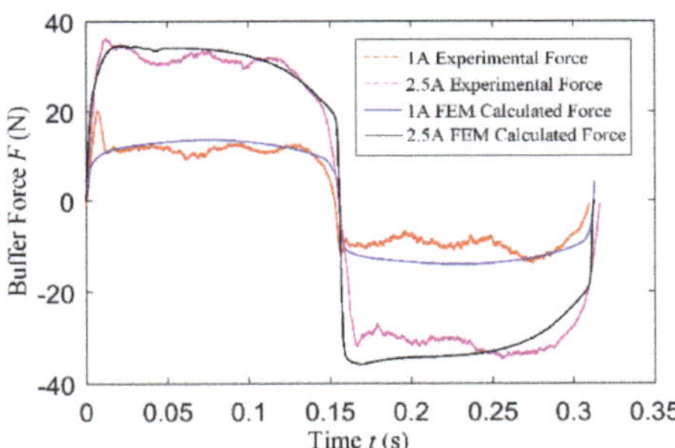

Figure 18. Buffer force curve at 3 Hz frequency.

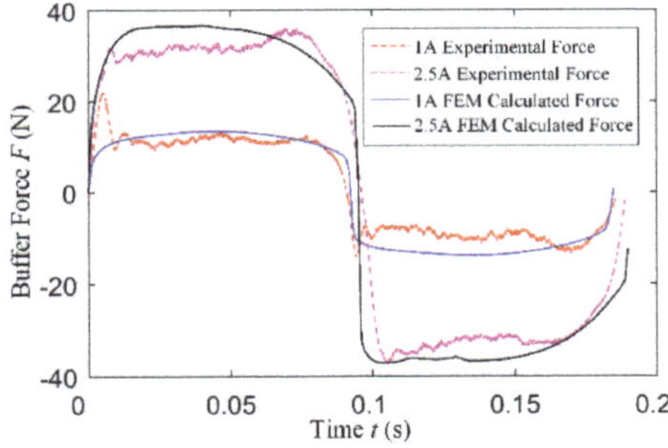

Figure 19. Buffer force curve at 5 Hz frequency.

5.3. Seating Performance Experiment

The main purpose of this experiment is to verify the seat buffering effect of the EMVT with the MR buffer. The experimental bench is shown in Figure 20. The experimental bench includes the EMVT with the MR buffer, displacement sensor, and valve components. Where the EMVT with the MR buffer is the control object, a displacement sensor is used to measure the displacement curve of the mover.

In this experiment, the EMLA is used to drive the valve opening and closing. The buffer effect of the system is verified by applying the buffering force before the valve seating.

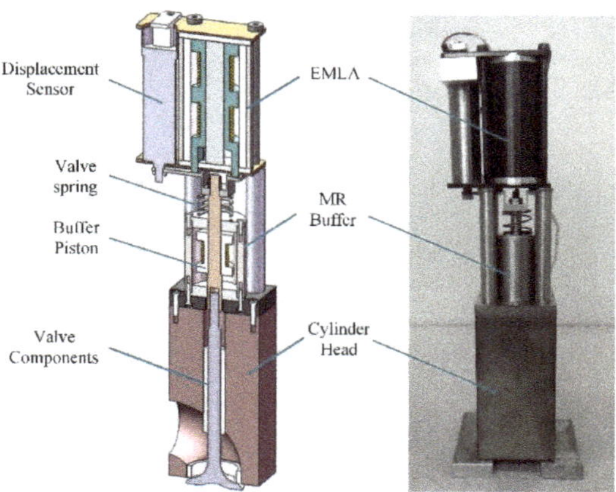

Figure 20. Valve seating test bench.

Figure 21 shows the experimental curve of the valve seat velocity and displacement when the valve lift is 8 mm. In this experiment, the coil in the buffer is energized before the valve seat to generate the buffering force, so as to reduce the valve seat velocity and realize the valve seat buffering. As can be seen from Figure 21, when the valve lift is 8 mm, the valve seat velocity decreases from 0.58 m/s to 0.14 m/s, and the valve rebound height decreases from 0.2 mm to 0.05 mm. The result shows that by applying buffering force before valve seat can reduce the valve seat velocity and restrain the valve rebound effectively. The seat buffer performance of the system is good.

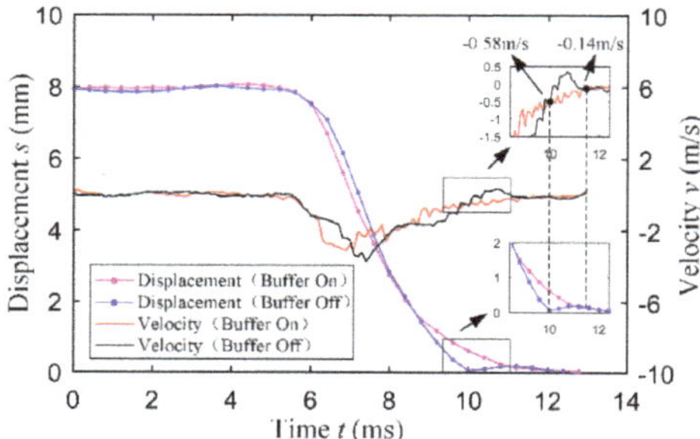

Figure 21. Experimental result of seating performance when the valve lift is 8 mm.

6. Conclusions

In this paper, an EMVT with an MR buffer is designed. The structure scheme of the system is put forward and the main components of the EMLA and MR buffer are designed. According to the structure of the main components, the magnetic field simulation of the EMLA and the MR buffer and the multi-physical field simulation of the MR buffer are carried out, respectively, allowing an accurate estimation of the magnetic force and buffer force. Electromagnetic simulation results show that the maximum electromagnetic force that EMLA can achieve is 320.3 N. The experimental bench of the system is built, and the measured buffer can achieve a 35 N buffer force when the piston motion frequency is 5 Hz and the buffer coil current is 2.5 A. The experimental results are in good agreement with the simulation results. The seating performance of the system is verified. When the lift of the valve is 8 mm, the seating velocity of the valve decreases from 0.58 m/s to 0.14 m/s, and the rebound height of the valve decreases from 0.2 mm to 0.05 mm. From the above results, the system can meet the electromagnetic force required when the valve is opened and the buffer force required when the valve is buffered, and it can reduce the valve seat velocity and restrain the seat rebound effectively. Therefore, the system has good seat buffering performance.

The requirements of the EMLA and MR buffer are higher for the EMVT with MR buffer. It not only requires the EMLA to have a large electromagnetic force, but also ensures that the buffering force can achieve seat buffering and quick response. Therefore, the EMVT with the MR buffer still needs further research on the buffering force response time of the MR buffer and the structure optimization of the system.

Author Contributions: Supervision: L.L., S.C., and Z.X.; research—original draft—editing: H.G.; system control: X.Z.

Funding: This work was supported by the National Natural Science Foundation of China (Grant No. 51975297, 51875290) and Shanghai Aerospace Science and Technology Innovation Fund (SAST2018-107).

Conflicts of Interest: The authors declare no conflict of interest.

References

1. Su, W.H.; Zhang, Z.J.; Liu, R.L. Development trend of automotive internal combustion engine technology. *China Eng. Sci.* **2018**, *20*, 97–103. [CrossRef]
2. Leone, T.; Pozar, M. Fuel economy benefit of cylinder deactivation—Sensitivity to vehicle application and operating constraints. *SAE Trans.* **2001**, *110*, 2039–2044. [CrossRef]
3. Millo, F.; Mirzaeian, M.; Luisi, S.; Doria, V.; Stroppiana, A. Engine displacement modularity for enhancing automotive s.i. engines efficiency at part load. *Fuel* **2016**, *180*, 645–652. [CrossRef]
4. Li, R.C.; Zhu, G.G. A real-time pressure wave model for knock prediction and control. *Int. J. Engine Res.* **2019**, 1–15. [CrossRef]
5. Xu, J.; Chang, S.; Fan, X.; Fan, A. Effects of electromagnetic intake valve train on gasoline engine intake charging. *Appl. Therm. Eng.* **2016**, *96*, 708–715. [CrossRef]
6. Fan, X.; Chang, S.; Liu, L.; Lu, J. Realization and optimization of high compression ratio engine with electromagnetic valve train. *Appl. Therm. Eng.* **2017**, *112*, 371–377. [CrossRef]
7. Fan, X.Y.; Chang, S.Q. Electromagnetic Exhaust Valve Event Optimization for Enhancing Gasoline Engine Performance. *MATEC Web Conf.* **2017**, *95*, 15009. [CrossRef]
8. Li, H. *Cam-Free Electro-Hydraulic Variable Valve Mechanism Control of Internal Combustion Engine Based on Model*; Beijing Institute of Technology: Beijing, China, 2018.
9. Yang, J.; Wang, Z.C.; Wang, Y. Design and analysis of electronically controlled hydraulic fully variable valve driving system. *J. Hunan Univ. Nat. Sci. Ed.* **2017**, *2017*. [CrossRef]
10. Chen, F. Simulation and experimental research of hydraulic pressure and intake valve lift on a fully hydraulic variable valve system for a spark-ignition engine. *Adv. Mech. Eng.* **2018**, *10*, 1–11. [CrossRef]
11. Tu, B.; Tian, H.; Wei, H.Q. Research on buffering process of electro hydraulic drive variable gas mechanism. *China Mech. Eng.* **2016**, *27*, 2652–2658. [CrossRef]
12. Paden, B.A.; Snyder, S.T.; Paden, B.E. Modeling and Control of an Electromagnetic Variable Valve Actuation System. *IEEE/ASME Trans. Mechatron.* **2015**, *20*, 2654–2665. [CrossRef]
13. Reinholz, B.; Seethaler, R. Experimental Validation of a Cogging Torque Assisted Valve Actuation System for Internal Combustion Engines. *IEEE/ASME Trans. Mechatron.* **2016**, *21*, 453–459. [CrossRef]
14. Gillella, P.K.; Song, X.; Sun, Z. Time-Varying Internal Model-Based Control of a Camless Engine Valve Actuation System. *IEEE Trans. Control Syst. Technol.* **2014**, *22*, 1498–1510. [CrossRef]
15. Reinholz, B.A.; Reinholz, L.; Seethaler, R.J. Optimal trajectory operation of a cogging torque assisted motor driven valve actuator for internal combustion engines. *Mechatronics* **2018**, *51*, 1–7. [CrossRef]

16. Haus, B.; Aschemann, H.; Mercorelli, P.; Werner, N. Nonlinear modelling and sliding mode control of a piezo-hydraulic valve system. In Proceedings of the 2016 21st International Conference on Methods and Models in Automation and Robotics (MMAR), Miedzyzdroje, Poland, 29 August–1 September 2016; pp. 442–447. [CrossRef]
17. Samani, R.; Khodadadi, H. A particle swarm optimization approach for sliding mode control of electromechanical valve actuator in camless internal combustion engines. In Proceedings of the IEEE International Conference on Environment & Electrical Engineering & IEEE Industrial & Commercial Power Systems Europe, Milan, Italy, 6–9 June 2017. [CrossRef]
18. Di Gaeta, A.; Hoyos Velasco, C.I.; Montanaro, U. Cycle-by-Cycle Adaptive Force Compensation for the Soft-Landing Control of an Electro-Mechanical Engine Valve Actuator. *Asian J. Control* **2014**, *17*, 1707–1724. [CrossRef]
19. Bernard, L.; Ferrari, A.; Micelli, D.; Perotto, A.; Rinolfi, R.; Vattaneo, F. Electrohydraulic Valve Control with MultiAir Technology. *MTZ Worldwide* **2009**, *70*, 4–10. [CrossRef]
20. Yang, Y.P.; Liu, J.J.; Ye, D.H. Multi-objective Optimal Design and Soft Landing Control of an Electromagnetic Valve Actuator for a Camless Engine. *IEEE/ASME Trans. Mechatron.* **2013**, *18*, 963–972. [CrossRef]
21. Fan, X.; Chang, S.; Lu, J.; Liu, L.; Yao, S.; Xiao, M. Energy consumption investigation of electromagnetic valve train at gas pressure conditions. *Appl. Therm. Eng.* **2019**, *146*, 768–774. [CrossRef]
22. Zhang, L.L. *Research on Engine Valve Load Characteristics and Design of Simulation Loading System*; Nanjing University of Science and Technology: Nanjing, China, 2019.
23. Case, D.; Taheri, B.; Richer, E. Design and Characterization of a Small-Scale Magneto-rheological Damper for Tremor Suppression. *IEEE/ASME Trans. Mechatron.* **2013**, *18*, 96–103. [CrossRef]
24. Sternberg, A.; Zemp, R.; De La Llera, J.C. Multi-physics behavior of a magneto-rheological damper and experimental validation. *Eng. Struct.* **2014**, *69*, 194–205. [CrossRef]
25. Case, D.; Taheri, B.; Richer, E. Dynamical Modeling and Experimental Study of a Small-Scale Magnetorheological Damper. *IEEE/ASME Trans. Mechatron.* **2014**, *19*, 1015–1024. [CrossRef]

Article

Comparison of the Emissions, Noise, and Fuel Consumption Comparison of Direct and Indirect Piezoelectric and Solenoid Injectors in a Low-Compression-Ratio Diesel Engine

Stefano d'Ambrosio *, Alessandro Ferrari, Alessandro Mancarella, Salvatore Mancò and Antonio Mittica

Energy Department, Politecnico di Torino, Corso Duca degli Abruzzi 24, 10129 Torino, Italy;
alessandro.ferrari@polito.it (A.F.); alessandro.mancarella@polito.it (A.M.); salvatore.manco@polito.it (S.M.);
antonio.mittica@polito.it (A.M.)
* Correspondence: stefano.dambrosio@polito.it; Tel.: +39-011-090-4415

Received: 19 September 2019; Accepted: 21 October 2019; Published: 23 October 2019

Abstract: An experimental investigation has been carried out to compare the performance and emissions of a low-compression-ratio Euro 5 diesel engine featuring high EGR rates, equipped with different injector technologies, i.e., solenoid, indirect-acting, and direct-acting piezoelectric. The comparisons, performed with reference to a state-of-the-art double fuel injection calibration, i.e., pilot-Main (*pM*), are presented in terms of engine-out exhaust emissions, combustion noise (CN), and fuel consumption, at low–medium engine speeds and loads. The differences in engine performance and emissions of the solenoidal, indirect-acting, and direct-acting piezoelectric injector setups have been found on the basis of experimental results to mainly depend on the specific features of their hydraulic circuits rather than on the considered injector driving system.

Keywords: solenoid injectors; indirect-acting piezoelectric injectors; direct-acting piezoelectric injectors; engine-out emissions; fuel consumption; combustion noise

---

1. Introduction

In the last few years, significant improvements have been made, by the automotive industry, in the development of innovative solutions that are able to comply with increasingly stringent international regulations, in terms of reduction of both CO_2 and pollutant emissions [1]. One of the main research fields in the diesel sector deals with the development of the Common Rail (CR) fuel injection systems [2–4] and the related fuel injection strategies [5].

Fuel spray formation plays a fundamental role in diesel engines, as the interaction between the injected fuel and the in-cylinder intake charge affects pollutant formation to a great extent [6]. For instance, a greater air entrainment in the spray can lead to a reduction in soot emissions, but can also increase NO_x as combustion temperatures increase [7]. In this regard, a major role is played by diesel injectors, which are one of the core components of CR systems [8,9]. Two main driving technologies exist to control the needle lift, and diesel injectors can thus be classified into solenoid- [10,11] or piezoelectric-driven [12,13]. In this latter category, it is possible to further distinguish between direct- and indirect-acting piezo injectors.

Solenoid injectors obtain the needle upstroke thanks to a magnetic field that is generated by means of an electric current circulating in a coil, while the piezoelectric ones use piezoelectric crystals that are able to elongate when subjected to an electric potential difference [14]. The solenoid technology has proved to be robust and simple, and it is the most widely adopted in diesel automotive engines. Nevertheless, it is often claimed that the piezoelectric alternatives have a wider range of potentialities than their conventional solenoid counterparts [15]. Ueda et al. [16], applying the same fuel injection calibration (featuring a double pilot injection), stated that a 4th generation indirect-acting piezoelectric injector could achieve lower soot emissions, thanks to its steeper fuel injection rate than a 4th generation solenoid reference measurement. Moreover, a potential fuel consumption reduction of up to 2.8% was also pointed out, thanks to the fast opening and closing of the needle which creates a short injection and combustion duration. Kim et al. [17] compared direct- and indirect-acting piezo injector performance, and pointed out that direct-acting piezo injectors could reduce NO_x emissions, with only slight HC penalties, in most of the analyzed tested conditions. In fact, most of the conditions considered in their study resulted in a higher heat release rate in the premixed combustion phase when an indirect piezo injector was used.

The fuel spray pattern and atomization characteristics of indirect-acting piezo injection systems may be different from those of the solenoid-driven technology, and may be a potential source of improvement for fuel consumption and reduction of exhaust pollutant emissions from the engine [18]. In this regard, not only can the particular driving technology (piezo or solenoid) play an important role, but also the hydraulic layout and the mechanical setup of each injector [19]. If a solenoid-driven injector features an innovative pressure-balanced pilot-valve layout, its static leakage can be reduced significantly, by up to 25% less than similar injectors featuring standard pilot valves [20,21]. Moreover, a further reduction, of up to 50%, may be obtained if the innovative pressure-balanced pilot-valve is coupled with an integrated Minirail [22]. The hydraulic setups commonly applied to modern indirect-acting piezo injectors may be endowed with 3-way pilot valves [23] and bypass-circuits [19], with both devices affecting the dynamics of the injector control chamber, which is placed at the rear of the needle. The 3-way pilot-valve has the potential to reduce the dynamic leakage of the injector, while the bypass-circuit can improve the dynamic response of the needle during the nozzle closure phase [23].

When the direct-acting piezoelectric technology is implemented, the injector needle is directly actuated by the piezoelectric driving system, without the need for any pilot-valve: in this way, a reduction of the injector leakage, major control of the injected quantity, flexible multiple injections and flowrate shaping capabilities are all claimed benefits of this solution [24].

The present work has the aim of comparing the performance of solenoid-driven, direct, and indirect piezo-driven injectors in terms of engine-out pollutant emissions, fuel consumption and combustion noise (CN), under different working conditions in a low-compression-ratio diesel engine for passenger cars. The considered solenoid-driven injector is equipped with a pressure-balanced pilot-valve and an integrated Minirail, while the indirect-acting piezo-driven injector is endowed with a bypass-circuit and the direct-acting one features a feedback control strategy of the injected fuel mass.

2. Experimental Setup and Engine Conditions

The experimental tests were performed on a 2.0 L, four-cylinder, four-stroke diesel engine, fueled with conventional diesel oil (EN 590), whose main technical specifications are reported in Table 1.

Table 1. Main technical specifications of the engine adopted for the experimental tests.

Engine Specifications	
Engine type	2.0 L diesel direct-injection Euro 5
Number of cylinders	4
Displacement	1956 cm^3
Bore × stroke	83.0 mm × 90.4 mm
Compression ratio	16.3
Valves per cylinder	4
Turbocharger	Twin-stage with valve actuators and wastegates
Fuel injection system	High-pressure (max. 2000 bar) Common Rail
EGR system	Short-route cooled EGR

A simplified scheme of the engine is shown in Figure 1. It is homologated for Euro 5 regulations and equipped with a high-pressure CR fuel injection system, a twin-stage turbocharger regulated by means of valve actuators and wastegate valves and a short-route high-pressure EGR system endowed with an EGR cooler. The EGR valve is installed downstream of the EGR cooler.

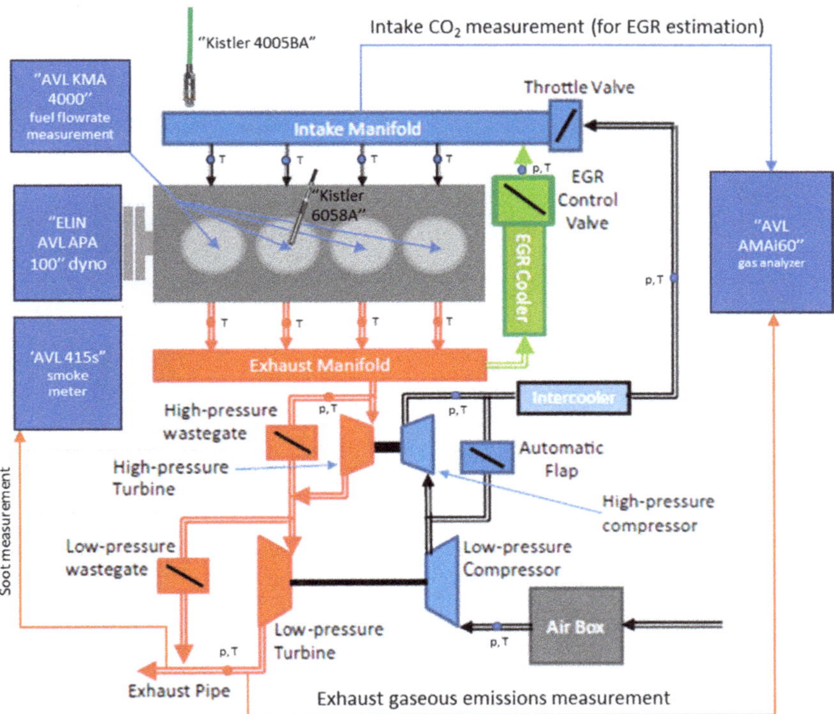

Figure 1. Scheme of the tested engine installed on the dynamic test bench at the Politecnico di Torino.

The experimental tests were carried out at the dynamic test cell of the Politecnico di Torino (ICEAL—Internal Combustion Engines Advanced Laboratories). This test cell is equipped with an "ELIN AVL APA 100" AC dynamometer and an "AVL KMA 4000" fuel flowrate measuring system. The brake specific fuel consumption (*bsfc*) values reported in the "Results and discussion" Section were calculated as the ratio of this fuel flowrate measurement to the brake power of the engine.

An "AVL AMAi60" was used to measure the raw gaseous emissions from the engine. It includes a complete analyzer train endowed with devices capable of simultaneously measuring gaseous concentrations of NO, NO_x, HC, CH_4, CO, CO_2, and O_2 chemical species, while another analyzer train is only equipped with a CO_2 measurement instrument, which was mounted at the intake manifold to estimate the EGR rate to the engine, defined as the ratio of the recirculated exhaust gas mass flowrate to the total mass flowrate inducted in the cylinders [25]:

$$X_{EGR} = \frac{\dot{m}_{EGR}}{\dot{m}_{EGR} + \dot{m}_a} \tag{1}$$

where $\dot{m}_{EGR}$ and $\dot{m}_a$ represent the EGR and fresh air mass flowrates, respectively. In the present investigation the calculation of X_{EGR} was performed considering the accurate expression developed in [26], which requires the evaluation of the volumetric concentrations of all the species at the exhaust and the knowledge of the combustion air composition at the inlet. An "AVL 415S" smokemeter was used to measure the soot engine-out emissions.

The test engine was instrumented with a high-frequency piezoelectric pressure transducer (Kistler 6058 A), which was installed on the cylinder head to provide measurements of the gas in-cylinder pressure temporal traces in cylinder #2. A high-frequency piezoresistive pressure transducer (Kistler 4005 BA) was installed in the corresponding intake runner of the same cylinder to obtain the absolute in-cylinder pressure measurement, which was used as a reference. Furthermore, slow-frequency piezoresistive pressure transducers and thermocouples were installed at different positions of the gas flow path (e.g., upstream and downstream of the turbocompressor, intercooler and turbine, in the intake manifold, in the cylinder runners and along the EGR circuit) to perform pressure and temperature measurements (cf. Figure 1).

Finally, all the above-mentioned measurement devices are controlled by AVL Puma Open 1.3.2 and IndiCom automation software. Data post-elaboration was performed with AVL CONCERTO 5 software. AVL CONCERTO 5 was in particular used to calculate the reported CN values, based on the measured in-cylinder pressure signals, which were filtered by means of a low-pass filter with a cut-off frequency of 5 kHz. Each in-cylinder pressure acquisition was averaged over 100 consecutive engine cycles.

Tables 2 and 3 report the available data to establish the accuracy of the measured pollutant emission values. Previous works [27] have shown that the expanded uncertainties of pollutant emission measurements taken at this engine test facility fall within a 2–4% range. As far as the extended uncertainties pertaining to the brake specific emissions are concerned, the fuel flowrate system accuracy (0.1% over a 0.28–110 kg/h fuel flowrate measurement range) and the maximum errors of the engine speed (1.50 rpm at full scale) and torque (0.30 Nm at full scale) also have to be taken into account.

Table 2. Composition of the gas calibration cylinder and extended uncertainty (95% confidence interval).

Composition of the Gas Calibration Cylinder and Extended Uncertainty	
NO (lower range) [ppm]	89.7 ± 1.7
NO (higher range) [ppm]	919 ± 18
CO (lower range) [ppm]	4030 ± 79
CO (higher range) [%]	8.370 ± 0.097
CO_2 (lower range) [ppm]	4.980 ± 0.067
CO_2 (higher range) [%]	16.78 ± 0.15
C_3H_8 (lower range) [ppm]	88.8 ± 1.8
C_3H_8 (higher range) [ppm]	1820 ± 36

Table 3. Manufacturer's data for the measurement errors of the CLD, NDIR, and HFID analyzers.

Measurement Errors of the CLD, NDIR, and HFID Analyzers	
Linearity	≤1% of full-scale range ≤2% of measured value whichever is smaller
Drift 24 h	≤1% of full-scale range
Reproducibility	≤0.5% of full-scale range

3. Injector Hydraulic Performance

The three considered injector types, that is a new-generation indirect-acting solenoid (for brevity here referred to as IAS), an indirect-acting piezoelectric (IAP) and a direct-acting piezoelectric (DAP) injector, all feature the same nozzle, conical angle, and geometrical sizes of the needle. However, the IAP and IAS injectors both have a ballistic needle (i.e., it does not reach its stroke end position for all the engine working conditions) with the same key features, while the DAP injector needle is not ballistic and reaches the stroke end at 0.1 mm [12]. Consequently, the capability of the injected fuel flowrate of the DAP injector differs from the other two solutions. In general, the injected mass results to be lower for medium to high ET and p_{rail} values, while it leads to a larger injected fuel mass for low ET and p_{rail} values, than for IAS and IAP types [22,23]. In the former case, this is due to the reduced needle lift of the DAP injector, which yields a smaller flow restriction in correspondence to the needle-seat passage. In the latter case, the DAP injector delivers more fuel than the other two solutions, because the elongation of its piezoelectric crystal is linked directly to the mechanical actuation of the needle, which is thus able to open more quickly than in the case of an indirect actuation for both IAP and IAS [22,23]. In fact, it has been verified that the injected fuel flowrate of the DAP injector increases with the highest time derivative, at least for lower p_{rail} levels than 1200 bar [23].

Figure 2 shows a schematic of the indirect-acting piezo-driven injector, endowed with the bypass-circuit [23], while Figures 3 and 4 represent the direct-acting piezo injector [12] and the solenoid-driven one, respectively; the latter highlights the pressure-balanced pilot-valve [22] and a standard pilot-valve.

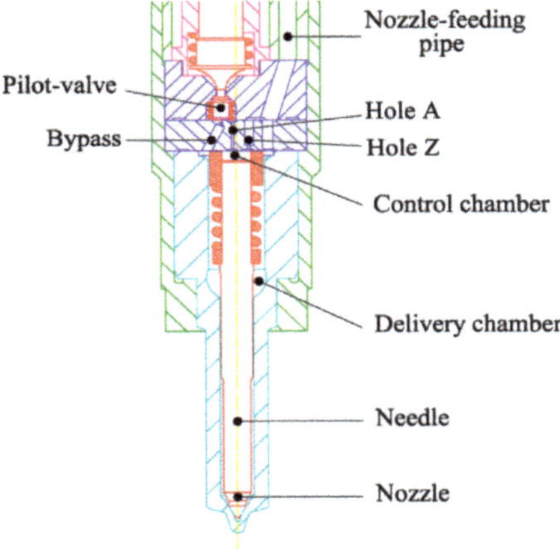

Figure 2. Schematic of the indirect-acting piezoelectric injector.

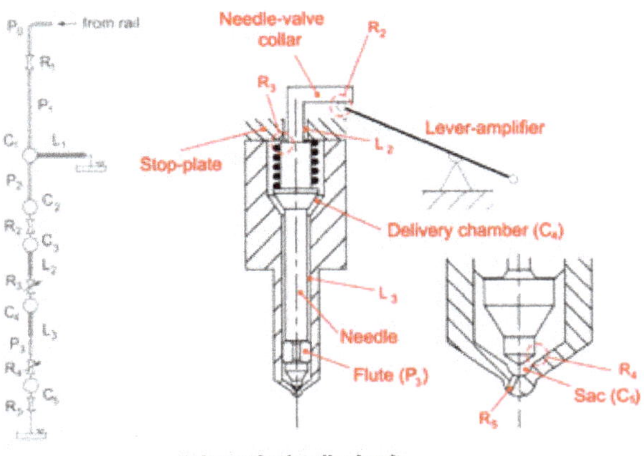

Figure 3. Schematic of the direct-acting piezoelectric injector.

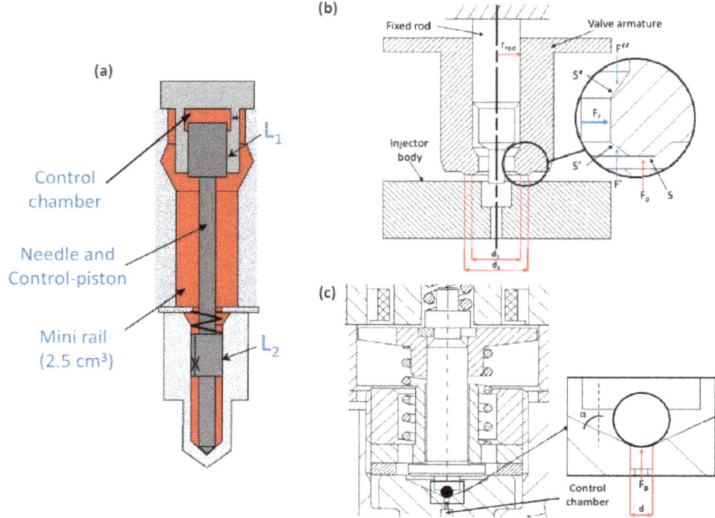

Figure 4. Solenoid injector. (**a**) Hydraulic circuit of the injector; (**b**) Pressure-balanced pilot-valve; (**c**) Standard pilot-valve.

4. Results and Discussion

The results pertaining to the experimental tests carried out at the dynamometer test cell are presented in what follows, and the performance of the engine is highlighted (in terms of engine-out pollutant emissions, *bsfc* and CN) when equipped with IAS, IAP, or DAP injectors. The engine was originally provided by the OEM with a state-of-the-art pilot-main (*pM*) fuel injection calibration referring to IAP injectors. Experimental tests were then carried out with DAP and IAS injectors instead of IAP ones, keeping the same fuel injection calibration, to perform a comparison and investigate their different performance.

Some engine operating points, which were representative of an application of the tested engine to a vehicle over the New European Driving Cycle (NEDC), were selected for the analysis. The resulting key-points (expressed in terms of speed n [rpm] $\times$ *bmep* [bar]) were in the 1.5 bar $\leq$ *bmep* $\leq$ 12 bar and 1000 rpm $\leq n \leq$ 2750 rpm ranges. For conciseness reasons, only three key-points are analyzed hereafter: a low load and speed condition, namely 1500 $\times$ 2, (whose results are presented in paragraph 4.1), and two medium–high speed and load points, namely 2000 $\times$ 5 and 2500 $\times$ 8 (whose results are presented in paragraph 4.2). For each working condition, results are presented for the three different injector technologies along some EGR sweeps, to have a wider comparison than that based on the single baseline points. The aim is to compare the performance of IAP, DAP and IAS injectors when the same fuel injection calibration is implemented and to highlight the impact of the particular injector technology on CN, *bsfc*, and engine-out pollutant emissions.

4.1. Low Load and Speed (1500 × 2)

Starting from the baseline double *pM* calibration (whose main parameters are reported in Table 4) EGR trade-offs were performed at a low speed and load engine point (i.e., 1500 $\times$ 2), for the three different injector typologies under investigation. The results of this comparison, in terms of CN, fuel consumption and engine-out pollutant emissions, are shown in Figure 5. Different colors and symbols refer to different injector typologies, while the contoured symbols refer to the baseline calibration results. The x-axis reports the NO_x brake specific emissions (in g/kWh) for each diagram, with the lowest values corresponding to the highest applied EGR fractions (X_{EGR}).

Table 4. Main parameters of the reference *pM* calibration at 1500×2, with IAP injectors. The main quantity is set directly by the test bench to obtain the desired *bmep*.

Quantity	Reference *pM* Injection Calibration by OEM, IAP Injector
SOI_{Pil} [°CA bTDC]	10.6
SOI_{Main} [°CA bTDC]	−2.4
ET_{Pil} [μs]	250
p_{Rail} [bar]	450
q_{Pil} [mm³/(stk·cyl)]	1.7
X_{EGR} [%]	47.5

The engine-out soot and NO_x emissions (Figure 5a) do not show the typical trade-off behavior of the conventional diesel combustion mode, for any of the three examined injector configurations. A PCCI-like trend (which means a simultaneous reduction of soot and NO_x emissions when X_{EGR} grows) is clearly achieved in each case, thanks to the realization of a high degree of premixed combustion. The high X_{EGR}, in addition to the somewhat delayed fuel injection timings (which make combustion occur mainly during the piston expansion stroke) and the reduced engine compression ratio, are all able to enhance the local mixing of the injected fuel plumes with the inducted charge, producing a highly premixed stratified charge and relatively low peak combustion flame temperatures [28]. Unfortunately, PCCI-like strategies lead to high incomplete combustion species (HC emissions are reported as a reference in Figure 5d, but significant levels were also obtained for CO emissions, which are not presented here for conciseness reasons) and fuel consumption (Figure 5c), which show rapidly worsening trends as the EGR fraction increases. If all the graphs in Figure 5 are compared, the DAP injector provides the worst results, in terms of CN, *bsfc*, and HC emissions, because it has the highest tendency toward premixed combustion of the three analyzed solutions.

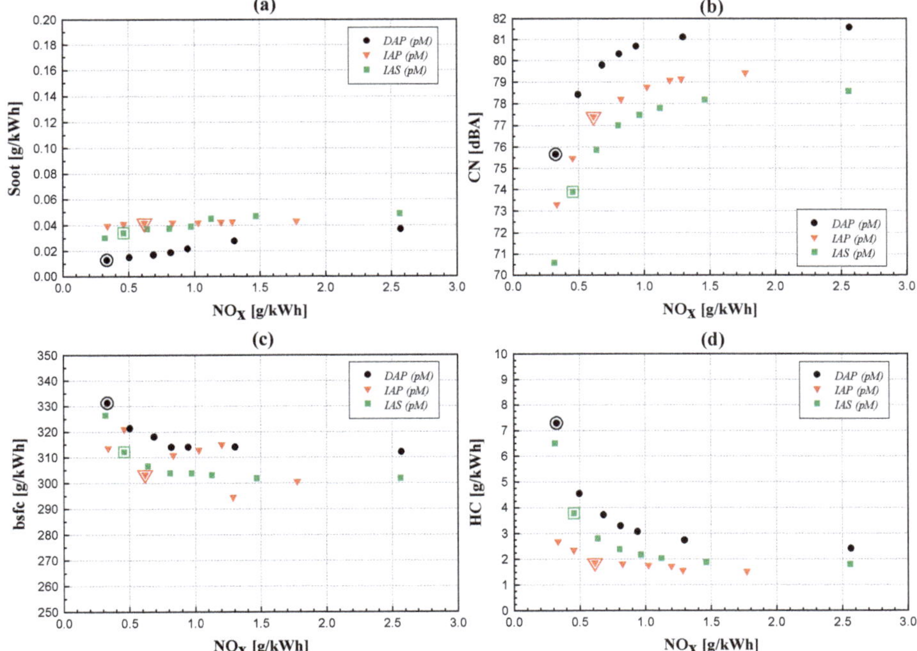

Figure 5. Engine-out soot emissions (**a**) CN; (**b**), *bsfc*; (**c**) and engine-out HC emissions; (**d**) vs. NO_x for the three injector typologies, at 1500×2.

The IAS and DAP injectors are both able to deliver fuel with an improved atomization compared to the IAP solution, especially when relatively low p_{rail} and ET values are implemented, as in this case. The improved atomization for the IAS injector can be mainly ascribed to the presence of the integrated Minirail in its hydraulic layout. The Minirail, in fact, is able to maintain a more stable (and thus, higher) injection pressure during both the pilot injection and the first part of the main shot, thereby counterbalancing the pressure reduction due to the injection in the injector delivery chamber, and allowing a faster needle opening [22]. The DAP injector features a faster needle upstroke because of its direct mechanical actuation on the needle, which gives rise to a steeper fuel injected flowrate, at least for p_{rail} values below 1200 bar [23]. This improved fuel atomization (with respect to the IAP solution) creates leaner equivalence ratios locally, increases the local oxygen concentration and improves the tendency toward low temperature premixed combustion through an increased autoignition delay and more delayed combustion developments. Unlike high temperature diesel ignition, which mostly arises through the chain-branching $H + O_2 = O + OH$ reaction (which means that a decrease in the local oxygen concentration enlarges the autoignition delay and slows down the reaction process), at lower temperatures typical of PCCI-like combustion fuel decomposition and ignition mainly proceed along a different chain-branching reaction (alkyl $\rightarrow$ alkyl peroxyl $\rightarrow$ hydroperoxyl alkyl $\rightarrow$ hydroperoxyl peroxyl $\rightarrow$ keto-hydroperoxide) [29,30]. In the latter case, unlike the former, the leaner local equivalence ratios (i.e., lower fuel concentrations) achievable with IAS and DAP injectors, thanks to the improved atomization, can reduce the concentrations of fuel radicals inside the air-fuel mixture, thus effectively slowing down the ignition process and increasing the premixed stage of combustion.

As already mentioned, the highest premixed degree is achieved using DAP injectors, thus justifying their worse HC, *bsfc*, and CN levels in Figure 5. The higher the premixed degree is, the more likely the occurrence of fuel over-mixing and flame quenching phenomena. These phenomena are thought to be the main mechanisms responsible for the emission of incomplete combustion species, i.e., unburned HC and CO, under low speed and low load conditions, together with wall wetting phenomena [31]. Moreover, these phenomena may also negatively affect the fuel economy.

The CN vs. NO_x (or the CN vs. EGR rate, as NO_x emissions have strong correlations with EGR fractions) trends reported in Figure 5b do not show any trade-off behavior. On the one hand, when the EGR rate increases, the combustion develops more in the premixed stage, primarily thanks to the dilution effect of the recirculated gases, which increase the autoignition delay [25], and this leads to higher HRR peaks, higher cylinder pressure derivatives during the main combustion and deteriorated CN. On the other hand, the reduced combustion velocity, which is induced by a decrease in the oxygen fraction [32], dampens the intensity of the premixed combustion. Furthermore, if the autoignition delay grows, the combustion process tends to develop further during the piston expansion stroke (a late PCCI mode is implemented) and is thus influenced more by the expansion cooling effect, which reduces the cylinder pressure rise. The latter effects tend to prevail over the dilution effect (cf. Figure 6), thus making CN decrease as the EGR rate increases.

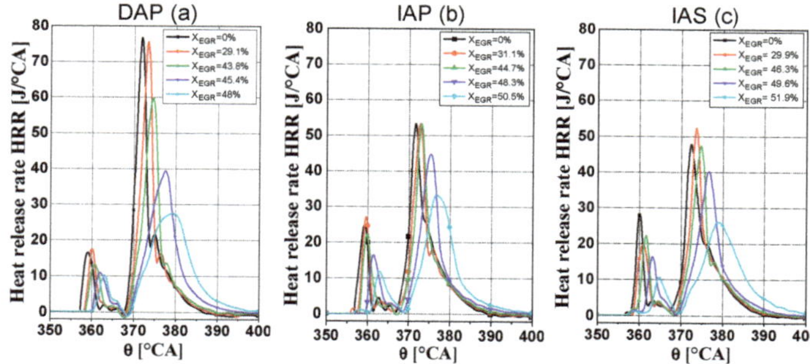

Figure 6. Comparison of the HRR traces featuring different EGR rates, at 1500 × 2, for the three injector typologies under investigation: (**a**) DAP; (**b**) IAP; (**c**) IAS.

The DAP injector provides a higher CN (and the worst CN-NO$_x$ trade-off) than the other two investigated solutions, if a similar EGR rate (i.e., similar intake oxygen concentrations) is applied, while the solenoid injector results to be the best one. Figure 7 shows the HRR traces related to the three injector typologies for almost the same intake oxygen concentration (around 17.5%, corresponding to an X_{EGR} of around 44%). The highest HRR peak pertaining to the DAP injector is likely due to the combination of a longer autoignition delay (as already explained), which increases the premixed stage of combustion, and a larger fuel injected quantity at the start of combustion, thanks to the faster upstroke of its needle than that of the IAP and IAS solutions. The IAS injector features a lower CN level than the IAP one, and its combustion process develops further during the piston expansion stroke with consequent lower HRR peaks. This is mostly due to a better atomization of the IAS injector, which is reached because of the presence of the Minirail [22].

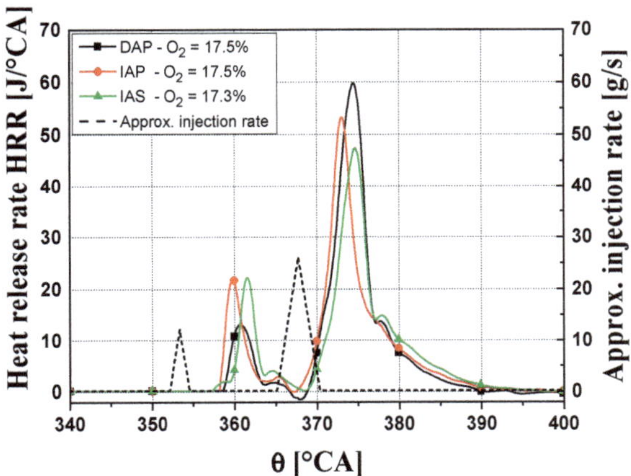

Figure 7. HRR traces featuring similar EGR rates, at 1500 × 2, for the three injector typologies under investigation.

4.2. Medium–High Speed and Load (2000 × 5, 2500 × 8)

Figures 8 and 9 report the soot, *bsfc*, CN and HC for the EGR trade-offs performed at two medium speeds and loads (engine points 2000 × 5 and 2500 × 8) for the three different injector types under investigation, in a similar way to the previous section. As has already been seen for the low load and speed engine conditions, the contoured symbols in each graph refer to the baseline tests obtained, for each injector type, when implementing the same baseline double *pM* calibration (whose main parameters are reported in Table 5) provided by the OEM and originally applied to the engine equipped with the IAP injectors.

Table 5. Main parameters of the reference *pM* calibration at 2000 × 5 and 2500 × 8, with IAP injectors.

Quantity	Reference *pM* Injection Calibration by OEM, IAP Injector, 2000 × 5	Reference *pM* Injection Calibration by OEM, IAP Injector, 2500 × 8
SOI$_{Pil}$ [°CA bTDC]	15.2	23.6
SOI$_{Main}$ [°CA bTDC]	−1.8	2.2
ET$_{Pil}$ [µs]	190	155
p$_{Rail}$ [bar]	750	1200
q$_{Pil}$ [mm^3/(stk·cyl)]	1.4	1.2
X$_{EGR}$ [%]	32.3	22.7

The engine-out soot and NO_x emissions at 2000×5 (Figure 8a) show a clear trade-off behavior, which is typical of a conventional diesel combustion mode, for each of the three examined injector typologies. Nevertheless, a low-NO_x combustion can also be achieved for these higher load and speed conditions, through the application of a relatively delayed fuel injection pattern and a high EGR rate (cf. the contoured symbols in Figure 8, which represent the results pertaining to the reference calibration for each injector type). The worst soot-NO_x trade-off can be observed for the IAP injector (cf. triangle symbols in Figure 8a), which features more than doubled soot emissions compared to the other two injector solutions for the highest EGR rates (lowest NO_x values). This is because, as already mentioned, the IAP injector gives the worst atomization and the lowest premixed degree, thus entraining the least air inside the injected fuel plumes and locally creating the richest mixture portions, which are prone to forming soot particles.

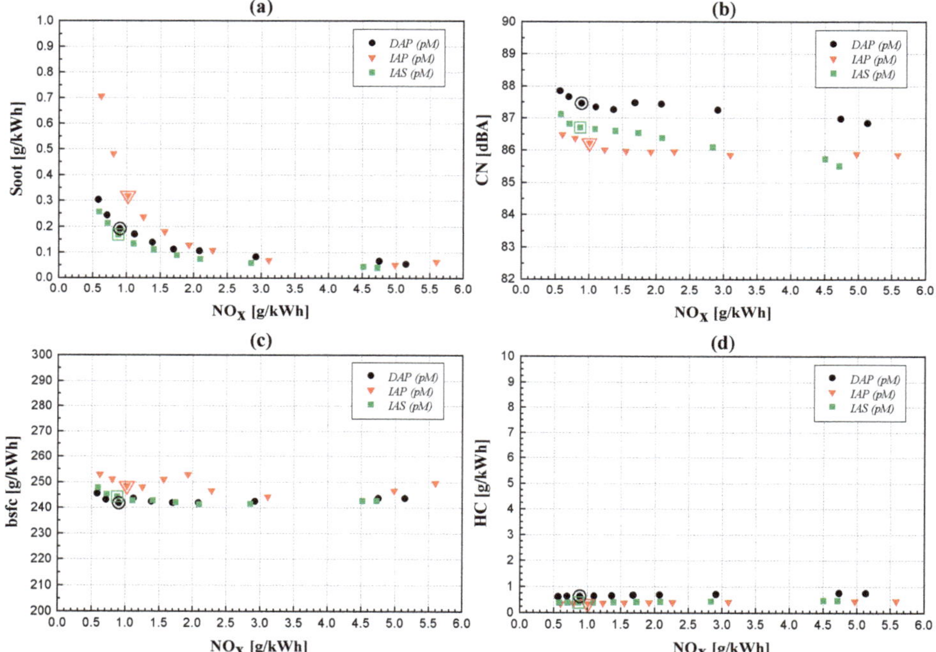

Figure 8. Engine-out soot emissions (**a**) CN; (**b**) *bsfc*; (**c**) and engine-out HC emissions; (**d**) vs. NO_x for the three injector typologies, at 2000×5.

Differences in fuel consumption (Figure 8c) have been highlighted between the three injectors under investigation. A mean penalty in *bsfc* of about 3–4% may be observed for the IAP solution, while DAP and IAS show almost the same behavior. However, the differences in fuel leakage between the IAP, DAP and IAS injectors should only play a negligible role at this engine operating point: the IAP and IAS injectors show comparable injector leakages, while the DAP injector, which features the lowest leakage values, does not show any appreciable *bsfc* reduction. The 3–4% reduction in *bsfc*, when IAS and DAP injectors are installed, is likely due to the enhancement in the thermal efficiency of the engine deriving from the higher premixed degree, which makes the combustion approach constant volume rather than constant pressure conditions and reduces its overall duration. Unlike the low load case, this gain in thermal efficiency is not offset by any severe CO (here not reported) or

HC penalties (Figure 8d). When load increases, concerns about incomplete combustion species are generally mitigated.

The CN vs. NO_x trends (cf. Figure 8b) at 2000 × 5 show a slight trade-off behavior, i.e., a slight increase in CN when NO_x emissions are reduced as a result of higher EGR rates, unlike what is highlighted at 1500 × 2. In this case, the larger premixed combustion phase prevails over the reduction in combustion velocity due to increased EGR and the damping effect resulting from the delayed combustion that develops along the expansion stroke. This leads to higher HRR peaks, higher cylinder pressure derivatives and increased CN as the EGR fractions grow. The lowest CN level of three injector types is achieved by the IAP injector, with improvement of up to 1–1.5 dBA with respect to the other solutions. The three HRR traces pertaining to the baseline tests, for each injector, are plotted in Figure 9 versus the crankshaft angle. It is clear that the IAP injector determines the lower ignition delay and, thus, less fuel to burn in the premixed stage, and this gives rise to the lowest HRR development and consequently to the lowest CN.

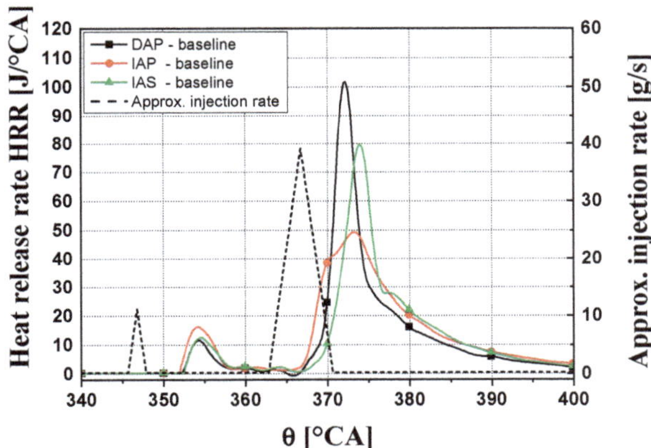

Figure 9. HRR traces of the reference calibrations, at 2000 × 5, for the three injector typologies under investigation.

The worst soot-NO_x trade-off at 2500 × 8 may be observed for the DAP injector (cf. the circle symbol curve in Figure 10a), while the IAS injector still gives the best results (cf. the square symbol curve), even though the differences between the three solutions are not big. In the case of the IAS injector, the lowest soot production at each EGR level is still due to its higher tendency toward premixed combustion, as a consequence of the presence of the Minirail in the hydraulic layout, which determines a faster needle upstroke and, thus, a better fuel atomization in the first part of the injection. The fact that the DAP injector does not give better results than the IAP, as observed at lower loads, is due to the higher rail pressure (as can be seen in Table 5, the rail pressure at 2500 × 8 is equal to 1200 bar). The needle velocity of the DAP injector for $p_{rail} \geq 1200$ bar during the upstroke tends to become similar to that of the IAP one, as the increased rail pressure level generates a higher pressure force that counteracts the needle lift.

Once again, the IAP injector shows an increased *bsfc* (about 3–4% more than the other injector typologies, cf. Figure 10c) compared to the other two injectors. The previously highlighted differences in the premixed combustion portion reflect on the fuel consumptions of the injectors, as occurred at 2000 × 5. Moreover, the considerations on the negligible levels of CO (here not reported) and HC emissions (Figure 10d) for all the considered injector types are even more valid at this higher considered load.

Finally, the CN (Figure 10b) has been confirmed to be higher for the DAP injector, even though it does not highlight the maximum HRR peak (cf. Figure 11) as it does at 2000 × 5. In fact, when double injection is performed, CN not only correlates with the heat release peak of the main injection (HRR_{pk2}), but also with that of the pilot shot (HRR$_{pk1}$) and with the timings of the two HRR bumps [33]. However, smaller differences are present at 2500 × 8 for the three injectors, than for the previously analyzed points.

Figure 11 reports the three HRR traces versus the crankshaft angle pertaining to the baseline tests at 2500 × 8 for all the injector types. Again, in this case, the IAP injector determines the lowest autoignition delay, even though the differences are smaller than in the previous cases.

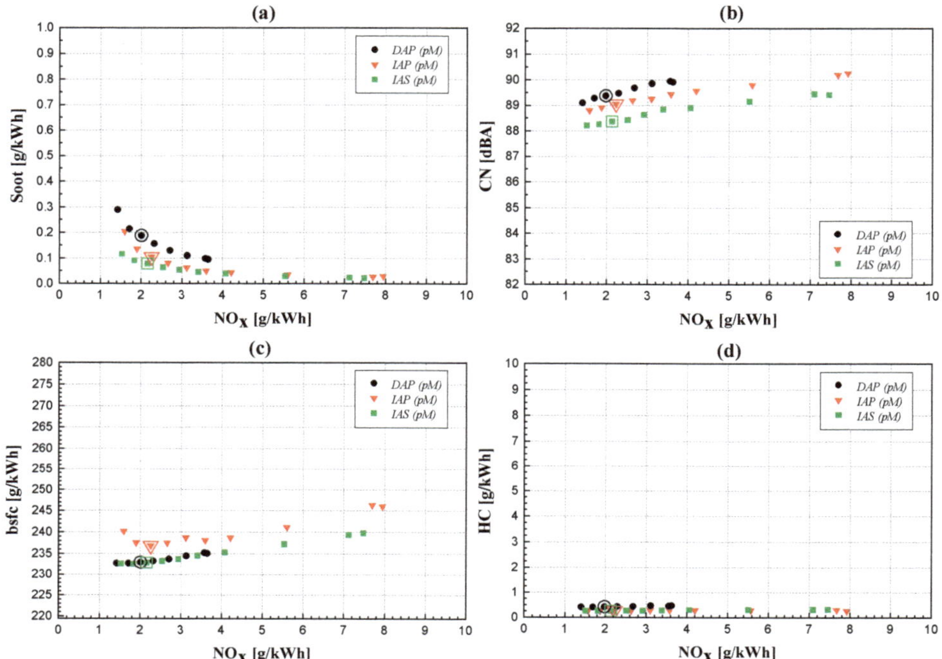

Figure 10. Engine-out soot emissions (**a**) CN; (**b**) *bsfc*; (**c**) and engine-out HC emissions; (**d**) vs. NO$_x$ for the three injector typologies, at 2500 × 8.

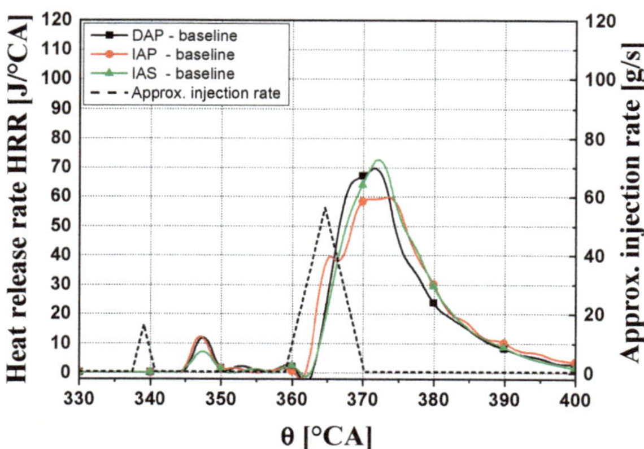

Figure 11. HRR traces of the reference calibrations, at 2500 × 8, for the three injector typologies under investigation.

5. Conclusions

The experimental investigation has concerned a comparison of the performance of indirect-acting solenoid-driven, direct, and indirect piezo-driven injectors in terms of engine-out pollutant emissions, *bsfc* and CN, under different working conditions in a low temperature combustion diesel engine for passenger cars (a pilot-Main *pM* fuel injection strategy was adopted in all of the tested calibrations).

It is possible to avoid the typical NO_x-soot trade-off of a conventional diesel combustion mode at low loads and speeds, i.e., at 1500×2, and thus to simultaneously obtain low-NO_x and low-soot PCCI-combustion for all of the three investigated injector types, even though the PCCI behavior is more pronounced for the DAP injector. In fact, the DAP injector provides penalties of about 2–3 dBA, in terms of CN, and the worst *bsfc* and HC trade-offs. In general, both the IAS injector (due to the presence of the Minirail integrated in its hydraulic layout) and DAP injector (due to its direct mechanical actuation on the needle) are able to deliver the fuel with an improved atomization compared to the IAP solution, and this leads to a more pronounced premixed combustion stage.

The engine equipped with each of the analyzed injector solutions shows a conventional diesel combustion mode at medium loads and speeds, i.e., at 2000×5 and at 2500×8. This can be inferred from the obvious presence of a NO_x-soot trade-off. The worst soot-NO_x trade-off at 2000×5 is shown for the IAP injector, due to its poorer fuel atomization and the lower premixed degree than the other two injector solutions. Instead, it is the DAP injector that gives the highest soot emissions at 2500×8, for any EGR rate applied. In fact, the needle velocity of the DAP injector tends to be slower during the nozzle opening phase at higher speed and load, as the accordingly increased rail pressure generates a higher pressure force that counteracts the needle lift.

A mean penalty in *bsfc* of about 3–4% was observed for the IAP injector, at both 2000×5 and 2500×8, whereas CN was confirmed to be higher for the DAP injector, although smaller differences between the injector types were present at 2500×8 than at the lower load points.

In general, the performance differences between the analyzed injector technologies can mainly be ascribed to the different layout solutions present in their internal hydraulic circuits (such as the bypass, the pressure-balanced pilot-valve and the Minirail), rather than to the injector driving system (solenoidal, indirect, or direct piezoelectric).

Author Contributions: The authors contributed equally to the preparation of the paper. Conceptualization, S.d., A.F.; Methodology, S.d., A.F.; Data Curation, S.d., A.F.; Writing-Original Draft Preparation, A.M. (Alessandro Mancarella); Writing-Review and Editing, S.d., A.F., A.M. (Alessandro Mancarella); Supervision, S.M., A.M. (Antonio Mittica).

Funding: This research received no external funding.

Conflicts of Interest: The authors declare no conflict of interest.

Abbreviations

bmep	brake mean effective pressure
bsfc	brake specific fuel consumption
bTDC	before Top Dead Center
CA	crank angle (deg)
CLD	Chemiluminescence Detector
CN	Combustion Noise
CR	Common Rail
DAP	Direct-Acting Piezoelectric
EGR	Exhaust Gas Recirculation
ET	Energizing Time
ET_{Pil}	Energizing Time of the pilot injection
HC	unburned hydrocarbons
HFID	Heated Flame Ionization Detector
HRR	Heat Release Rate

HRR_{pk1} Peak of the Heat Release rate for pilot combustion
HRR_{pk2} Peak of the Heat Release rate for main combustion
IAP   Indirect-Acting Piezoelectric
IAS   Indirect-Acting Solenoid
ID   Ignition Delay
n Rotational speed of the engine
NDIR   Non-Dispersive Infrared detector
NO_x nitrogen oxides
OEM   Original Equipment Manufacturer
p   pressure [bar]
PCCI   Premixed Charge Compression Ignition
pM pilot + main double fuel injection strategy
P_{rail} fuel rail pressure
q   fuel injected volume per stroke and per cylinder
q_{Pil} fuel injected volume of the pilot injection
SOI   electric Start of Injection
SOI_{Main} electric Start of Injection of the main injection
SOI_{Pil} electric Start of Injection of the pilot injection
TDC   Top Dead Center
X_{EGR} EGR rate
Greek symbols
θ Rotational angle of the crankshaft

References

1. Finesso, R.; Hardy, G.; Mancarella, A.; Marello, O.; Mittica, A.; Spessa, E. Real-time simulation of torque and nitrogen oxide emissions in an 11.0 L heavy-duty diesel engine for model-based combustion control. *Energies* **2019**, *12*, 460. [CrossRef]
2. Catania, A.; Ferrari, A.; Mittica, A.; Spessa, E. Common rail without accumulator: Development, theoretical-experimental analysis and performance enhancement at DI-HCCI level of a new generation FIS. In *SAE Technical Paper 2007-01-1258*; SAE International: Warrendale, PA, USA, 2007. [CrossRef]
3. Catania, A.; Ferrari, A.; Mittica, A. High-pressure rotary pump performance in multi-jet common rail systems. In Proceedings of 8th Biennial ASME Conference on Engineering Systems Design and Analysis, Torino, Italy, 4–7 July 2006. [CrossRef]
4. Ferrari, A.; Mittica, A.; Pizzo, P.; Jin, Z. PID Controller Modelling and Optimization in CR Systems with Standard and Reduced Accumulators. *Int. J. Automot. Technol.* **2018**, *19*, 771–781. [CrossRef]
5. Nikzadfar, K.; Shamekhi, A.H. More than one decade with development of common-rail diesel engine management systems: A literature review on modelling, control, estimation and calibration. *Proc. Inst. Mech. Eng. Part D J. Automob. Eng.* **2015**, *229*, 1110–1142. [CrossRef]
6. Musculus, M.P.B.; Pickett, L.M. 17—In-cylinder spray, mixing, combustion, and pollutant-formation processes in conventional and low-temperature-combustion diesel engines. In *Advanced Direct Injection Combustion Engine Technologies and Development*; Woodhead Publishing: Cambridge, UK, 2010; Volume 2, pp. 644–675. ISBN 9781845697440. [CrossRef]
7. Leach, F.; Ismail, R.; Davy, M. Engine-out emissions from a modern high speed diesel engine—The importance of Nozzle Tip Protrusion. *Appl. Energy* **2018**, *226*, 340–352. [CrossRef]
8. Ferrari, A.; Mittica, A. Response of different injector typologies to dwell time variations and a hydraulic analysis of closely-coupled and continuous rate shaping injection schedules. *Appl. Energy* **2016**, *169*, 899–911. [CrossRef]
9. Ferrari, A.; Mittica, A.; Pizzo, P.; Wu, X.; Zhou, H. New methodology for the identification of the leakage paths and guidelines for the design of common rail injectors with reduced leakage. *J. Eng. Gas Turbines Power* **2018**, *140*, 022801. [CrossRef]
10. Ferrari, A.; Mittica, A.; Paolicelli, F.; Pizzo, P. Hydraulic Characterization of Solenoid-actuated Injectors for Diesel Engine Common Rail Systems. *Energy Procedia* **2016**, *101*, 878–885. [CrossRef]

11. Huber, B.; Ulbrich, H. Modeling and Experimental Validation of the Solenoid Valve of a Common Rail Diesel Injector. SAE Technical Paper 2014-01-0195. 2014. Available online: https://www.sae.org/publications/technical-papers/content/2014-01-0195/ (accessed on 23 October 2019). [CrossRef]

12. Ferrari, A.; Mittica, A. FEM modeling of the piezoelectric driving system in the design of direct-acting diesel injectors. *Appl. Energy* **2012**, *99*, 471–483. [CrossRef]

13. Fettes, C.; Leipertz, A. Potential of a Piezo-Driven Passenger Car Common-Rail System to Meet Future Emission Legislations—An Evaluation by Means of in-Cylinder Analysis of Injection and Combustion. SAE Technical Paper 2001-01-3499. 2001. Available online: https://www.sae.org/publications/technical-papers/content/2001-01-3499/ (accessed on 23 October 2019). [CrossRef]

14. Magno, A.; Mancaruso, E.; Sequino, L.; Vaglieco, M.B. Analysis of spray evolution from both piezo and solenoid injectors in single cylinder research engine. In Proceedings of the 25th European Conference on Liquid Atomization and Spray Systems ILASS-2013, Chania, Greece, 1–4 September 2013.

15. Koyanagi, K.; Oing, H.; Renner, G.; Maly, R. Optimizing Common Rail Injection by Optical Diagnostics in a Transparent Production Type Diesel Engine. SAE Technical Paper 1999-01-3646. 1999. Available online: https://www.sae.org/publications/technical-papers/content/1999-01-3646/ (accessed on 23 October 2019). [CrossRef]

16. Ueda, D.; Tanada, H.; Utsunomiya, A.; Kawamura, J.; Weber, J. 4th Generation Diesel Piezo Injector (Realizing Enhanced High Response Injector. SAE Technical Paper 2016-01-0846. 2016. Available online: https://www.sae.org/publications/technical-papers/content/2016-01-0846/ (accessed on 23 October 2019). [CrossRef]

17. Kim, J.; Kim, J.; Jeong, S.; Han, S.; Lee, J. Effects of different piezo-acting mechanism on two-stage fuel injection and CI combustion in a CRDi engine. *J. Mech. Sci. Technol.* **2016**, *30*, 5727–5737. [CrossRef]

18. Suh, K.H.; Lee, C.S. Experimental and Numerical Analysis of Diesel Fuel Atomization Characteristics of a Piezo Injection System. *Oil Gas Sci. Technol. Rev. IFP* **2008**, *63*, 239–250. [CrossRef]

19. Ferrari, A.; Mittica, A.; Spessa, E. Benefits of hydraulic layout over driving system in piezo-injectors and proposal of a new-concept CR injector with an integrated Minirail. *Appl. Energy* **2012**, *103*, 243–255. [CrossRef]

20. Jorach, R.W.; Bercher, P.; Meissonnier, G.; Milanovic, N. Common rail system from Delphi with solenoid valves and single plunger pump. *Auto Tech Rev.* **2012**, *1*, 38–43. [CrossRef]

21. Matsumoto, S.; Date, K.; Taguchi, T.; Herrmann, O.E. The new Denso common rail diesel solenoid injector. *Auto Tech Rev.* **2013**, *2*, 24–29. [CrossRef]

22. D'Ambrosio, S.; Ferrari, A. Diesel engines equipped with piezoelectric and solenoid injectors: Hydraulic performance of the injectors and comparison of the emissions, noise and fuel consumption. *Appl. Energy* **2018**, *211*, 1324–1342. [CrossRef]

23. D'Ambrosio, S.; Ferrari, A. Direct Versus Indirect Acting Piezoelectric CR Injectors: Comparison of Hydraulic Performance, Pollutant Emissions, Combustion Noise, and Fuel Consumption. *SAE Int. J. Engines* **2018**, *11*, 585–612. [CrossRef]

24. Delphi Automotive. *Direct Acting Light-Duty Diesel CR System*; Technical Report; Delphi Automotive: Dublin, Ireland, 2008.

25. D'Ambrosio, S.; Ferrari, A. Effects of exhaust gas recirculation in diesel engines featuring late PCCI type combustion strategies. *Energy Convers. Manag.* **2015**, *105*, 1269–1280. [CrossRef]

26. D'Ambrosio, S.; Finesso, R.; Spessa, E. Calculation of mass emissions, oxygen mass fraction and thermal capacity of the inducted charge in SI and diesel engines from exhaust and intake gas analysis. *Fuel* **2011**, *90*, 152–166. [CrossRef]

27. D'Ambrosio, S.; Finesso, R.; Lezhong, F.; Mittica, A.; Spessa, E. A control-oriented real-time semi-empirical model for the prediction of NOx emissions in diesel engines. *Appl. Energy* **2014**, *130*, 265–279. [CrossRef]

28. D'Ambrosio, S.; Ferrari, A. Potential of double pilot injection strategies optimized with the design of experiments procedure to improve diesel engine emissions and performance. *Appl. Energy* **2015**, *155*, 918–932. [CrossRef]

29. Curran, H.J.; Gaffuri, P.; Pitz, W.J.; Westbrook, C.K. A Comprehensive Modeling Study of n-Heptane Oxidation. *Combust. Flame* **1998**, *114*, 149–177. [CrossRef]

30. Goldsborough, S.S. A chemical kinetically based ignition delay correlation for iso-octane covering a wide range of conditions including the NTC region. *Combust. Flame* **2009**, *156*, 1248–1262. [CrossRef]

31. Kiplimo, R.; Tomita, E.; Kawahara, N.; Yokobe, S. Effects of spray impingement, injection parameters, and EGR on the combustion and emission characteristics of a PCCI diesel engine. *Appl. Therm. Eng.* **2011**, *37*, 165–175. [CrossRef]
32. Selim, M. Effect of exhaust gas recirculation on some combustion characteristics of dual fuel engine. *Energy Convers. Manag.* **2003**, *44*, 707–721. [CrossRef]
33. D'Ambrosio, S.; Ferrari, A.; Iemmolo, D.; Mittica, A. Dependence of combustion noise on engine calibration parameters by means of the response surface methodology in passenger car diesel engines. *Appl. Therm. Eng.* **2019**, *163*, 114209. [CrossRef]

Article

Combustion Analysis of Homogeneous Charge Compression Ignition in a Motorcycle Engine Using a Dual-Fuel with Exhaust Gas Recirculation

Yuh-Yih Wu [1,*] and Ching-Tzan Jang [1,2]

[1] Department of Vehicle Engineering, National Taipei University of Technology, Taipei 10608, Taiwan; edward.ct.jang@gmail.com
[2] Giant Lion Know-How Co., Ltd., Taipei 10692, Taiwan
* Correspondence: cyywu@ntut.edu.tw; Tel.: +886-2-2771-2171 (ext. 3620)

Received: 25 January 2019; Accepted: 25 February 2019; Published: 4 March 2019

Abstract: Exhaust emissions from the large population of motorcycles are a major issue in Asian countries. The regulation of exhaust emissions is therefore becoming increasingly stringent, with those relating to nitrogen oxides (NO_x) the most difficult to pass. The homogeneous charge compression ignition (HCCI) has special combustion characteristics and hence produces low NO_x emissions and exhibits high thermal efficiency. This study developed an HCCI system for a 150 cc motorcycle engine. The target engine was modified using a dual-fuel of dimethyl ether (DME) and gasoline with exhaust gas recirculation (EGR). It was tested at 2000–4000 rpm and the analysis was focused on 2000 rpm. The DME was supplied continuously at an injection pressure of $1.5 \, \text{kg/cm}^2$. The gasoline injection rate was adjusted at a pressure of $2.5 \, \text{kg/cm}^2$. A brake-specific fuel consumption of <250 g/kW·h was achieved under a condition of air–fuel equivalence ratio (λ) < 2 and an EGR of 25%. The nitric oxide concentration was too low to measure. The brake mean effective pressure (BMEP) increased by 65.8% from 2.93 to 4.86 bar when the EGR was 0% to 25%. The combustion efficiency was close to 100% when the BMEP was >3 bar.

Keywords: homogeneous charge compression ignition (HCCI); exhaust gas recirculation (EGR); dual-fuel; dimethyl ether (DME); exhaust emission

---

1. Introduction

The number of motorcycles in Asia is extremely large. This is because motorcycles are low in cost and small in size, allowing freedom of movement in crowded areas, easy parking, and high mobility. For example, the density of motorcycles in Taiwan in 2018 was $384/\text{km}^2$ [1], the highest in the world. Motorcycles contribute to air pollution more than other vehicles [2,3]. They not only damage human health but also contribute to global warming. Consequently, regulations on their exhaust emissions and fuel consumption are becoming more stringent.

For instance, the Economic Commission of Europe (ECE) established long-term stage-by-stage emission regulations for motorcycles from 1999, as shown in Figure 1. The Taiwan Environmental Protection Administration (EPA) implemented motorcycle emission regulations from 1988, in a stage-by-stage approach. Taiwan's regulations were harmonized with ECE regulations from their fifth stage, which is similar to EURO 3 (i.e., the third stage of the ECE). The double-arrow red lines in Figure 1 link the corresponding stages between Taiwan and ECE. The implementation of Taiwanese regulations occurred approximately 1 year later than ECE regulations.

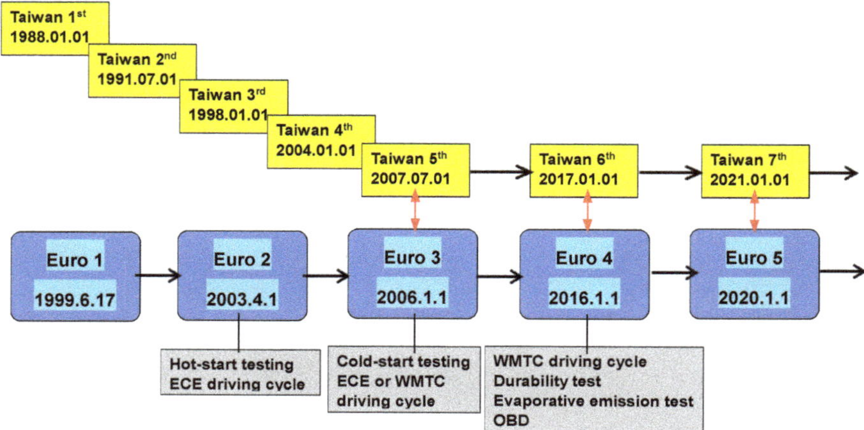

Figure 1. Evolution of motorcycle emission regulations by the ECE and Taiwan (WMTC: worldwide motorcycle test cycle; OBD: on-board diagnostics).

Table 1 shows the current and next-stage emission standards for motorcycles in Taiwan [4]. This encompasses carbon monoxide (CO), hydrocarbons (HC), non-methane hydrocarbons (NMHC), nitrogen oxides (NO_x), and particulate matter (PM). The emission standards of the sixth and seventh stages in Table 1 are the same as those in EURO 4 and 5, respectively. Numerous Asian countries follow the ECE regulations. To comply with emission standards, all the motorcycles produced in Taiwan now use an electronic fuel injection system. The NO_x standard is the most difficult to pass. In-use motorcycles emission controlled by Taiwan's EPA have even been recalled because of high NO_x emissions. Although lean-burn can improve fuel economy, motorcycle manufacturers do not use this approach. This is because NO_x emissions are difficult to reduce in a lean-burn system. The seventh-stage standard will be implemented from 1 January 2021. This will be extremely strict, which means that new technologies must be developed.

Table 1. Emission standards of motorcycles in Taiwan.

Stage (Implemented Date)	Maxi. Speed (km/h)	CO (mg/km)	HC (mg/km)	NMHC (mg/km)	NO_x (mg/km)	PM [1] (mg/km)
6th stage (1 Jan. 2017)	<130	1140	380	-	70	-
	≥130	1140	170	-	90	-
7th stage (1 Jan. 2021)	-	1000	100	68	60	4.5

[1] Particulate matter (PM) is for gasoline direct injection engines only.

In a spark ignition (SI) engine, a three-way catalytic converter efficiently reduces HC, CO, and NO_x, although the mixture must be maintained within a narrow air-fuel-ratio window. Other technologies for reducing NO_x include exhaust gas recirculation (EGR) [5], selective catalytic reduction (SCR) [6], a NO_x trap [7], a plasma reactor [8], water injection [9], water/fuel emulsions [10], and homogeneous charge compression ignition (HCCI) [11–13]. SCR and lean NO_x-trap have been applied in diesel vehicles to meet standard emission limits [14]. However, these systems increase production costs and cannot be used in motorcycles. An HCCI engine has special characteristics, as it constitutes a combination of an SI engine and a diesel engine. The homogeneous charge of fuel–air mixture is similar to that in an SI engine, whereas the compression ignition is similar to that in a diesel engine. Such combustion characteristics produce very low amounts of NO_x emissions and exhibit high thermal efficiency.

HCCI is an advanced low-temperature combustion concept that has attracted global attention in recent years [12]. However, the operational range of HCCI combustion is restricted due to the absence of direct control of the ignition timing and heat release rate [15]. In short, the key issues affecting HCCI combustion are thermal effects and chemical kinetics.

The factors that cause thermal effects include negative valve overlap (NVO), intake heating, glow plug, spark-assisted combustion, and compression ratio. NVO engines retain hot residual gas in the cylinder to heat the intake charge, extending the engine load range [16]. The intake heating improves autoignition, advances the combustion phase, and shortens combustion duration [15]. Spark-assisted HCCI reduces combustion noise under high-load conditions [17,18] and extends the operating load up to 750 kPa indicated mean effective pressure (IMEP) [18]. The glow plug can be used to control the heat release rate, and the temperature distribution is broadened to heat the charge unevenly [19]. A suitable compression ratio is necessary, which is from 12 to 18 according to most studies on HCCI engines.

The factors that cause the chemical kinetic effect include fuel composition, fuel properties, dual-fuel, external EGR, EGR stratification, and fuel stratification. One of the vital properties of fuel used for HCCI is its autoignition temperature. Fuel with a low autoignition temperature is termed diesel-like fuel and fuel with a high autoignition temperature is called gasoline-like fuel. To ensure suitable ignition timing and combustion phasing, an additive fuel is used. Diesel-like fuel can be used as an additive to enhance autoignition in a gasoline-like fuel system. Gasoline-like fuel can also be used as an additive to inhibit autoignition in a diesel-like fuel system.

Ji et al. [20] used a diesel-like fuel of 2-ethylhexyl nitrate to enhance the autoignition of an E10 gasoline HCCI engine. A maximum indicated thermal efficiency of 50.1% was found at 1800 rpm and an intake pressure of 180 kPa. Wang et al. [21] developed a DME-diesel blend fuel system using the gasoline-like fuel of liquefied petroleum gas as an ignition inhibitor. Pedersen et al. [22] used the port fuel injection (PFI) of methanol in the direct injection (DI) of a DME engine. The added methanol increased the BMEP and slowed the combustion so that it was after top dead center (aTDC). Mohebbi et al. [14] applied a PFI of ethanol and diethyl-ether blend fuel in a DI diesel engine: this resulted in a 14% increase in IMEP and a 33% reduction in the maximum rate of pressure rise (MRPR). Li et al. [23] investigated a blend of n-butanol and n-heptane in HCCI combustion and found that the knock tendency decreased as the n-butanol volume fraction increased. Khandal et al. [24] used a PFI of hydrogen in the DI of a biodiesel engine, which resulted in 65%–67% less smoke and 98%–99% lower NO_x emissions. Finally, Zheng et al. [25] developed a dual-fuel system using the DI of biodiesel and PFI of several gasoline-like fuels. Among the gasoline-like fuels, the biodiesel-ethanol produced low NO_x and soot emissions (soot < 0.3 FSN and NO_x < 1.5 g/kW·h).

Regarding the mechanism of the effect of EGR on combustion, the addition of EGR slows the decomposition of hydrogen peroxide (H_2O_2) by reducing the rate of hydroxyl radicals (OH) [26]. This results in the suppression of autoignition and delay of combustion phasing, thus allowing high load operation. Researchers [27,28] have reported that in-cylinder EGR stratification reduced the MRPR of HCCI engines. Other researchers [13,29] have reported that the effects of external EGR on HCCI engines were a delay in combustion phasing, decrease in maximum temperature and maximum HRR (heat release rate), prolonged high-temperature heat release, and decreased MRPR. Lee et al. [30] investigated the optimization of EGR (0–25%) and two-stage injection with compression ratios of 15 and 17.8.

Superior HCCI operation can be achieved using a combination of several strategies, such as the stratification of external EGR, fuel stratification using PFI and DI injectors including asymmetric injection, open valve injection using a PFI injector, and NVO injection [31]. Researchers [16,32] have developed an HCCI engine with intake boosting, NVO, and external EGR. The engine load was extended to an IMEP of 8 bar. This type of system can achieve a thermal efficiency of 47%, NO_x of $\leq$0.1 g/kW·h, and combustion efficiency of $\geq$96.5% [16].

The best elements of previous research have been developed for automobile engines and cannot be used for motorcycles. Most motorcycle engines are small-scale, with a displacement $\leq$150 cc, while the surface-volume ratio is large compared to automobile engines and hence causes high heat loss [33]. This condition makes HCCI operation in motorcycles difficult. Additionally, the combustion characteristics of a small-scale air-cooled HCCI engine are insufficient. Therefore, a combination of DME-gasoline dual-fuel and external EGR with a suitable compression ratio for HCCI operation was developed. This is useful for low-cost motorcycle engines with only minor modifications. The purpose of this study was to therefore investigate the combustion characteristics of this HCCI system.

2. Experimental Methodology

2.1. Experimental Setup

A commercial motorcycle engine (SYM, Taipei, Taiwan) with an electronic fuel injection system was retrofitted for HCCI operation. The engine was a 150 cc air-cooled SI engine. Detailed engine specifications are listed in Table 2. The compression ratio was increased from 10.5 to 12.4 by replacing the cylinder head with a smaller clearance volume. The increased compression ratio raises the temperature of the compressed mixture, which easily achieves compression ignition.

Table 2. Engine specifications.

Items	Specifications	Units
Engine type	4-stroke, 1-cylinder, SI	-
Valve system	4-valve, overhead cam	-
Cooling system	Forced air cooling	-
Displacement	150	cc
Bore × stroke	57.4 × 57.8	mm
Compression ratio	10.5 changed to 12.4	-
Fuel system	Electronic port fuel injection	-
Intake valve open [1]	10° bTDC	CA
Intake valve close [1]	20° aBDC	CA
Exhaust valve open [1]	30° bBDC	CA
Exhaust valve close [1]	10° aTDC	CA

[1] Valve timing is defined at 1 mm of valve lift. aTDC: after top dead center; bTDC: before top dead center; aBDC: after bottom dead center; bBDC: before bottom dead center.

The fuel for HCCI operation in an SI engine should have a low autoignition temperature. Previous research [34] has concluded that DME, the properties of which are listed in Table 3, is a favorable choice for this. The cetane number of DME is 60, higher than that of diesel fuel (i.e., 40–60). It can autoignite in the target engine. Therefore, DME was selected as the main fuel. Increasing the fuel amount of DME to increase engine load will cause high MRPR or engine knocking.

Table 3. Fuel properties.

Properties	DME	Gasoline
Chemical structure	C_2H_6O	-
Lower heating value (MJ/kg)	28.9	44.0
Octane number	35	92
Cetane number	60	5–12
Autoignition temperature [1] (K)	508	553–729
Stoichiometric air-fuel ratio	9.0	14.7
Viscosity at 20 °C (cP)	0.224	0.74

[1] Autoignition temperatures were obtained from Material Safety Data Sheets.

Gasoline with a research octane number (RON) 92 and external EGR were added to extend the operating range of the engine. The autoignition temperature of gasoline is much higher than that of DME, as shown in Table 3. The addition of gasoline can therefore increase the engine load without knocking.

The experimental setup for the proposed HCCI engine is shown in Figure 2. The external EGR system was built on the target engine with a small EGR pump and a control valve, which adjusted the EGR ratio. A surge tank was installed between the air flow meter and the throttle; it was used for attenuating the pressure pulsation in the intake system of the engine. So the intake air flow rate can be measured stably. The volume of the surge tank is 40 L, which is larger than the minimum requirement calculated according to SAE J244. A dual-fuel supplying system was then built into the target engine (Figure 3). The original fuel and ignition systems of the target engine were retained to start the engine. The spark plug was used only for starting the engine and the spark timing was the same as the original engine. The original gasoline injector was also used for the addition of gasoline in the HCCI engine. The additional DME supplying system, which includes a DME tank, pressure regulator, filter, and a flow meter, was attached to the target engine with a DME supply tube installed near the intake port.

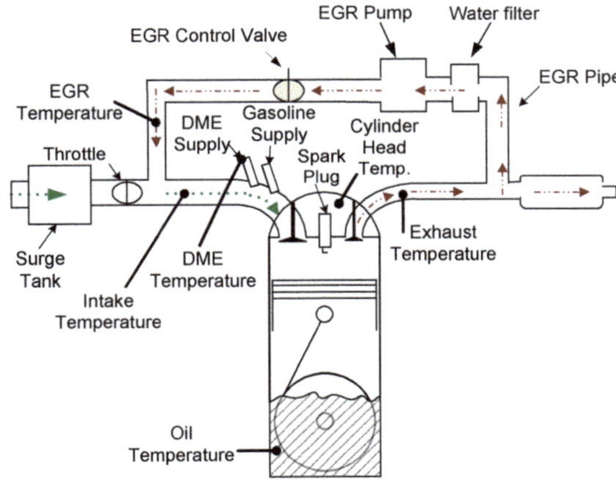

Figure 2. Experimental setup for the proposed dual-fuel HCCI engine with external EGR.

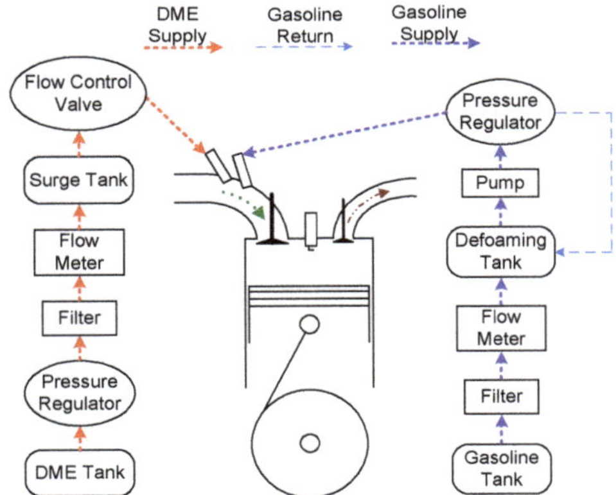

Figure 3. Experimental setup for the fuel supply system.

For the engine test, an eddy-current engine dynamometer (FE150-S, Borghi & Saveri S.R.L., Bologna, Italy) was used to measure the engine speed and brake torque. The gasoline flow rate was measured using a mass burette flow detector (FX-1110, Ono Sokki, Yokohama, Japan). The DME flow rate was measured using a thermal mass flow controller (NM-2100, Tokyo Keiso, Tokyo, Japan). The exhaust emissions of CO, HC, NO, carbon dioxide (CO_2), oxygen, and the lambda were measured using an emission analyzer (MEXA-584L, Horiba, Kyoto, Japan). The lambda is calculated by the carbon balance method in standard configuration using the MEXA-584L emission analyzer. The hydrogen/carbon ratio and oxygen/carbon ratio of the fuel must be input to this analyzer. Additionally, another emission analyzer (MEXA-584L, Horiba) was installed in the intake system to measure the CO_2 concentration and thus calculate the EGR ratio. K-type thermocouples were installed on the engine to measure the temperature of the intake gas, exhaust gas, cylinder head, and lubricating oil, as shown in Figure 2.

A piezoelectric pressure transducer (Kistler 6051B, Winterthur, Switzerland) coupled to a charge amplifier (Kistler 5018A) was used to record in-cylinder gas pressure. A shaft encoder (BEI H25, Thousand Oaks, CA, USA) was used to detect the crank angle (CA). The pressure signal was transmitted to a data acquisition system (IndiCom 619, AVL, Graz, Austria) for every 1 °CA of 100 continuous cycles. The number of cylinder pressure data of 100 cycles is good enough for the combustion analysis. The pressure data were used to analyze engine combustion parameters, such as the in-cylinder gas temperature, coefficient of variation (COV), and HRR.

The engine control was replaced by a controller (MotoHawk ECU 555-80, Woodward, CO, USA) which controlled the spark timing (for SI engine starting), gasoline injector, DME flow control valve, and EGR control valve.

Temperature is an important factor in HCCI engines [35]. Reference [36] has shown that, for stable HCCI operation in a small engine the cylinder head and oil temperatures should be maintained above 120 °C and 65 °C, respectively. The intake charge temperature of all tests in this research ranges from 21.2 °C to 23.2 °C, its mean value is 22.4 °C, and the standard deviation is 0.6 °C. The range of intake temperature is not large, so the influence of intake temperature on combustion can be neglected.

The engine was started using the original ignition and fuel systems. When the cylinder head and oil temperatures reached 120 °C and 65 °C, respectively, the controller switched the engine operation to HCCI mode by interrupting the ignition system. Simultaneously, the dual-fuel of DME and gasoline was injected and the throttle was opened further. The throttle was used only for SI operation of the original engine. For HCCI operation, it was fully opened. The fuel supply was adjusted for stable HCCI operation; the ratio of these two fuels was therefore not fixed. The engine was tested at 2000–4000 rpm and with a wide-open throttle. Most of the analyses were focused on 2000 rpm because its operating load range was large. The DME fuel was supplied continuously; its injection pressure was adjusted before engine test for stable combustion and it was found to be 1.5 kg/cm^2. The gasoline was injected at a pressure of 2.5 kg/cm^2, which was the same as the original engine. The gasoline fuel injection pulse width was adjusted to set the fuel flow rate. The EGR ratio was adjusted from 0% to 30%.

The engine speed, brake torque, DME flow rate, gasoline flow rate, intake air mass flow rate, exhaust emissions, and EGR ratio were measured and recorded during the engine tests. The EGR ratio was calculated using the CO_2 percentages measured in the intake and exhaust systems, as formulated in Equation (1) [37]:

$$EGR = \frac{[CO_2]_{in}}{[CO_2]_{exh}} \times 100\%,\tag{1}$$

where $[CO_2]$ is the CO_2 concentration.

The maximum error of the experimental data was calculated with an engine test at 2000 rpm and repeated five times using the Kline and McClintock method [38]. The results are listed in Table 4.

Table 4. Maximum error in the experimental results of the engine test.

Item	Maximum Error ($\pm$ %)
Engine Speed	0.68
Engine Torque	1.75
IMEP	1.92
Fuel Rate	5.57

2.2. Calculation of Combustion Parameters

Combustion parameters, such as IMEP, COV, combustion efficiency, cylinder gas temperature, HRR, and MFB, are used to study the combustion characteristics of an engine.

IMEP is defined as the work divided by the engine displacement volume, which can be calculated during compression and expansion strokes without pumping work. Because this study focused on combustion effect, the elimination of intake and exhaust strokes is better for analyzing. This is expressed as:

$$IMEP = \frac{\int_{-180CAD}^{180CAD} P\, dV}{V_d},$$ (2)

where CAD denotes CA degrees.

The COV is expressed as:

$$COV = \frac{IMEP_{std}}{IMEP_{avg}},$$ (3)

The combustion efficiency η_c can be defined as the fraction of carbon emitted as CO_2 in relation to the total carbon emitted (CO_2, CO, HC and PM). The modified combustion efficiency is defined as the ratio between CO_2 and CO_2 plus CO. This study used the modified combustion efficiency for convenient calculation. This is expressed as:

$$\eta_c = \frac{[CO_2]}{[CO_2] + [CO]},$$ (4)

where square bracket [ ] represents exhaust species concentration as a percentage.

The in-cylinder gas temperature is obtained using the state equation of ideal gas. The HRR equation can be derived from the first law of thermodynamics and is thus expressed as:

$$\frac{dQ_{hr}}{d\theta} = \frac{\gamma}{\gamma - 1} p \frac{dV}{d\theta} + \frac{1}{\gamma - 1} V \frac{dp}{d\theta} + \frac{dQ_{ht}}{d\theta},$$ (5)

where $dQ_{hr}/d\theta$ is the HRR, $dQ_{ht}/d\theta$ is the heat transfer rate between cylinder gas and the wall. The specific heat ratio γ is a function of temperature, which can be obtained using the empirical equation presented in [39].

The heat transfer rate $dQ_{ht}/d\theta$ is expressed as:

$$\frac{dQ_{ht}}{d\theta} = hA(T_g - T_w),$$ (6)

where h is the heat transfer coefficient, as shown in Equation (7):

$$h = S_t \rho_g c_p (0.5 C_m) \text{ and}$$ (7)

$$S_t = 0.718 \exp(-0.145 C_m),$$ (8)

where S_t is the Stanton number and C_m the average piston speed. Equations (7) and (8) were proposed by [33] for small engines.

The MFB at any CA is calculated from HRR and is expressed as:

$$MFB = \frac{\int \left(\frac{dQ_{hr}}{d\theta}\right) d\theta}{m_f Q_{LHV}}.$$ (9)

3. Results and Discussion

The experimental results are presented in terms of the performances, efficiencies, exhaust emissions, and combustion characteristics of the target engine. The data in Sections 3.1–3.4 are all at 2000 rpm of engine speed.

3.1. Engine Performances

The BMEP, IMEP, and brake-specific fuel consumption (BSFC) are discussed in this section. HCCI engines are always run with a lean mixture. Generally, the output of a HCCI engine increases with a richer mixture but is limited by engine knocking or MRPR [40]. The EGR is a kind of diluent in air–fuel mixture which suppresses the combustion rate. Previous research [41] has reported that autoignition timing is delayed and burn duration prolonged by applying EGR in a HCCI engine, which suppresses engine knocking. The test results for engine output with various EGR ratios are shown in Figure 4.

The BMEP and IMEP indicate the work output per cycle divided by the engine displacement volume. As shown in Figure 4, the engine output increases when the air–fuel equivalence ratio (lambda, λ) decreases. Each curve in Figure 4 indicates the operational range of each EGR ratio. The maximum engine output on each curve increases in line with the EGR. However, the EGR of 30% seems too high for this HCCI system because the operational range is very small and the engine output cannot be extended. In this study, an EGR of 25% yields the best engine output. When the EGR was increased from 0% to 25%, the maximum BMEP increased by 65.8% from 2.93 to 4.86 bar. In Figure 4, the first and last point of each EGR line were determined by the limit of stable operation, which is COV < 10% for low load and MRPR < 6 bar/deg for high load.

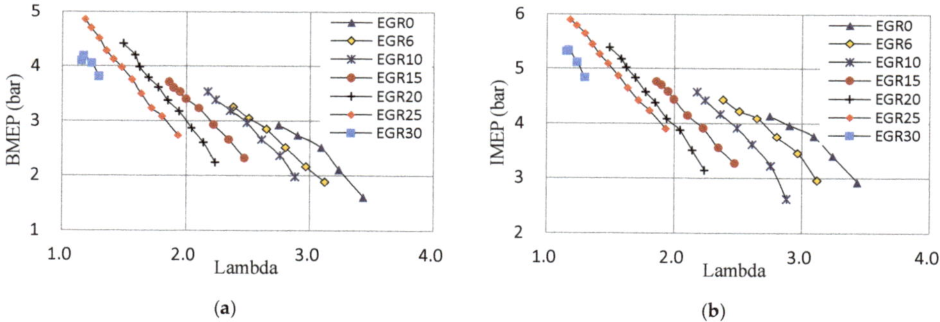

Figure 4. Effect of lambda on engine output with various EGR ratios: (**a**) BMEP; (**b**) IMEP.

Gasoline and DME were used in the dual-fuel system. DME was selected as the main fuel and was supplied at an almost constant flow rate. Increasing the gasoline mass ratio in the dual-fuel causes λ to decrease, as shown in Figure 5a. The BMEP increases in line with the amount of gasoline, as shown in Figure 5b. At an EGR of 25%, BMEP increases by 77.4% from 2.74 to 4.86 bar when the gasoline ratio rises from 0.13 to 0.35.

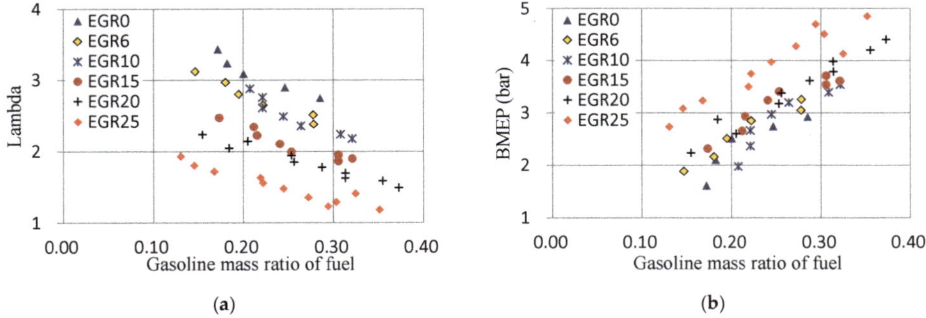

Figure 5. Effect of gasoline mass ratio in the dual-fuel on: (**a**) Lambda; (**b**) BMEP.

The fuel consumed contains gasoline and DME; therefore, an equivalent fuel mass was used. The equivalent fuel flow rate was calculated based on the heating value of the fuels, as shown in Equation (10), and the *BSFC* was calculated using Equation (11):

$$\dot{m}_e = \frac{\dot{m}_{DME} Q_{LHV\ DME} + \dot{m}_{gasoline} Q_{LHV}}{Q_{LHV}} \quad \text{and} \tag{10}$$

$$BSFC = \frac{\dot{m}_e}{bhp}, \tag{11}$$

where $\dot{m}_e$ is the fuel flow rate equivalent to gasoline, $\dot{m}_{DME}$ is DME flow rate, $\dot{m}_{gasoline}$ is gasoline flow rate, $Q_{LHV\ DME}$ is the low heating value of DME, Q_{LHV} is the low heating value of gasoline, and bhp is the brake horsepower.

Figure 6 shows the BSFC of all test points. Most of the operation conditions have a BSFC ranging from 230 to 260 g/kW·h. The BSFC of a HCCI engine is much lower than that of a conventional motorcycle engine, which is approximately 350 g/kW·h [42]. When λ is >2 or BMEP is <3 bar, the BSFC clearly increases. This might be caused by the low engine output and low combustion efficiency. A BSFC of ≤250 g/kW·h was achieved under conditions of λ < 2 and an EGR of 25%.

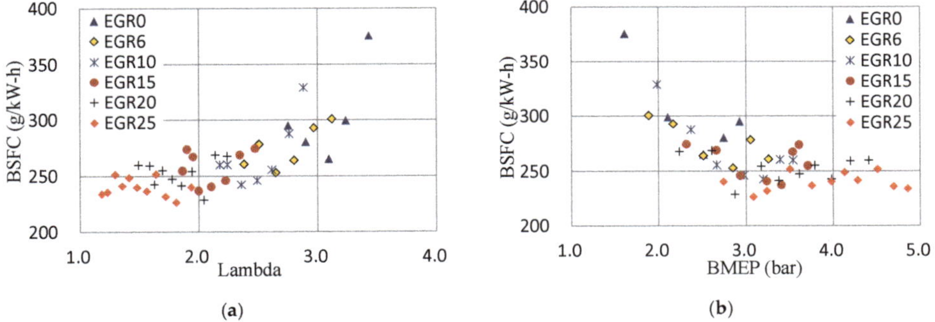

Figure 6. Brake specific fuel consumption with respect to: (**a**) Lambda; (**b**) BMEP.

3.2. Engine Efficiencies

The combustion efficiency, η_c, is the ratio of the fuel mass burned and the fuel mass delivered into the engine. It can be calculated using Equation (4). The fuel efficiency η_f is the ratio of the power developed by the engine to the rate of fuel energy in the engine. It is calculated using Equation (12):

$$\eta_f = \frac{bhp}{\dot{m}_f Q_{LHV}} = \frac{3600(s/h)}{BSFC(g/kW \cdot h) \cdot Q_{LHV}(kJ/g)}, \tag{12}$$

where bhp is the brake horsepower, $\dot{m}_f$ is the fuel flow rate, and Q_{LHV} the lower heating value of fuel. The brake thermal efficiency η_{th} is calculated using Equation (13):

$$\eta_{th} = \frac{bhp}{\eta_c \dot{m}_f Q_{LHV}} = \frac{\eta_f}{\eta_c}, \tag{13}$$

Figure 7 shows that the combustion efficiency decreases with an increase in λ or decrease in engine load (BMEP) because a high λ or low BMEP leads to a leaner mixture and incomplete combustion. In Figure 7a, the effect of λ on combustion efficiency for each EGR ratio is separated each other and does not coincide together. However, the effect of BMEP on combustion efficiency is evident (Figure 7b), irrespective of the amount of EGR. The combustion efficiency was close to 100% when BMEP > 3 bar.

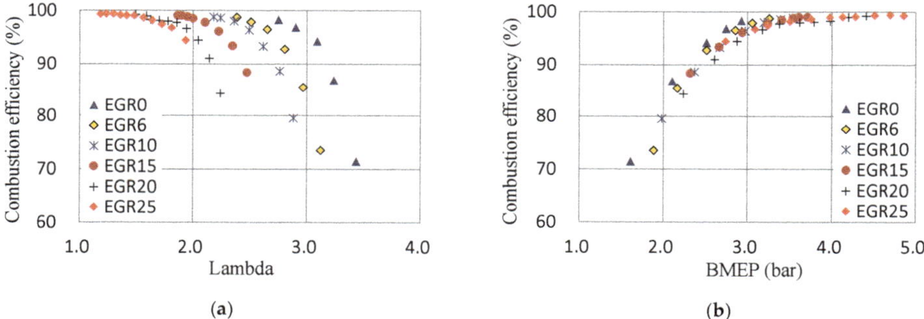

Figure 7. Combustion efficiency with respect to: (**a**) Lambda; (**b**) BMEP.

Figure 8 shows that the fuel efficiency decreases with increasing λ or decreasing engine load (BMEP) because high λ or low BMEP leads to incomplete combustion. When λ is >2 or BMEP is <3 bar, the fuel efficiency clearly drops.

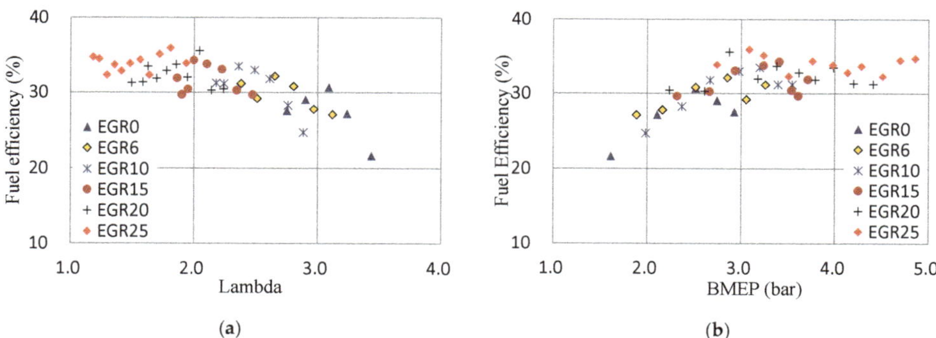

Figure 8. Fuel efficiency with respect to: (**a**) Lambda; (**b**) BMEP.

The brake thermal efficiency is shown in Figure 9. Most of the brake thermal efficiencies are clustered within the range of 30%–35%. They have no clear relationship with λ or engine load (BMEP). The thermal efficiency is not very high because the target engine is small in size and heat loss is relatively high [33]. Furthermore, the HCCI engines do not require as much cooling as a conventional engine due to the low combustion temperature of HCCI. Therefore, the modification of a cooling system to reduce cooling capacity will be better for HCCI operation and will improve thermal efficiency.

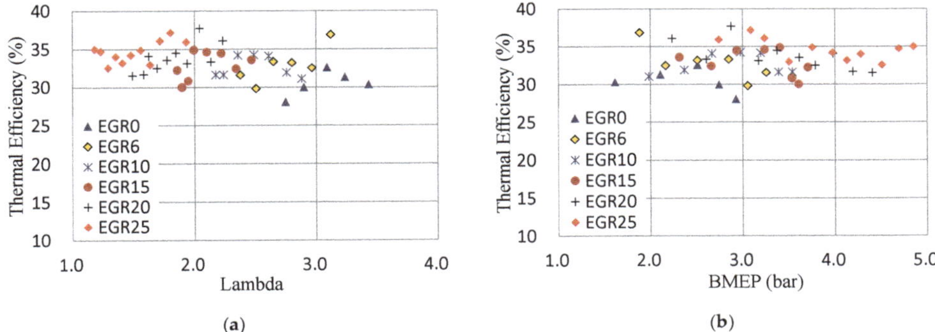

Figure 9. Brake thermal efficiency with respect to: (**a**) Lambda; (**b**) BMEP.

3.3. Exhaust Emissions

Emission data for NO is lacking because the concentration was too low to be measured during the engine test. In general, when the combustion temperature is lower than 1800 K, much less NO will be emitted [43]. The maximum cylinder temperatures of all test points in this study were less than 1600 K, which is much lower than 1800 K, so the NO emission was close to zero ppm.

The exhaust emissions of CO and HC are depicted through the brake-specific emissions as BSCO and BSHC. The BSCO increases with increasing λ or decreasing exhaust temperature, as shown in Figure 10. The leaner mixture causes higher CO emission in the HCCI engine (Figure 10a), which is the opposite of that in a conventional SI engine. The results obtained are in agreement with previous research [44,45]. This is because the combustion temperature of HCCI is lower and, again, will decrease with increasing λ. Sjöberg and Dec [44] reported that CO oxidation does not reach completion with a peak temperature below 1500 K because the OH level becomes too low.

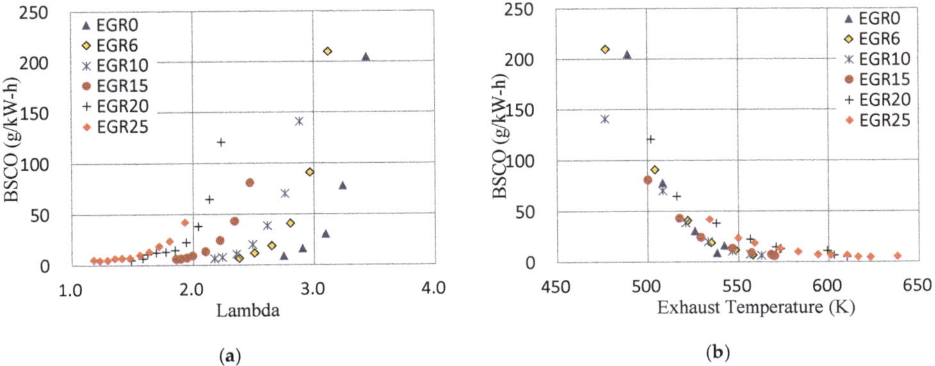

Figure 10. Brake specific CO emission with respect to: (**a**) Lambda; (**b**) Exhaust temperature.

In Figure 10a, the effect of λ on BSCO for each EGR ratio is separated each other and does not coincide together. However, the effect of exhaust temperature on BSCO is evident (Figure 10b), irrespective of the amount of EGR. The BSCO is close to 0 g/kW·h when the exhaust temperature is > 550 K.

The BSHC increases with increasing λ or decreasing exhaust temperature, as shown in Figure 11. A higher λ indicates a leaner mixture and leads to a lower combustion temperature, which causes an increase in HC emissions. Both BSCO and BSHC show good correlation with the exhaust temperature, but they do not correlate with the maximum cylinder gas temperature (the Figure for which is not included here). Thus, the high temperature of the exhaust gas results in the oxidation of CO and HC.

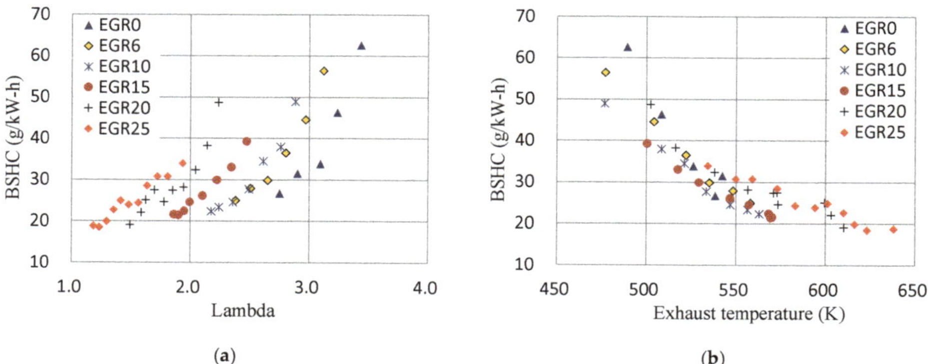

Figure 11. Brake specific HC emission with respect to: (**a**) Lambda; (**b**) Exhaust temperature.

The exhaust temperature increases in line with BMEP. The curve fitting of Figure 12a shows a good correlation between exhaust temperature and BMEP, with an R^2 of 0.9713. The oxygen concentration in the exhaust gas increases with λ as a quadratic equation, the R^2 of which is 0.9899 (Figure 12b).

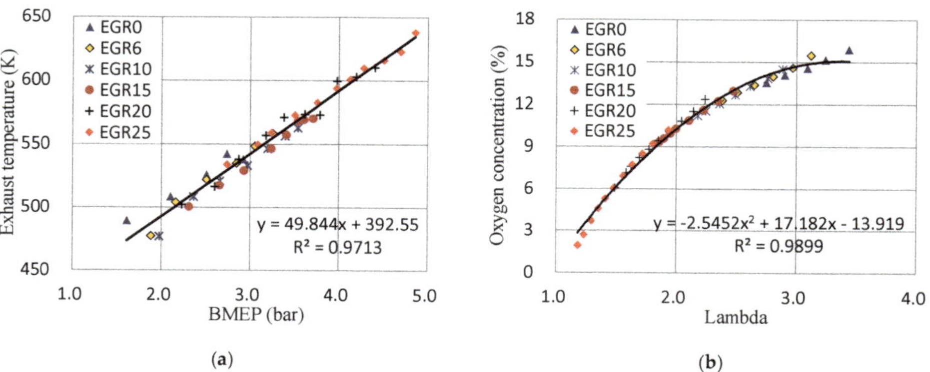

Figure 12. Curve fitting of exhaust properties: (**a**) Exhaust gas temperature versus BMEP; (**b**) Oxygen concentration in exhaust gas versus lambda.

3.4. Combustion Characteristics

The combustion characteristics were calculated from cylinder pressure values. The cylinder pressure and temperature under a condition of 25% EGR and 20% EGR with various λ values are illustrated in Figure 13. A lower λ means a large amount of gasoline is added, which delays the combustion phase and slows the onset of maximum cylinder pressure (Figure 13a,c). The cylinder temperature is very low (Figure 13b,d) compared with an SI engine because HCCI involves low-temperature combustion [12].

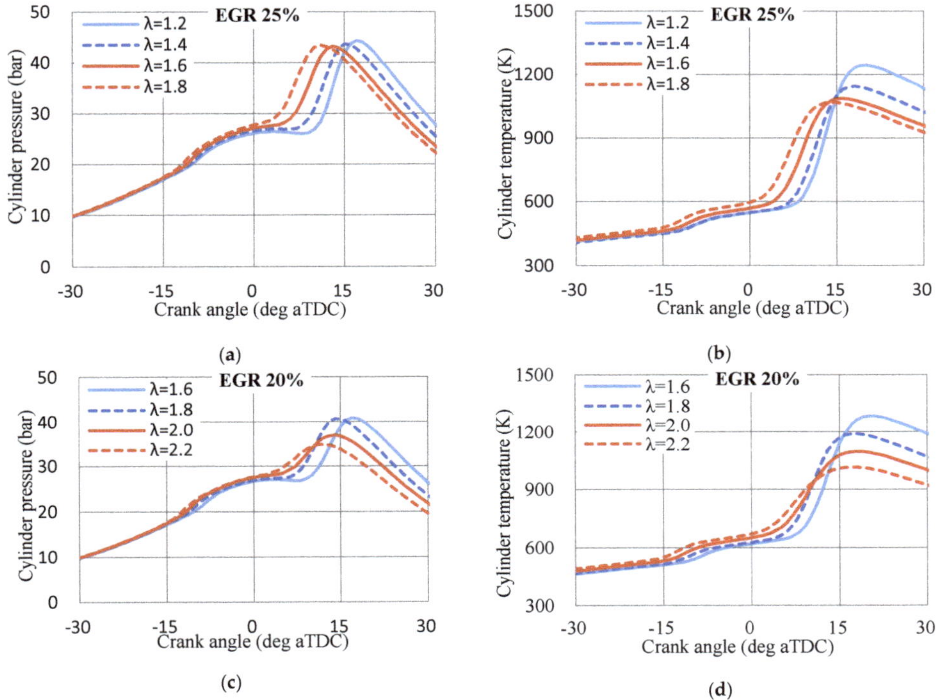

Figure 13. Cylinder pressure and temperature: (**a**) Cylinder pressure with EGR 25%; (**b**) Cylinder temperature with EGR 25%; (**c**) Cylinder pressure with EGR 20%; (**d**) Cylinder temperature with EGR 20%.

HRR and MFB under a condition of 25% EGR and 20% EGR with various λ values are illustrated in Figure 14. It is clear that the addition of more gasoline (less λ) causes higher HRR values and delays HRR and MFB. The beginning of the increase in HRR indicates the start of combustion. Figure 14a shows that the start of combustion is delayed by adding gasoline.

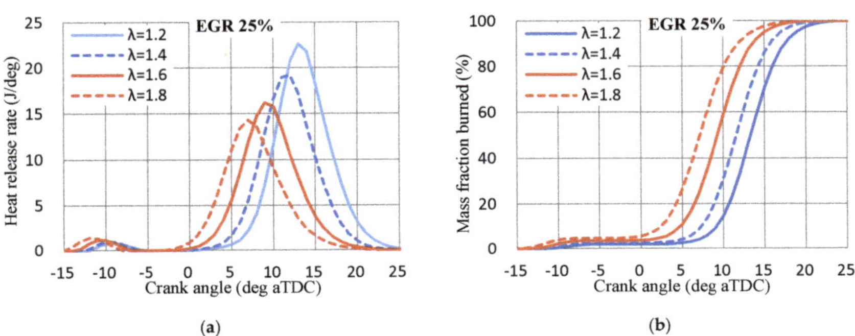

Figure 14. *Cont.*

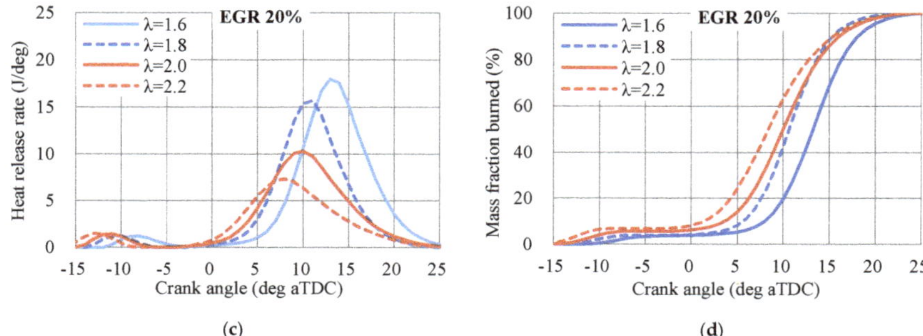

(c)

(d)

Figure 14. Heat release rate and mass fraction burned: (**a**) Heat release rate with EGR 25%; (**b**) Mass fraction burned with EGR 25%; (**c**) Heat release rate with EGR 20%; (**d**) Mass fraction burned with EGR 20%.

Figure 15 shows that COV decreases and MRPR increases with an increase in BMEP. The COV increases rapidly as BMEP is <3 bar (Figure 15a). In general, a low load causes unstable combustion and a high load causes a high rate of pressure rise or knocking [11,26]. Almost all the MRPR values are less than 6 bar/deg in Figure 15b, which is a criterion for avoiding engine damage [40]. This is because the addition of gasoline and EGR delays the ignition and suppresses the combustion reaction.

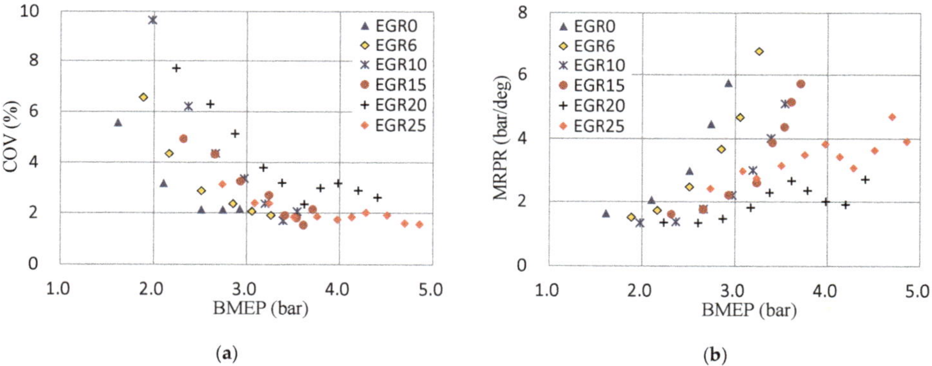

(a)

(b)

Figure 15. The coefficient of variation of IMEP (COV) and the maximum rate of pressure rise (MRPR) with respect to: (**a**) COV; (**b**) MRPR.

In Figure 16, the maximum HRR1 (first-stage HRR) increases with increasing λ, whereas maximum HRR2 (second-stage HRR) decreases with increasing λ. The first stage of combustion (Figure 16a) expresses the ignition property, whereas the second stage of combustion (Figure 16b) expresses the combustion property. The first stage of combustion is the result of cool-combustion chemistry and negative temperature coefficient behavior [46]. In the first stage of combustion, the reaction rate decreases after the temperature reaches a certain value. Therefore, the maximum HRR1 depends on λ only (Figure 16a). By contrast, the maximum HRR2 depends on both the λ value and EGR ratio (Figure 16b).

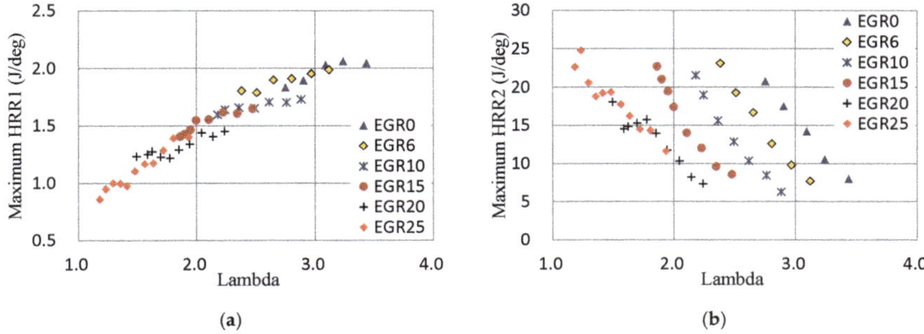

Figure 16. Effects of lambda on maximum heat release rate: (**a**) First stage combustion, HRR1; (**b**) Second stage combustion, HRR2.

CA10, CA50, and CA90 are the CA values when MFB is 10%, 50%, and 90%, respectively. These indicate the combustion phase. Both CA10 and CA50 are delayed when λ is lower, as shown in Figure 17. This is caused by the effect of dual-fuel on combustion. A richer mixture (i.e., less λ) has much more gasoline (Figure 5a), which delays combustion. CA50 is a combination of first- and second-stage combustion, whereas CA10 represents first-stage combustion only. Therefore, the correlation between CA10 and λ (Figure 17a) is clearer than that of CA50 (Figure 17b).

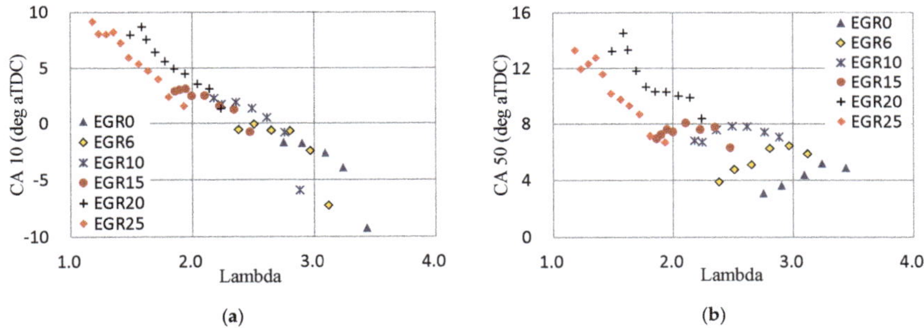

Figure 17. Effects of lambda on combustion phasing: (**a**) CA10; (**b**) CA50.

Figure 18a shows that the burn duration (period between CA10 and CA90) increases with an increase in λ, because a leaner mixture causes slower combustion. Figure 18b shows that the burn duration increases rapidly when BMEP is <3 bar.

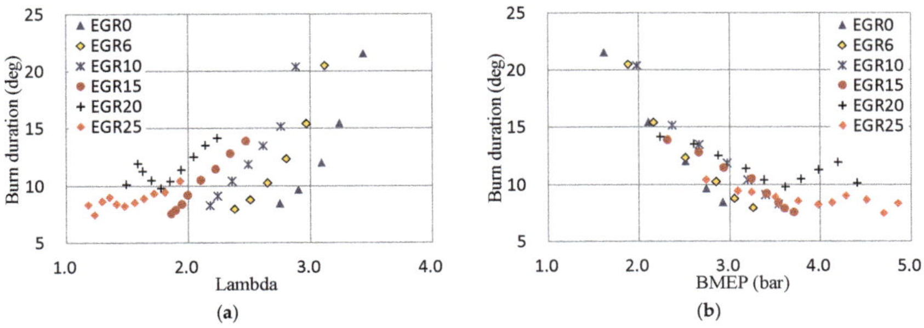

Figure 18. Burn duration with respect to: (**a**) Lambda; (**b**) BMEP.

The combustion efficiency is close to 100% at a burn duration of <10 deg CA (Figure 19a), which equates to a BMEP of >3 bar (Figure 18b). The burn duration indicates the combustion rate, whereas CA10 indicates the start of combustion. Therefore, the influence of burn duration on the combustion efficiency is more obvious (Figure 19a) than in CA10 (Figure 19b).

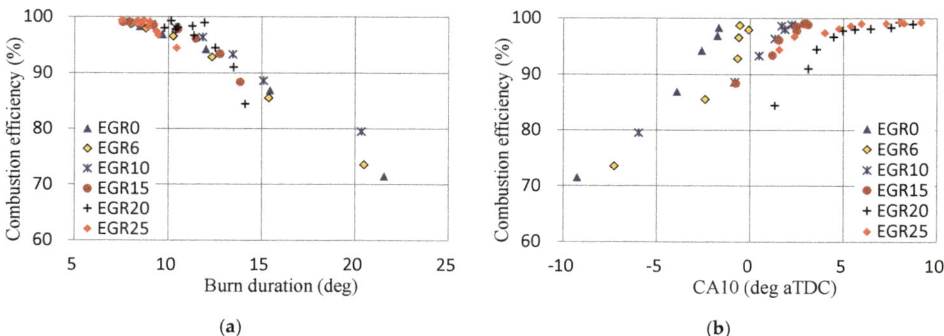

Figure 19. Combustion efficiency with respect to: (**a**) burn duration; (**b**) CA10.

3.5. Comparison between HCCI and SI

The engine speeds operated with HCCI contained 2000, 2600, 3000, 3300, 3500, and 4000 rpm. The experimental results of HCCI are compared with that of original SI engine as shown in Figures 20 and 21.

Figure 20a shows that the operating range of HCCI engine is much smaller than that of the original SI engine, so the proposed HCCI engine cannot be used in a conventional motorcycle. However, it could be used as a range extender for an electric motorcycle. The BSFC of HCCI engine is much better than that of original SI engine as shown in Figure 20b. The lower BSFC of HCCI engine might be caused by several reasons: (1) lower heat transfer loss from cylinder gas to the wall, (2) short combustion duration, and (3) lean mixture. The combustion duration of HCCI engine is much shorter than that of conventional SI engine. When the load increases (i.e., λ decreases in Figure 14a), the combustion phasing has to be retarded to avoid engine knocking as shown in Figure 14a.

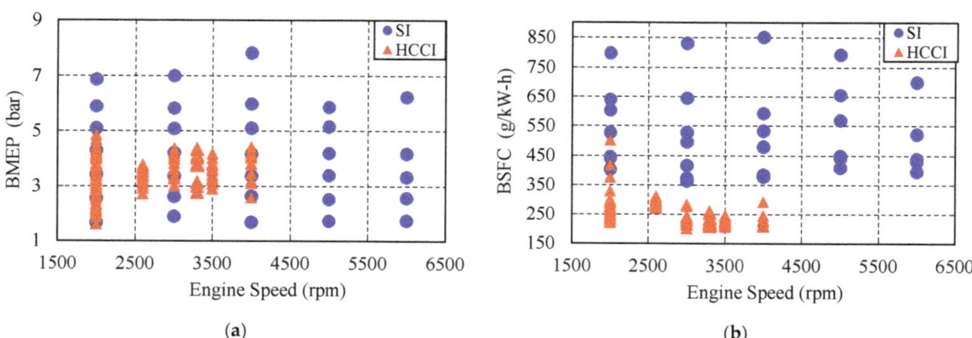

Figure 20. Comparison of engine characteristics between proposed HCCI and original SI engine: (**a**) BMEP; (**b**) BSFC.

The CO emission of HCCI engine is smaller than that of original SI engine (Figure 21a) due to the lean combustion. However, the HC emission is higher (Figure 21b) because of the low combustion temperature.

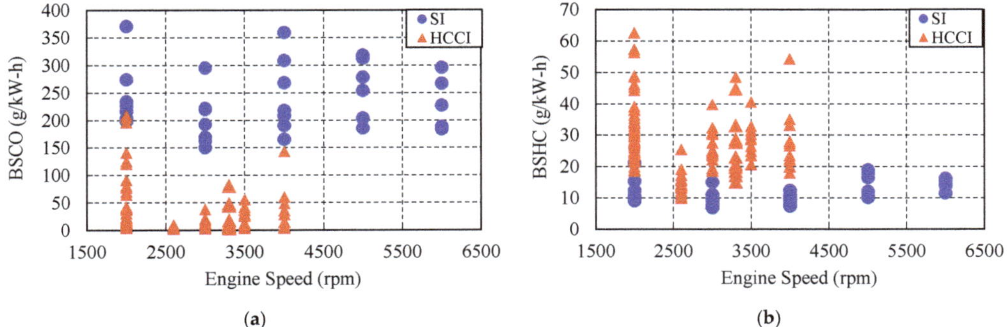

Figure 21. Comparison of exhaust emissions between proposed HCCI and original SI engine: (a) BSCO; (b) BSHC.

4. Conclusions

The proposed HCCI engine was operated with DME-gasoline dual-fuel in a conventional motorcycle engine. The engine test results and combustion analysis led to the following conclusions:

(1) To pursue both high engine output and low BSFC, the proposed HCCI system for a motorcycle engine is DME-gasoline dual-fuel with 25% EGR and $\lambda < 2$. Therefore, the design guide for HCCI engine obtaining high output and low BSFC can be led to a DME-gasoline dual-fuel system with suitable EGR ratio and air-fuel mixture not too lean.

(2) The maximum BMEP increase was from 2.93 to 4.86 bar, an increase of 65.8%, when the EGR was 0% to 25%. At 25% EGR, BMEP increased by 77.4% from 2.74 to 4.86 bar when the gasoline ratio increased from 0.13 to 0.35.

(3) The BSFC was improved great as compared with the original SI engine and NO emision was too small to measure.

(4) The thermal efficiency ranged from 30%–35% and had no clear relationship with λ or BMEP.

(5) Both BSCO and BSHC decreased when the exhaust temperature increased, whereas the exhaust temperature increased linearly with BMEP. When the exhaust temperature was > 550 K or BMEP was > 3.16 bar, the amount of CO emitted was very small.

(6) Both CA10 and CA50 were delayed by a decrease in λ. This is caused by the addition of more gasoline fuel, which delays combustion.

(7) The burn duration increased in line with λ because a leaner mixture causes slower combustion. The combustion efficiency was close to 100% when the burn duration was <10 deg CA.

To further improve the thermal efficiency of the proposed HCCI engine, future studies should modify the cooling system to reduce the cooling capacity.

Author Contributions: Y.-Y.W. designed the experiments; C.-T.J. performed the experiments and data reduction; Y.-Y.W. and C.-T.J. analyzed the data; Y.-Y.W. wrote the paper.

Funding: This research was funded by Ministry of Science and Technology (MOST, Taipei, Taiwan), grant number MOST 106-3113-E-027-002-CC2.X.

Acknowledgments: The authors would like to thank the Ministry of Science and Technology (MOST, Taipei, Taiwan) for their financial support. This manuscript was edited by Wallace Academic Editing.

Nomenclature

A	area of combustion chamber surface
BMEP	brake mean effective pressure
BSCO	brake-specific CO
BSFC	brake-specific fuel consumption
BSHC	brake-specific HC
BDC	bottom dead center
C_m	average piston speed
c_p	specific heat for constant pressure
CA	crank angle
CAD	crank angle degrees
CA10	crank angle at which the mass fraction burned is 10%
CA50	crank angle at which the mass fraction burned is 50%
CA90	crank angle at which the mass fraction burned is 90%
CO	carbon monoxide
CO_2	carbon dioxide
COV	coefficient of variation
DI	direct injection
DME	dimethyl ether
ECE	Economic Commission of Europe
EGR	exhaust gas recirculation
EPA	Environmental Protection Administration
FSN	filter smoke number
HC	hydrocarbons
HCCI	homogeneous charge compression ignition
HRR	heat release rate
H_2O_2	hydrogen peroxide
h	heat transfer coefficient
IMEP	indicated mean effective pressure
$IMEP_{avg}$	average IMEP
$IMEP_{std}$	standard deviation of IMEP
MFB	mass fraction burned
MHRR	maximum heat release rate
MRPR	maximum rate of pressure rise
NMHC	non-methane hydrocarbons
NO	nitric oxide
NO_x	nitrogen oxides
NVO	negative valve overlap
OBD	on-board diagnostics
OH	hydroxyl radical
P	cylinder gas pressure (bar)
PFI	port fuel injection
PM	particulate matter
Q_{LHV}	low heating value of fuel
RON	research octane number
S_t	Stanton number
SCR	selective catalytic reduction
SI	spark ignition
T	cylinder gas temperature
TDC	top dead center
V	cylinder volume
WMTC	worldwide motorcycle test cycle
bhp	brake horsepower

m_f	fuel mass supplied per cycle
λ	air–fuel equivalence ratio
θ	crank angle (degree)
γ	specific heat ratio
η_c	combustion efficiency (%)
η_f	fuel efficiency (%)
η_{th}	brake thermal efficiency (%)

References

1. Taiwan Ministry of Transportation and Communications (MOTC). Motor Vehicle Registration (Sept. 2018). Available online: http://stat.motc.gov.tw/mocdb/stmain.jsp?sys=100&funid=e3301 (accessed on 6 November 2018).
2. Lin, C.Y.; Wu, S.Y.; Liang, H.J.; Liu, Y.C.; Ueng, T.H. Metabolomic analysis of the effects of motorcycle exhaust on rat testes and liver. *Aerosol Air Qual. Res.* **2014**, *14*, 1714–1725. [CrossRef]
3. Chuang, S.C.; Chen, S.J.; Huang, K.L.; Chang-Chien, G.P.; Wang, L.C.; Huang, Y.C. Emissions of polychlorinated dibenzo-p-dioxin and polychlorinated dibenzofuran from motorcycles. *Aerosol Air Qual. Res.* **2010**, *10*, 533–539. [CrossRef]
4. Taiwan Environmental Protection Administration (EPA). Air Pollutant Emission Regulations for Transportation Vehicles. Available online: https://oaout.epa.gov.tw/law/LawContent.aspx?id=FL015347 (accessed on 6 November 2018).
5. Finesso, R.; Spessa, E.; Venditti, M.; Yang, Y. Offline and Real-Time Optimization of EGR ratio and Injection Timing in Diesel Engines. *SAE Int. J. Engines* **2015**, *8*, 2099–2119. [CrossRef]
6. Theis, J.; Kim, J.; Cavataio, G. TWC+LNT/SCR Systems for Satisfying Tier 2, Bin 2 Emission Standards on Lean-Burn Gasoline Engines. *SAE Int. J. Fuels Lubr.* **2015**, *8*, 474–486. [CrossRef]
7. De Abreu Goes, J.; Olsson, L.; Berggrund, M.; Kristoffersson, A.; Gustafson, L.; Hicks, M. Performance Studies and Correlation between Vehicle- and Rapid- Aged Commercial Lean NOx Trap Catalysts. *SAE Int. J. Engines* **2017**, *10*, 1613–1626. [CrossRef]
8. Lin, Y.C.; Yang, P.M.; Chen, C.B. Reducing emissions of polycyclic aromatic hydrocarbons and greenhouse gases from engines using a novel plasma-enhanced combustion system. *Aerosol Air Qual. Res.* **2013**, *13*, 1107–1115. [CrossRef]
9. Wei, M.; Nguyen, T.S.; Turkson, R.F.; Liu, J.; Guo, G. Water injection for higher engine performance and lower emissions. *J. Energy Inst.* **2017**, *90*, 285–299. [CrossRef]
10. Wu, Y.Y.; Chen, B.C.; Hwang, J.J.; Chen, C.Y. Performance and emissions of motorcycle engines using water–fuel emulsions. *Int. J. Veh. Des.* **2009**, *49*, 91–110. [CrossRef]
11. Hasan, M.M.; Rahman, M.M. Homogeneous charge compression ignition combustion: Advantages over compression ignition combustion, challenges and solutions. *Renew. Sustain. Energy Rev.* **2016**, *57*, 282–291. [CrossRef]
12. Agarwal, A.K.; Singh, A.P.; Maurya, R.K. Evolution, challenges and path forward for low temperature combustion engines. *Prog. Energy Combust. Sci.* **2017**, *61*, 1–56. [CrossRef]
13. Jung, D.; Iida, N. Thermal and chemical effects of the in-cylinder charge at IVC on cycle-to-cycle variations of DME HCCI combustion with combustion-phasing retard by external and rebreathed EGR. *Appl. Therm. Eng.* **2017**, *113*, 132–149. [CrossRef]
14. Mohebbi, M.; Reyhanian, M.; Hosseini, V.; Muhamad Said, M.F.; Aziz, A.A. The effect of diethyl ether addition on performance and emission of a reactivity controlled compression ignition engine fueled with ethanol and diesel. *Energy Convers. Manag.* **2018**, *174*, 779–792. [CrossRef]
15. Hasan, M.M.; Rahman, M.M.; Kadirgama, K.; Ramasamy, D. Numerical study of engine parameters on combustion and performance characteristics in an n-heptane fueled HCCI engine. *Appl. Therm. Eng.* **2018**, *128*, 1464–1475. [CrossRef]
16. Kalaskar, V.; Splitter, D.; Szybist, J. Gasoline-Like Fuel Effects on High-Load, Boosted HCCI Combustion Employing Negative Valve Overlap Strategy. *SAE Int. J. Fuels Lubr.* **2014**, *7*, 82–93. [CrossRef]
17. Yun, H.; Wermuth, N.; Najt, P. Extending the High Load Operating Limit of a Naturally-Aspirated Gasoline HCCI Combustion Engine. *SAE Int. J. Engines* **2010**, *3*, 681–699. [CrossRef]

18. Szybist, J.P.; Nafziger, E.; Weall, A. Load Expansion of Stoichiometric HCCI Using Spark Assist and Hydraulic Valve Actuation. *SAE Int. J. Engines* **2010**, *3*, 244–258. [CrossRef]
19. Lawler, B.; Lacey, J.; Güralp, O.; Najtc, P.; Filipi, Z. HCCI combustion with an actively controlled glow plug: The effects on heat release, thermal stratification, efficiency, and emissions. *Appl. Energy* **2018**, *211*, 809–819. [CrossRef]
20. Ji, C.; Dec, J.; Dernotte, J.; Cannella, W. Boosted Premixed-LTGC/HCCI Combustion of EHN-doped Gasoline for Engine Speeds Up to 2400 rpm. *SAE Int. J. Engines* **2016**, *9*, 2166–2184. [CrossRef]
21. Wang, Y.; Liu, H.; Huang, Z.; Liu, Z. Study on combustion and emission of a dimethyl ether-diesel dualfuel premixed charge compression ignition combustion engine with LPG (liquefied petroleum gas) as ignition inhibitor. *Energy* **2016**, *96*, 278–285. [CrossRef]
22. Pedersen, T.D.; Schramm, J.; Yanai, T.; Sato, Y. *Controlling the Heat Release in HCCI Combustion of DME with Methanol and EGR*; SAE Technical Paper; SAE International: Warrendale, PA, USA, 2010.
23. Li, G.; Zhang, C.; Zhou, J. Study on the knock tendency and cyclical variations of a HCCI engine fueled with n-butanol/n-heptane blends. *Energy Convers. Manag.* **2017**, *133*, 548–557. [CrossRef]
24. Khandal, S.V.; Banapurmath, N.R.; Gaitonde, V.N. Performance studies on homogeneous charge compression ignition (HCCI) engine powered with alternative fuels. *Renew. Energy* **2019**, *132*, 683–693. [CrossRef]
25. Zheng, Z.; Xia, M.; Liu, H.; Wang, X.; Yao, M. Experimental study on combustion and emissions of dual fuel RCCI mode fueled with biodiesel/n-butanol, biodiesel/2,5-dimethylfuran and biodiesel/ethanol. *Energy* **2018**, *148*, 824–838. [CrossRef]
26. Putrasari, Y.; Jamsran, N.; Lim, O. An investigation on the DME HCCI autoignition under EGR and boosted operation. *Fuel* **2017**, *200*, 447–457. [CrossRef]
27. Ito, S.; Ikeda, H.; Jung, D.W.; Iida, N. *Potential of Stratification Charge for Reducing Pressure-Rise Rate in HCCI Engines Based on Multi-Zone Modeling and Experiments by Using RCM*; SAE Technical Paper; SAE International: Warrendale, PA, USA, 2013.
28. Jamsran, N.; Lim, O.; Iida, N. *An Investigation on DME HCCI Engine about Combustion Phase Control Using EGR Stratification by Numerical Analysis*; SAE Technical Paper; SAE International: Warrendale, PA, USA, 2012.
29. Nishi, M.; Kanehara, M.; Iida, N. Assessment for innovative combustion on HCCI engine by controlling EGR ratio and engine speed. *Appl. Therm. Eng.* **2016**, *99*, 42–60. [CrossRef]
30. Lee, C.; Chung, J.; Lee, K. Emission Characteristics for a Homogeneous Charged Compression Ignition Diesel Engine with Exhaust Gas Recirculation Using Split Injection Methodology. *Energies* **2017**, *10*, 2146. [CrossRef]
31. Lee, K.; Cho, S.; Kim, N.; Min, K. A study on combustion control and operating range expansion of gasoline HCCI. *Energy* **2015**, *91*, 1038–1048. [CrossRef]
32. Szybist, J.P.; Edwards, K.D.; Foster, M.; Confer, K.; Moore, W. *Characterization of Engine Control Authority on HCCI Combustion as the High Load Limit is Approached*; SAE Technical Paper; SAE International: Warrendale, PA, USA, 2013.
33. Wu, Y.Y.; Chen, B.C.; Hsieh, F.C. Heat transfer model for small-scale air-cooled spark ignition four-stroke engines. *Int. J. Heat Mass Transf.* **2006**, *49*, 3895–3905. [CrossRef]
34. Wu, Y.Y.; Chen, B.C.; Wang, J.H. Experimental study on HCCI combustion in a small engine with various fuels and EGR. *Aerosol Air Qual. Res.* **2016**, *16*, 3338–3348. [CrossRef]
35. Mack, J.H.; Flowers, D.L.; Buchholz, B.A.; Dibble, R.W. Investigation of HCCI combustion of diethyl ether and ethanol mixtures using carbon 14 tracing and numerical simulations. *Proc. Combust. Inst.* **2005**, *30*, 2693–2700. [CrossRef]
36. Wu, Y.Y.; Jang, C.T.; Chen, B.L. Combustion characteristics of HCCI in motorcycle engine. *J. Eng. Gas Turbines Power-Trans. ASME* **2010**, *132*. [CrossRef]
37. Yao, M.; Chen, Z.; Zheng, Z.; Zhang, B.; Xing, Y. Study on the controlling strategies of homogeneous charge compression ignition combustion with fuel of dimethyl ether and methanol. *Fuel* **2006**, *85*, 2046–2056. [CrossRef]
38. Holman, J.P. *Experimental Methods for Engineers*, 5th ed.; McGraw-Hill: New York, NY, USA, 1989; ISBN 9780070296220.
39. Ceviz, M.A.; Kaymaz, I. Temperature and air-fuel ratio dependent specific heat ratio functions for lean burned and unburned mixture. *Energy Convers. Manag.* **2005**, *46*, 2387–2404. [CrossRef]

40. Sun, R.; Thomas, R.; Gray, C.L., Jr. *An HCCI Engine: Power Plant for a Hybrid Vehicle*; SAE Technical Paper; SAE International: Warrendale, PA, USA, 2004.
41. Fathi, M.; Saray, R.K.; Checkel, M.D. The Influence of Exhaust Gas Recirculation (EGR) on Combustion and Emissions of n-heptane/Natural Gas Fuelled Homogeneous Charge Compression Ignition (HCCI) Engines. *Appl. Energy* **2011**, *88*, 4719–4724. [CrossRef]
42. Wu, Y.Y.; Chen, B.C.; Wang, J.H. Application of HCCI Engine in Motorcycle for Emission Reduction and Energy Saving. *Aerosol Air Qual. Res.* **2015**, *15*, 2140–2149. [CrossRef]
43. Turns, S.R. *An Introduction to Combustion*, 3rd ed.; McGraw-Hill: Singapore, 2012; p. 172. ISBN 978-007-108687-5.
44. Sjöberg, M.; Dec, J.E. An investigation into lowest acceptable combustion temperatures for hydrocarbon fuels in HCCI engines. *Proc. Combust. Inst.* **2005**, *30*, 2719–2726. [CrossRef]
45. Machrafi, H.; Cavadias, S.; Amouroux, J. A parametric study on the emissions from an HCCI alternative combustion engine resulting from the auto-ignition of primary reference fuels. *Appl. Energy* **2008**, *85*, 755–764. [CrossRef]
46. Kelly-Zion, P.L.; Dec, J.E. A computational study of the effect of fuel type on ignition time in homogeneous charge compression ignition engines. *Proc. Combust. Inst.* **2000**, *28*, 1187–1194. [CrossRef]

 *energies*

Article

Comparative Analysis of the Combustion Stability of Diesel-Methanol and Diesel-Ethanol in a Dual Fuel Engine

Arkadiusz Jamrozik *, Wojciech Tutak, Renata Gnatowska and Łukasz Nowak

Faculty of Mechanical Engineering and Computer Science, Czestochowa University of Technology,
42-201 Czestochowa, Poland; tutak@imc.pcz.pl (W.T.); gnatowska@imc.pcz.pl (R.G.); luknowak@imc.pcz.pl (Ł.N.)
* Correspondence: jamrozik@imc.pcz.pl

Received: 1 February 2019; Accepted: 6 March 2019; Published: 13 March 2019

Abstract: The co-combustion of diesel with alcohol fuels in a compression ignition dual fuel engine is one of the ways of using alternative fuels to power combustion engines. Scientific explorations in this respect should not only concern the combustion process in one engine cycle, which is most often not representative for a longer engine life, but should also include an analysis of multiple cycles, which would allow for indicating reliable parameters of engine operation and its stability. This paper presents experimental examinations of a CI engine with a dual fuel system, in which co-combustion was performed for diesel and two alcohol fuels (methanol and ethanol) with energy contents of 20%, 30%, 40% and 50%. The research included the analysis of the combustion process and the analysis of cycle-by-cycle variation of the 200 subsequent engine operation cycles. It was shown that the presence and increase in the share of methanol and ethanol used for co-combustion with diesel fuel causes an increase in ignition delay and increases the heat release rate and maximum combustion pressure values. A larger ignition delay is observed for co-combustion with methanol. Based on changes in the coefficient of variation of the indicated mean effective pressure (COV_{IMEP}) and the function of probability density of the indicated mean effective pressure (f(IMEP)), prepared for a series of engine operation cycles, it can be stated that the increase in the percentage of alcohol fuel used for co-combustion with diesel fuel does not impair combustion stability. For the highest percentage of alcohol fuel (50%), the co-combustion of diesel with methanol shows a better stability.

Keywords: co-combustion; dual fuel; combustion stability; coefficient of variation of IMEP; probability density of IMEP

---

1. Introduction

Compression ignition engines (CI) are commonly used in transport, industrial machines and in the agricultural economy due to their long life time durability. The CI engines are a subject of great criticism due to their emissions output. The biggest problem in these engines is the simultaneous reduction of NO_x and soot emissions [1,2]. Emissions of soot and NO are opposed to each other. In paper [3] authors presented results of an investigation of the emissions of dual fuel CI engine. Authors stated that in all analyzed cases, conditions were achieved in which soot emission starts to increase again with lower NO emissions. Emissions of NO_x and soot should be reduced because these are harmful to human health and environment as well [4]. Another motivation for these activities is the European Union Directive 2009/28/EC obliging the use of a 20% share of renewable biofuels in overall transport and diesel fuel consumption by 2020 [5]. One of the reasons for using biofuels is their smaller negative impact on the environment through lower greenhouse gas emissions, while another is in order to develop diversification, which can increase energy independence. Increased consumption of biofuels causes reduced consumer reliance on imported fossil fuels and the depletion of crude oil reserves [6,7]. Fuelling of compression ignition internal combustion engine with unconventional fuels seems to be an interesting alternative that allows for increasing the use of biofuels in the transport and energy sectors. Biofuels such as alcohols and biodiesel have been proposed as alternatives to fossil fuels for internal combustion engines [8]. Combined with diesel fuel or biodiesel, alcohol fuels may represent a valuable fuel for compression ignition engines. Alternative liquid fuels are used for powering compression ignition engines in two modes. The first one is powering engines by using a diesel-alternative fuel blend and second one is a co-combustion of fuels using dual fuel technology [9–12]. The main disadvantage of using a fuel mixture is the limited share of alcohol due to the phase separation [13,14]. In a dual-fuel engine, additional fuel is supplied to the intake system by low-pressure injection. Such produced air-alcohol mixture enters into cylinders during the inlet valve opening period. The engine in this mode of operation can burn alcohol to its 90% energy share [15,16]. Recently, there has been a growing interest in using organic oxygen compounds in engines of this type, especially alcohols [17]. Some disadvantages of alcohols compared to diesel fuel include a relatively high compression ignition temperature and a tendency for the decreasing of charge in-cylinder temperature due to the high value of heat of vaporization. Furthermore, the heating value of ethanol and methanol is lower compared to diesel. Thus, to provide the same energy of delivered fuel to the engine cycle, higher amounts of alcohol are needed [15,16]. The reduction of exhaust emissions of the compression ignition engine can be obtained by controlling the combustion process and by using alternative fuels [18]. Therefore, the substitution of diesel fuel with alternative fuels such as alcohol fuels represents a good approach to solve the environmental problems. With its properties, alcohol can also have a positive effect on NO_x and particulate matter emissions [19]. The diesel fuel consists of various hydrocarbons, the ratio of carbon to hydrogen (C/H) is high, and there is a tendency to form soot under fuel rich combustion conditions. The particulate matter is created in the internal combustion engine (mainly in the self-ignition engine), when combustion causes local or temporary lack of air, and more specifically oxygen. The additional oxygen delivered in the alcohol fuel promotes the burning of coal and reduces the formation of PM. Due to their organic nature, alcohols contain up to 50% oxygen in their structure, allowing for widespread oxidation of fuel, resulting in more complete combustion. One of the alcohols that can be used for co-combustion with diesel is ethanol. Ethanol can be produced from raw materials of plant origin and therefore can be considered a fully renewable fuel [17]. Methanol can be produced from such resources as natural gas, coal and various renewable biomasses. The main disadvantages of using ethanol in diesel engines include a low cetane number and problems with solubility in the petroleum diesel [20]. Among these alternative fuels, methanol also represents a promising alternative fuel [21]. Since methanol has a higher latent heat of vaporization than ethanol, methanol has the potential to reduce NO_x due to a cooling effect on the charge, leading to lower in-cylinder temperatures [20]. Another advantage of using alcohol is an oxygen content in methanol structure which reduces exhaust fume emissions. Ethanol or ethanol mixed with diesel leads to the occurrence of phase separation between these fuels. Phase separation occurs each time because diesel fuel and methanol or ethanol are immiscible [22]. Therefore, a better solution is to use alcohols in dual fuel mode. In the available literature, there are many publications about the co-combustion of ethanol or methanol with diesel fuel in internal combustion engine. The papers [14,23] presented the results of examinations of compression ignition engine powered by blends of diesel fuel with various alcohols. The authors demonstrated that diesel-ethanol blends could be used in the compression ignition engines with quite a large fraction. Using ethanol fuel in a blend with diesel fuel is supplied additional oxygen for the combustion, which can cause improvements in the overall combustion process. Another statement was that the combustion process of alcohol-diesel fuels blend in IC diesel engine takes place in a shorter time than in the diesel engine fuelled by pure diesel fuel. With an increase of the ethanol fuel fraction, the ignition delay increased. With the increase in the proportion of ethanol, the combustion duration decreases [14,20,24]. The paper [25] presented results of investigation of the combustion process in compression ignition engine powered by a diesel-methanol blend. They stated that up to a 30% methanol fraction, the combustion process occurred as it would in engines powered by pure diesel. A further increase in the alcohol content led to a significant decline in cylinder pressure and the abnormal combustion process. For mixtures up to 40% of methanol fraction, it did not exceed the value of COV_{IMEP} equal to 10% [25]. In another paper [24] results are presented which show that the use of dual fuel technology gave a satisfactory performance in a standard diesel engine without any modification in the engine. In addition, emissions of carbon monoxide, unburned hydrocarbon and smoke were lower, whereas nitrogen oxides emissions were almost equivalent with diesel. In paper [24], results are presented of varying the ethanol ratio on engine performance and emissions of the dual fuel combustion with various load conditions. The test engine was a single-cylinder diesel engine equipped with two direct injectors operated with constant speed. The authors stated, inter alia, that NO_x and PM emissions decreased with increasing ethanol fraction. Additionally, they stated that for low and high loads conditions, the ethanol fraction is limited due to insufficient ignition energy at a low load and at full load due to a high in-cylinder pressure rise. The engine can also burn so-called hydrated ethanol [26]. It seems that this is a good solution because the removal of water from fermented products of ethanol consumes lots of energy. In the paper [27], results are presented of hydrous ethanol combustion in a dual fuel engine, including impacts on the combustion and emissions. The thermal efficiency reduced significantly at the ethanol purity of 60%. The reduction of ethanol purity can reduce NO_x emissions but CO and THC emissions increase. In paper [4] results are presented of investigation of cylinder-to-cylinder pressure variation in a diesel engine fuelled with diesel/methanol using dual fuel technology. Research was carried out on the 4-cylinder turbocharged direct injection diesel engine. In the experiment, methanol was injected into the inlet duct by two low-pressure injectors, and directly mixed with intake air before entering the engine. The diesel fuel was directly injected into cylinder. The experimental results showed that the unevenness degree of engine increases with the increase of methanol percent under various engine loads. The uniqueness of pressure in subsequent cycles increases with the methanol share under a low engine load, and the COV_{pp} (coefficient of variation of peak pressure) curves vary very little under the low methanol percentage under 75% and 100% engine loads, but a significant rise appeared when the methanol substitution percent further increased [4]. Co-combustion of alternative fuels such as alcohols causes lowering in cylinder temperature which then affects the value of the ignition delay due to higher value of heat of evaporation of alcohols. In the paper [28] results are presented of investigation of in-cycle pressure fluctuation intensity under long ignition delay conditions. Authors stated that under low temperature charge conditions, the difference in peak pressure between the average cycle and cycles showed low fluctuation intensity. The average intensity of the pressure fluctuations showed an increase with increasing amounts of the premixed combustion phase. This resulted in higher cycle-to-cycle variations under these conditions [28]. Other works presented comparative analysis of the impact of fuel on the uniqueness of engine cycles. In the paper [29], results are presented of investigation of cycle-to-cycle variations of peak pressure in compression ignition engine powered by diesel and biodiesel fuel. The results show that at a lower load, in-cylinder pressure variations for biodiesel were lower compared to mineral diesel fuel. On the other hand, at medium and high loads, biodiesel dominated the peak n-cylinder pressure variations [29]. The cycle of unrepeatability influences emissions variation as well. In the paper [30], results are presented of assessment of cyclic variations on NO emissions of a direct injection diesel engine. Ono of statements was that the intensity of the pressure fluctuations increases with increasing ignition delay. Additionally, the authors stated that in cases with a significant diffusion combustion phase, cycles with pressure fluctuations showed higher-than average NO concentrations, with the intensity of fluctuation determining the increase in NO. Cycle-to-cycle variations in compression ignition engines are undesirable due to negative impacts on efficiency and higher emission. This undesirable phenomenon causes power output limitations.

This work aims to increase knowledge of influence of type of alternative fuel co-combusted with mineral diesel fuel on operation parameters and cycle-to-cycle variations of the CI engine. The results of experiments of co-combustion of diesel fuel with alcohols in the compression ignition engine working in the system of dual fuel are presented in these studies. This paper presents experimental examinations of a CI engine with dual fuel system, in which co-combustion was performed for diesel and two alcohol fuels (methanol and ethanol) with energy contents of 20%, 30%, 40% and 50%. The research concerned the analysis of the combustion process and the analysis of non-repeatability for 200 subsequent engine operation cycles. In the available literature there is not enough information about the influence of the share of alcohol on the stability of the combustion process. However, a lot of information is provided about the deterioration of the combustion by the share of alcohol. In this work, we showed the effect of alcohol on combustion process stability using indicators such as COV_{IMEP} and the probability density of IMEP.

2. Materials and Methods

2.1. Cycle-to-Cycle Variability Determination

During the observation of the course of the indicator pressures in successive engine operation cycles, it is easy to notice differences in the values of maximum pressure and indicated mean effective pressure in subsequent cycles. Cycle variability is an undesirable feature of combustion in internal combustion engines. Cyclic variations in diesel engines are undesirable since they are understood to lead to lower efficiency and higher emissions and to power output limitations [28]. They are caused by varied conditions in the cylinder from one combustion cycle to the other, under nominally identical operating conditions. Due to air pollution and increasingly stringent emission standards around the world, mitigation of the effects of cycle variability can reduce fuel consumption and minimize the emissions of harmful compounds in IC engines. Furthermore, cycle variability limitation allows for better engine optimization, i.e., it is possible to achieve efficiency close to optimal conditions. The main sources of the non-repeatability of engine cycles include:

-   flow processes occurring in the engine cylinder,
-   variability of air-fuel ratio in individual cycles,
-   uneven composition of the combustible mixture in the cylinder due to incomplete mixing of air, fuel and residual gas,
-   ignition non-repeatability.

The assessment of the of non-repeatability degree of the maximum pressure and indicated mean effective pressure of the thermal cycle occurring in the cylinder of the internal combustion engine can be conducted by analysing several dozen or several hundred consecutive engine cycles.

The obtained set of maximum pressures can be divided into classes k, from the interval:

$$k \in \langle p_{max\,A}, p_{max\,B} \rangle, \tag{1}$$

where $p_{max\,A}$ is the smallest value of maximum pressure in set of cycles, $p_{max\,B}$ is the maximum value of the maximum pressure in set of cycles.

The average value of maximum pressure pmax, determined on the set of pressure:

$$\overline{P}_{max} = \frac{1}{N} \sum_{i=1}^{N} P_{max\,i}, \tag{2}$$

where N is the cycle index, $p_{max\,I}$ is maximum pressure in individual cycles.

One of the most popular criteria for the evaluation of the correct operation of combustion engines is non-repeatability of the work cycles. The coefficient of non-repeatability of maximum pressure COV_{pp}, expressed in percentages and calculated as a ratio of standard deviation of the maximal pressure to its mean value from multiple engine operation cycles, was adopted as a measure of non-repeatability [4,31].

The coefficient of variation of peak pressure (COV_{pp}) is defined as:

$$COV_{pp} = \frac{\sigma_{pmax}}{\overline{P_{max}}} \cdot 100\%,$$

(3)

The standard deviation of peak pressure:

$$\sigma_{pmax} = \sqrt{\frac{1}{N}\sum_{i=1}^{N}(P_{max\,i} - \overline{P_{max}})^2},$$

(4)

where $\overline{P_{max}}$ is the mean value of peak pressure of N cycle $P_{max\,i}$.

The indicated mean effective pressure is evaluated based on the recorded changes in the cylinder pressure and represents one of the indices that characterize operation of combustion engines in terms of the opportunities to ensure high and expected functional performance.

Indicated mean effective pressure for a single engine cycle:

$$IMEP_i = \frac{1}{V_d}\int_{0}^{720} pdV,$$

(5)

The average value of IMEP:

$$\overline{IMEP} = \frac{1}{N}\sum_{i=1}^{N} IMEP_i,$$

(6)

where $IMEP_i$ is indicated mean effective pressure in individual cycles.

The uniqueness of IMEP can be determined in the same way as the uniqueness of the maximum pressure value.

$$COV_{IMEP} = \frac{\sigma_{IMEP}}{\overline{IMEP}} \cdot 100\%,$$

(7)

The standard deviation of IMEP:

$$\sigma_{IMEP} = \sqrt{\frac{1}{N}\sum_{i=1}^{N}(IMEP_i - \overline{IMEP})^2},$$

(8)

The probability density function allows for expressing the probability of obtaining or occurrence of a concrete value of the analyzed parameter. Distribution of probability density of the indicated mean effective pressure represents a repeatability index (index of probability of occurrence) of individual IMEP values obtained for multiple analyzed cycles of the test engine operation. This function can also be used as an indicator to assess the stability of operation of the internal combustion engine. Among other things, it indicates the frequency of occurrence of the most frequently repeated IMEP value, which is close to the mean value, indicating the repeatability of subsequent engine operation cycles. The IMEP probability density function has the profile of a normal distribution (Gaussian distribution), in which the density function is symmetrical in relation to the mean value of the distribution [32].

The probability density of the indicated the mean effective pressure:

$$f(IMEP) = \frac{1}{\sigma_{IMEP}\sqrt{2\pi}}\exp\left(\frac{-(IMEP_i - \overline{IMEP})^2}{2\sigma_{IMEP}^2}\right),$$

(9)

2.2. Test Stand

The experimental examinations of the process of co-combustion of diesel with alcohol fuels used the test engine based on a single-cylinder, four-stroke diesel engine 1CA90 manufactured by Andoria (Poland), with compression rate of 17 and engine displacement of 573 cm^3. This engine is a stationary two-valve unit with a vertical arrangement of cylinders and an air cooling system with an axial fan. The engine worked at constant rotational speed of 1500 rpm and constant injection angle of 17° crank angle before TDC. The engine was with the one single injection and was running at a steady-state full load. In order to change and ensure the optimal position of the initial injection angle, necessary changes were made in the design, which included changing the design of the camshaft. The engine was equipped in a high-pressure installation of fuel supply with direct injection of diesel fuel to the cylinder. The test engine was modernized and equipped with an additional installation of alcohol fuel supply to the intake manifold. The research stand, with its main component being the engine 1CA90, was equipped in the necessary control and measurement apparatus that allowed e.g., for indication and recording of the necessary parameters of the engine work. Engine indication included measurements and digital recording and analysis of fast-changing pressure in the combustion chamber of the research engine. The engine indication system included Kistler 6061B piezoelectric pressure sensor, charge amplifier Kistler 5011B, crank angle sensor and the module for data acquisition with the A/D converter (Measurement Computing USB-1608HS) (Figure 1). Table 1 presents the most basic technical data for this engine.

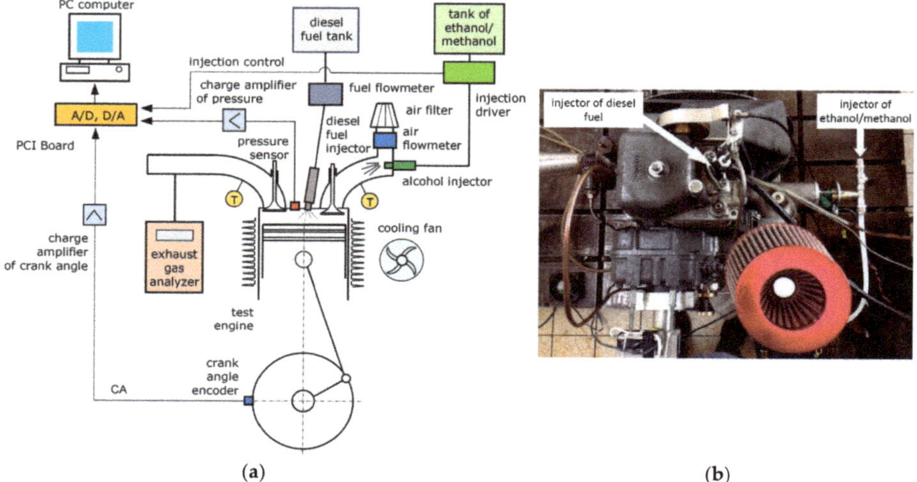

Figure 1. Diagram of the experimental setup (**a**) and view from the top of the test engine (**b**).

Table 1. Main engine parameters.

Parameter	Value	Unit
Displacement volume	0.000573	m^3
Bore	90	mm
Stroke	90	mm
Compression ratio	17:1	-
Rated power	7	kW
Crankshaft rotational speed	1500	rpm
Diesel injection pressure	21	MPa
Alcohol injection pressure	0.3	MPa
Injection timing	17	deg bTDC

The test bench consisted of:

- compression ignition engine 1CA90,
- digital acquisition system for analysis of fast changing data consisted of: pressure transducer, Kistler 6061B of sensitivity: ±0.5%, charge amplifier, Kistler 5011B of linearity of FS < ±0.05%,

data acquisition module, crank angle encoder, resolution of 360 pulses/rev, Measurement Computing USB-1608HS–16 bits resolution of sampling frequency 20 kHz,

- exhaust gas analyzer Bosch BEA 350 for:

THC:	range 0–9999 ppm vol,	accuracy: 12 ppm vol,
CO:	range 0–10% vol,	accuracy: 0.06% vol,
CO_2:	range 0–18% vol,	accuracy: 0.4% vol,
O_2:	range 0–22% vol,	accuracy: 0.1% vol,
λ:	range 0.5–9.999	accuracy: 0.01,
NO_x:	range: 0–5000 ppm,	According to OIML R99 Ed. 1998.

- Roots flowmeter Common CGR-01, range 0.2–650 m^3/h, accuracy < ±1%.
- temperature sensor TP-204K-1b-100-1.5, range +200 °C, resolution 0.1 °C.

2.3. Fuels

Diesel fuel and technical alcohols (methanol, ethanol), were used in the study. Due to the low value of the cetane number and difficulties with compression ignition, alcoholic fuels, i.e., methanol CH_3OH and ethanol C_2H_5OH, viewed as unconventional engine fuels, cannot be used as individual fuels in CI engines and, therefore, have been used only as fuels for SI engines. Nowadays, although methanol and ethanol belong to strongly oxygenated alcohols (with high content of oxygen in a fuel particle) and are characterized by a lower heating value compared to gasoline or diesel fuel, a continuous increase in interest in these alcohols as potential replacement fuels is being observed compared to conventional fuels, as well as in compression-ignition engines. Mass percentage of oxygen in structure of these alcohols is ca. 50% for methanol and ca. 35% for ethanol. High content of oxygen can have a positive effect on combustion in the engine's combustion chamber since this oxygen is characterized by higher activity compared to the molecular oxygen contained in air. It is conducive to intensification of combustion and contributes to the reduction in exhaust gas smokiness. Similar to ethanol, methanol is characterized by insignificant heating value compared to conventional fuels. The heating value of pure methyl alcohol (19.5 MJ/kg) is only 45%, whereas for ethyl alcohol (26.9 MJ/kg), this means 60% of the heating value of the diesel fuel or gasoline. Both alcohols are characterized by a high heat of vaporization, equal to 1100 kJ/kg for methanol and 840 kJ/kg for ethanol (Table 2). In compression ignition engines, this property reduces mixture temperature and improves the coefficient of cylinder filling, whereas in engines with compression ignition, it contributes to the ignition delay and impacts on the increased rate of pressure rise and can magnify the so-called hard operation of the engine. Hard operation is defined as engine work at a large pressure increase. For diesel engines, a limit of 1 MPa/deg is usually assumed. Alcohol fuels offer opportunities to reduce the use of fossil fuels in CI engines and to increase percentage of biofuels in the transport and energy sectors, where combustion engines are often used. Methyl and ethyl alcohols, due to relatively low costs and easy production, belong to the most popular fuels for combustion engines. Methanol can be manufactured on the industrial scale from broadly available resources, including both non-renewable (natural gas, coal) and renewable (gas generated from biomass or municipal waste) sources. Ethanol is typically obtained during alcohol fermentation of sugars, with sugar sources being mainly cereals (wheat, maize, barley) and potatoes.

Figure 2a shows the two most important features of the analyzed fuels from the standpoint of combustion engine operation. These are the calorific value, which determines engine efficiency, and heat of vaporization, which, in the case of compression ignition engines, impacts on the processes that initiate the ignition of the combustible mixture, causing a change in compression ignition delay and determining engine performance and levels of emissions. Figure 2b shows the molecular composition of diesel, methyl and ethyl alcohol. Hydrogen content in all the fuels is comparable. The main difference between them results from oxygen and carbon content. With its molecular structure, diesel fuel does not contain oxygen, whereas the most oxygenated alcohol (methanol) has a 50% oxygen content (Table 2).

The study examined the processes of co-combustion of diesel with methanol and diesel with ethanol. The percentages of alcohol fuels were 20%, 30%, 40% and 50%. The following symbols were used during the analysis: for pure diesel fuel: D100, for co-combustion of diesel fuel with methanol: DM20, DM30, DM40, DM50 and, for co-combustion of diesel with ethanol: DE20, DE30, DE40, DE50. The numbers in the symbols concern the percentage of alcohol energy dose in total fuel energy (alcohol and diesel fuel) supplied to the engine.

Table 2. Fuels properties [33–36].

Properties	Unit	Diesel	Methanol	Ethanol
Molecular formula	-	$C_{14}H_{30}$	CH_3OH	C_2H_5OH
Molecular weight	g	198.4	32.04	46.068
Cetane number	-	51	3	8
Research octane number	-	15–25	136	129
Boiling point	K	453–643	338	351
Liquid density	kg/m^3	840	796	789
Lower heating value (LHV)	MJ/kg	42.5	19.5	26.9
LHV of stoichiometric mixture	MJ/kg	2.85	2.68	2.69
Heat of evaporation	MJ/kg	243	1100	840
Auto-ignition temperature	K	503	736	698
Stoichiometric air-fuel ratio	-	14.6	6.45	9.06
Viscosity (at 40 °C)	cSt	4.59	0.65	1.52
Carbon content	%	85	37.5	52.2
Hydrogen content	%	15	12.5	13
Oxygen content	%	0	50	34.8

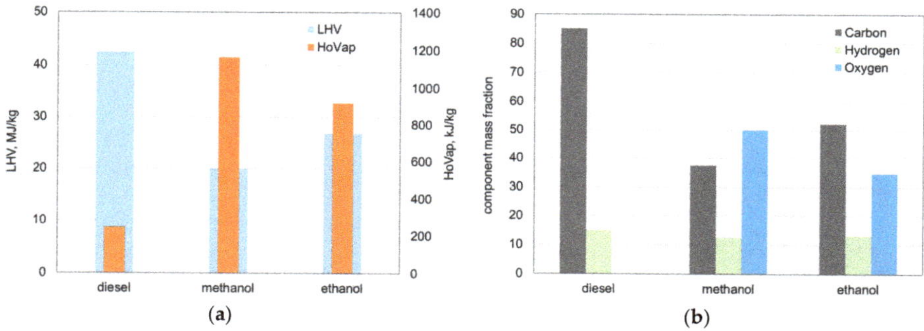

Figure 2. Lower heating value and heat of evaporation of fuels (**a**) and fuels component mass fraction (**b**).

3. Results and Discussion

3.1. Combustion Characteristics

The research was based on the process of indicating the engine, which consists of measuring the pressure in the engine cylinder vs. crank angle. The examinations were started from indication of the engine fuelled with pure diesel as a reference. Figure 3 shows the pressure profiles in the cylinder for the engine fuelled by pure diesel and the engine for co-combustion of diesel with methanol and ethanol. The pressure profiles were used to determine the heat release rate (HRR), which characterizes the combustion process in the engine and allows for evaluation of the operation of the test engine.

The presence and the increase in the percentage of methanol and ethanol used for co-combustion with diesel lead to an increase in ignition delay of the charge caused by a higher value of latent heat of vaporization of the alcohol fuel compared to diesel [37]. Figure 3 shows that a greater ignition delay is observed for co-combustion with methanol. The increase in ignition delay leads to the rise in heat release rate and the maximum combustion pressure values. The paper [38] showed that with an increasing methanol substitution ratio in diesel engine with diesel/methanol compound combustion mode, the combustion starting point is delayed, both for the peak heat release rate and the peak cylinder pressure increase. The increase in ignition delay with methanol fraction results in more fuel burnt in the premixed combustion (kinetic combustion phase), which, coupled with the combustion of diesel fuel in a richer methanol–air mixture, leads to the increase of the maximum cylinder pressure and heat release rate. The objective of paper [39] was to investigate the thermal performance, exhaust emissions and combustion behaviour of small capacity compression ignition dual-fuel engine using fumigated ethanol. It was found that with the induction of ethanol in the CI engine, an increased rate of heat release and increased peak pressure occurs, thus enhancing the thermal efficiency. Cycle to cycle variations were very small with the diesel fuel alone. Distinct engine cycle variations as indicated by higher value of COV of IMEP were evident with various ethanol flow rates. Figure 3 was supplemented with the computed values of the indicated mean effective pressure, which show that co-combustion of diesel with alcohol fuel does not cause significant changes in the values of this parameter. Similar values of IMEP were obtained for both pure diesel oil and for all analyzed percentages of alcohol, at the level of ca. 0.66 MPa. Figure 3 presents the value of IMEP measurement error. The IMEP measurement error was ±3.1%.

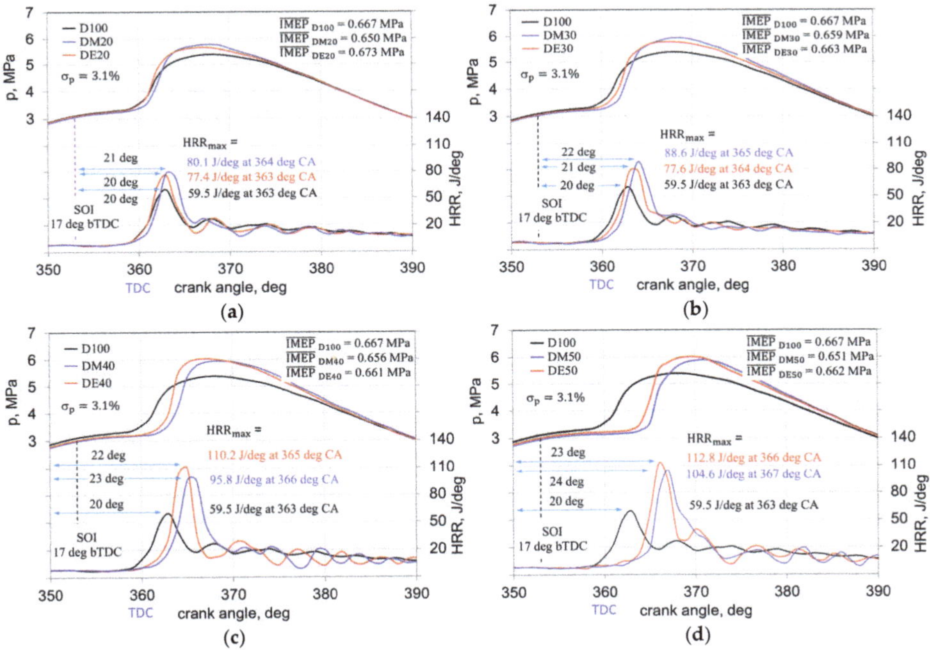

Figure 3. In cylinder pressure and heat release rate (HRR) traces of the engine for co-combustion of diesel fuel with methanol and diesel fuel with ethanol for 20% (**a**), 30% (**b**), 40% (**c**) and 50% (**d**) alcohol share.

In the conventional compression ignition engine, the combustion process is performed in two stages. The first, initial stage, is a period of kinetic combustion, which relates to combustion of the combustible mixture around the jet of injected fuel, whereas the second stage is diffusion combustion which characterizes the process of gradually vaporized fuel combustion. Depending on the conditions of work, the contribution of these stages changes, thus affecting engine performance and emissions. The stage of the kinetic combustion, which characterizes high values of temperature, is conducive to creation of the nitrogen oxides, whereas the stage of the diffusion combustion may be the source of non-burned hydrocarbons and soot particles. Co-combustion of fuels in the dual fuel engine differs from classical fuel combustion in the conventional engine. However, in most conditions, it is found that the combustion characteristics and heat release rate curves of dual-fuel engines can also be divided into two phases, and its shape is very similar to the typical curves [40]. In the dual fuel engine, the injection of additional fuel to the intake manifold, a homogeneous mixture is formed during the inlet stroke, which is then compressed in the cylinder during the compression stroke. With this prepared mixture, the injection of the dose of the diesel fuel occurs at the end of the compression stoke. This dose evaporates, ignites and leads to the ignition of the homogeneous mixture in the cylinder. Due to the reduced content of diesel fuel in total dose of the fuel supplied to the engine, its time of vaporization and combustion is shorter. This leads to shortening of the diffusion combustion stage and increases in the content of the stage of kinetic combustion. Co-combustion of fuels using the idea of dual fuel engine with two independent fuel supply systems, allows for control of individual stages of the combustion process through control of the moment of compression ignition through injection of the dose that initiates ignition and offers opportunities for smooth changes in the contents of individual fuels, adjusted to varied conditions of engine work.

Furthermore, Figure 4 shows the effect of alcohol energy percentage on the kinetic and diffusive combustion periods during co-combustion of diesel with methanol and ethanol. The division into two combustion phases was made arbitrarily on the basis of HRR analysis. The kinetic phase was defined as the period from the beginning of combustion to the local minimum on the HRR curve before the diffusion phase. The diffusion phase was defined as the period from the end of the kinetic phase to the end of combustion. 10% of the heat release was considered as the start of combustion and 90% of heat release was considered as the end of combustion. As the share of alcohol increased, the kinetic combustion period was extended due to the increased heat release rate and intensification of combustion, whereas the diffusion combustion period was reduced. Studies [40] showed that the increased in the percentage ratio of methanol energy to the total energy, from 0% to about 50%, enlarges the kinetic combustion phase, called the rapid burning duration. It is also shown that the proportion of rapid burning heat release in the whole combustion heat release increases dramatically with the increase of the alcohol energy in total fuel energy (alcohol and diesel fuel) supplied to the engine. In the case of DM50, the kinetic combustion stage accounted for 56% of the entire period of combustion of the load in the engine cylinder and was 31% longer than in the case of pure diesel. In the case of DE50, the kinetic combustion phase was 43% and was 18% longer than the kinetic phase accompanying the combustion of pure diesel.

For high energy contents, methanol, compared to ethyl alcohol, turned out to be a fuel that increased the rate of charge combustion, increasing the contribution of kinetic combustion and decreasing the contribution of diffusion combustion period. For 50% of alcohol, the use of methanol compared to ethanol increased the kinetic combustion period and led to a reduction in the diffusion combustion period by 13%.

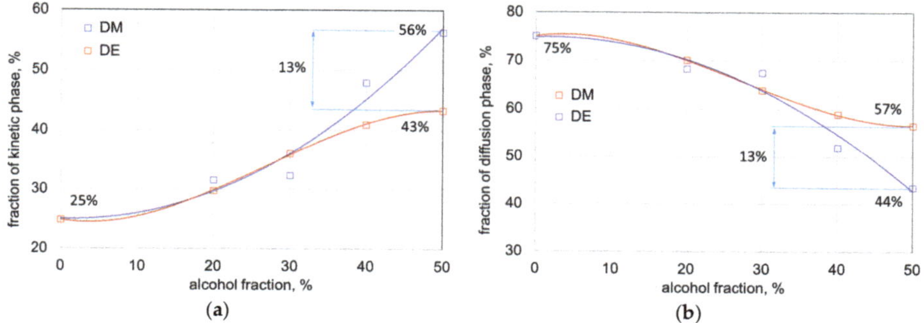

Figure 4. Kinetic (**a**) and diffusion (**b**) phase of the co-combustion process of diesel oil with methanol and ethanol for the analyzed energy shares of an alcohol fuel.

3.2. Cycle-to-Cycle Variation of the Engine

Combustion engine operation is characterized by the non-repeatability of subsequent operation cycles caused by the continuous changes in the conditions present in the engine cylinder, both during filling and combustion processes. Instantaneous conditions in the combustion chamber depend, among other things, on the wave phenomena in the intake channels during the suction stroke, instability of the ignition system and the ignition misfiring (SI engines) and the change in ignition delay (CI engines) associated with the heterogeneity of the combustible mixture. Furthermore, the increased non-repeatability of engine operation is affected by the instability of the rotational speed and engine temperature. The profiles of pressure changes in the cylinder obtained during engine indication for several hundred cycles differ from each other and always deviate to a greater or lesser extent from a single profile usually adopted as a representative profile. Figure 5 illustrates in cylinder pressure profiles and heat release rate during combustion of D100 and DM50 for 200 individual test engine operation cycles.

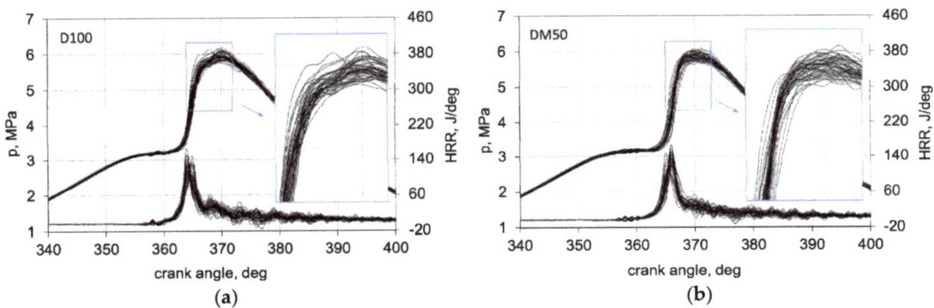

Figure 5. In cylinder pressure and heat release rate during combustion of D100 (**a**) and DM50 (**b**) for 200 individual cycles of a test engine.

The indicator diagrams can be used to determine characteristic values that are representative for many individual engine operation cycles, such as maximum combustion pressure or indicated mean effective pressure. Changes in these values can be computed using statistical analysis methods and presented in the form of the coefficient of variation of peak pressure (COV_{pp}) or of the coefficient of variation of the indicated mean effective pressure (COV_{IMEP}). This parameter can be used to evaluate stability of the internal combustion engine related to the non-repeatability of subsequent engine operation cycles and ignition misfiring. In paper [41], the COV of IMEP and p_{max} were both used to evaluate cyclic variations in multi-cylinder diesel engine fueled with diesel methanol dual fuel. Three different methanol injection positions were chosen to investigate their effect on cylinder-to-cylinder variation. According to literature, the threshold of correct operation of combustion engine, expressed by the maximal value of the coefficient of variation of the indicated mean effective pressure is 2–5% [35,42].

Figure 6 shows non-repeatability of the value of indicated mean effective pressure and maximum pressure, for a conventional diesel engine and a dual fuel engine for co-combustion of diesel with methanol and ethanol. It can be observed that in the case of both alcohols, the increase in their energy content leads to a reduction in the COV_{IMEP} coefficient, which suggests an improvement in engine operation stability. This may result from an increase in the intensification and speed of the combustion process that improved process repeatability due to the presence of active oxygen supplied in the alcohol fuel. The highest oxygen content is observed in methanol, with the lowest COV_{IMEP} = 1.63% values obtained for DM50. In paper [43] COV_{IMEP} coefficient was analyzed for a dual-fuel engine supplied with diesel fuel and methanol, at 60% load. The COV_{IMEP} of the dual-fuel engine remained smaller than 2% in all test conditions. This means the combustion of dual-fuel is very stable without large cycle-by-cycle variations. In paper [44] results are presented of investigation of the diesel/methanol dual-fuel combustion mode as a new combustion mode for NO_x and PM reduction from the diesel engine. The aim of this research was to investigate the effect of diesel fuel injection pressure on the performance and emissions characteristics in a 6-cylinder common-rail diesel engine. It showed that the COV_{IMEP} of the dual-fuel combustion mode remains under 2.1% among all the tests. Experimental investigations [31] of the engine with the ethanol-diesel dual-fuel combustion mode showed that, in the dual-fuel mode, more premixed combustion yielded higher levels of COV_{IMEP} than the conventional diesel combustion mode case for higher engine loads. The increase in COV_{IMEP} was proportional to the ethanol energy fraction. In addition, the dual-fuel operation resulted in higher COV_{pp} than the conventional diesel combustion mode at all engine loads except a lower load. Nevertheless, the COV_{IMEP} and COV_{pp} could be controlled between 1.0% and 3.0%. In the case of the analysis of non-repeatability of the maximal combustion pressure, no unambiguous changes in the COV_{pp} values were obtained. A low percentage of alcohol, both in the case of methanol and ethanol, led to a decrease of the COV_{pp} coefficient, which indicated the improved engine stability. After exceeding the level of 20% alcohol, an insignificant increase was observed in non-repeatability of engine operation, eventually reaching the value comparable to D100 for DM50 and DE50. It should be emphasized that both the engine powered by pure diesel and the engine for co-combustion of diesel with alcohol fuels were characterized by stable and even operation, for which the coefficient of non-repeatability was much lower than the acceptable limits for the reciprocating combustion engine (2–5%).

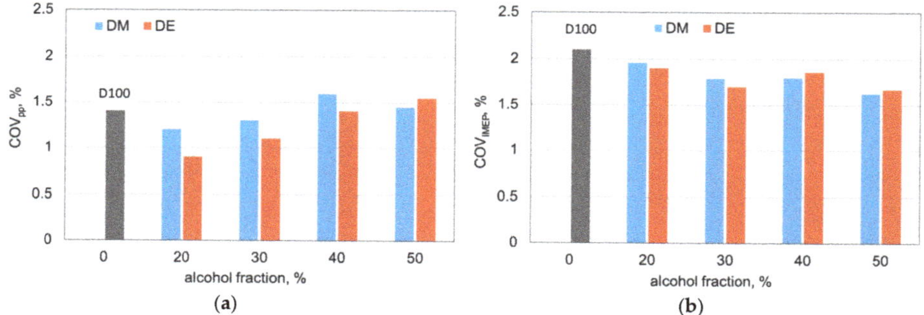

Figure 6. Coefficient of variation of peak pressure (**a**) and coefficient of variation of the indicated mean effective pressure (**b**) for a conventional engine and a dual fuel engine co-combustion diesel with ethanol and methanol.

One of the indicators that allow for the assessment of the operation stability of an internal combustion engine is probability density for the occurrence of selected parameters of engine operation. Figure 7 presents the probability density for the occurrence of IMEP in a conventional engine and a dual fuel engine for co-combustion of diesel with 20%, 30%, 40% and 50% of ethanol and methanol. The distribution of probability density of the indicated mean effective pressure, based on normal distribution, represents a repeatability index for individual IMEP values obtained for 200 analyzed cycles of the test engine operation. The mean values of IMEP presented in the drawings show insignificant differences in performance between combustion of pure diesel and co-combustion of diesel with methanol and ethanol. The largest differences were found for the lowest percentage of alcohol fuel, equal to 20% (Figure 7a).

The probability density curves of the indicated mean effective pressure show, among other things, the repeatability of the mean IMEP value, which may be considered a determinant of the stability of operation of the test engine. The probability density for IMEP only during combustion with the lowest percentage of alcohol fuel (20%) was lower compared to combustion of pure diesel (Figure 7a). In the case of higher percentage of both methanol and ethanol, higher values of f(IMEP) were recorded compared to D100. For the highest percentage of alcoholic fuel (50%) (Figure 7d), this value was 37.5% for methanol and 36.2% for ethanol, respectively.

Analysis of the profile of the probability density function in the case of DM50 for 200 engine operation cycles reveals that the individual IMEP values for the largest number of cycles were close to the mean value of IMEP. This suggests the best repeatability of subsequent cycles of the test engine operation for co-combustion of diesel with 50% energy from methyl alcohol. These findings are consistent with previous results obtained for the analysis of the stability of a dual fuel engine operation based on the changes in the coefficient of variation of the indicated mean effective pressure.

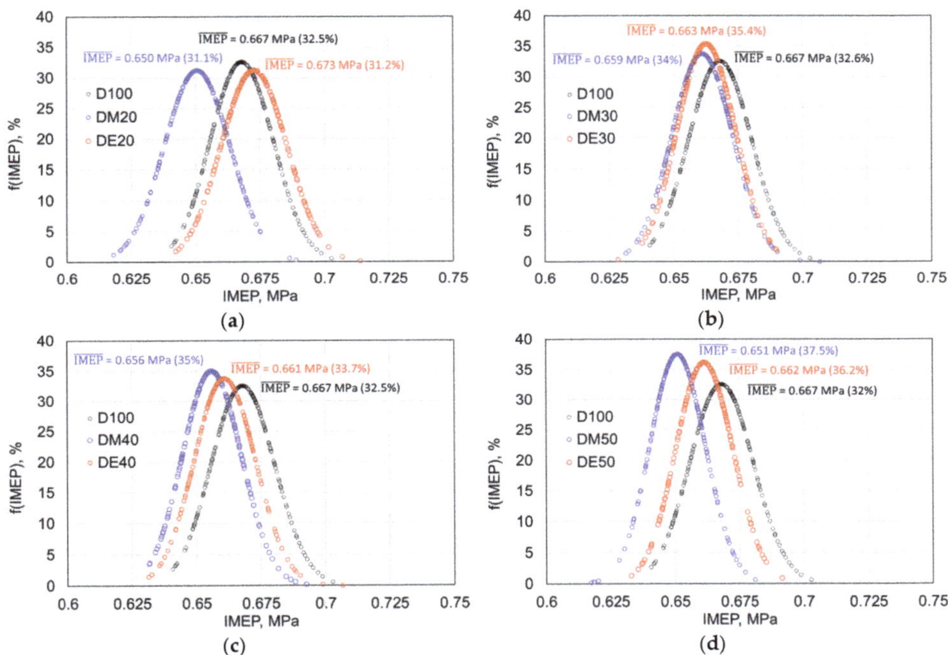

Figure 7. The probability density of the indicated mean effective pressure for co-combustion of diesel fuel with methanol and diesel fuel with ethanol for 20% (**a**), 30% (**b**), 40% (**c**) and 50% (**d**) alcohol share.

The correlation between the maximum combustion pressure and the indicated mean effective pressure for a conventional engine and a dual fuel engine for co-combustion of diesel with ethanol and methanol is presented in Figure 8. The figure illustrates the scattering of the pmax and IMEP values obtained for 200 engine cycles and the mean values of these parameters adopted as representative. The analysis of the scattering in subsequent cycles reveals that the increase in the indicated mean effective pressure was associated with the increase in the maximum combustion pressure for each case. Due to the fact that the tests were performed at constant engine load, insignificant changes in IMEP were obtained for each case, which did not differ much from the value of 0.66 MPa. Larger differences are observed in p_{max} values. Analysis of this parameter shows that for the lowest alcohol percentage (20%), the differences between co-combustion of methanol and ethanol are the greatest. The increase in the alcohol content to over 20% yielded p_{max} values similar for DM and DE while the differences were increased compared to D100. It is possible to obtain the expected IMEP in an engine for co-combustion of diesel with alcohol, even though the engine has a wider range of p_{max} changes compared to the combustion of pure diesel.

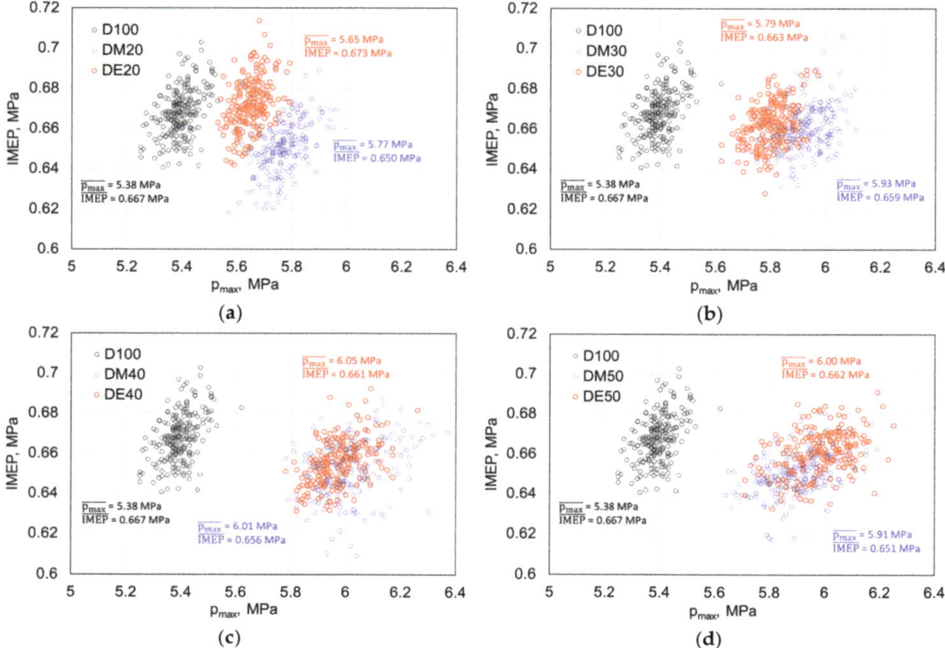

Figure 8. The correlation between the maximum combustion pressure and the indicated mean effective pressure for co-combustion of diesel fuel with methanol and diesel fuel with ethanol for 20% (**a**), 30% (**b**), 40% (**c**) and 50% (**d**) alcohol share.

4. Conclusions

Alcohol fuels offer opportunities to reduce the use of fossil fuels in CI engines and to increase the percentage of biofuels in the transport and energy sectors, where combustion engines are often used. This study presents experimental examinations of a CI engine with dual-fuel system, in which co-combustion of diesel with two alcohol fuels (methanol and ethanol) with energy content of 20%, 30%, 40% and 50% was performed. The research included the analysis of the combustion process and the analysis of the non-repeatability of the 200 subsequent engine operation cycles. Analysis of the results obtained in the study leads to the following conclusions:

-   presence and increase in the share of methanol and ethanol used for co-combustion with diesel causes an increase in compression ignition delay and increases the heat release rate and maximum combustion pressure values,

-   for high energy percentages of over 30%, methanol, compared to ethyl alcohol, turned out to be a fuel that is accompanied by a higher rate of charge combustion, contributing to intensification of the process and the increase in the contribution of the kinetic combustion period while decreasing the contribution of the diffusion combustion stage,

-   the analysis of the operation of a dual fuel engine based on the value of the coefficient of variation of the indicated mean effective pressure (COV_{IMEP}) for 200 operation cycles revealed an improvement in the stability of its operation with an increase in the percentage of alcohol. With the largest share of alcohol (50%), methanol proved to be a better fuel than ethanol. For DM50, the lowest COV_{IMEP} value of 1.63% was obtained.

-   based on the function of probability density of the indicated mean effective pressure (f(IMEP)) prepared for 200 engine operation cycles, it can be concluded that the increase in the percentage of alcohol fuel used for co-combustion with diesel is conducive to stability of operation of the dual-fuel engine. For a 50% share of alcoholic fuel, the higher value of f(IMEP), which demonstrates better engine stability, was characterized by methanol and was equal to 37.5%.

- assessment of engine operation stability based on the probability density function of selected engine performance parameters such as indicated mean effective pressure, leads to results similar to those obtained from the assessment based on the analysis of changes in the coefficient of variation of the indicated mean effective pressure,

- with the increase in the percentage of methanol and ethanol used for co-combustion with diesel, the dependence of the indicated mean effective pressure on the maximum combustion pressure becomes lower. Although combustion in a dual fuel engine was characterized by greater differences in maximum pressure compared to combustion in a conventional engine, the values of the indicated mean effective pressure (that reflect engine performance) were similar for both engines.

Author Contributions: Conceptualization, A.J. and W.T.; Data curation, A.J., W.T., R.G. and Ł.N.; Formal analysis, R.G.; Investigation, A.J., W.T., R.G. and Ł.N.; Methodology, A.J. and W.T.; Writing–original draft, A.J.; Writing–review & editing, W.T.

Funding: This research was funded by the Ministry of Science and Higher Education of Poland from the funds dedicated to scientific research No. BS/PB 1-103-3030/2018/P.

Conflicts of Interest: The authors declare no conflict of interest.

Abbreviations

CI	compression ignition engine
IMEP	indicated mean effective pressure, MPa
p_{max}	maximum pressure, MPa
COV_{IMEP}	coefficient of variation of the indicated mean effective pressure, %
COV_{PP}	coefficient of variation of peak pressure, %
σ_{IMEP}	standard deviation of the IMEP, MPa
σ_{pmax}	standard deviation of peak pressure, MPa
f(IMEP)	probability density of the indicated mean effective pressure, %
HRR	heat release rate, J/deg
V_d	displaced cylinder volume, m^3
p	pressure, bar
n	engine speed, rpm
TDC	top dead centre
λ	excess air ratio
SOI	start of injection
NO_x	nitrogen oxides
THC	total hydrocarbons
CO	carbon monoxide
CO_2	carbon dioxide
O_2	oxygen
PM	particulate matter

References

1. Yoon, S.K.; Kim, M.S.; Kim, H.J.; Choi, N.J. Effects of canola oil biodiesel fuel blends on combustion, performance, and emissions reduction in a common rail diesel engine. *Energies* **2014**, *7*, 8132–8149. [CrossRef]

2. Qasim, M.; Ansari, T.M.; Hussain, M. Combustion, performance, and emission evaluation of a diesel engine with biodiesel like fuel blends derived from a mixture of Pakistani waste canola and waste transformer oils. *Energies* **2017**, *10*, 1023. [CrossRef]

3. Tutak, W.; Jamrozik, A.; Bereczky, Á.; Lukács, K. Effects of injection timing of diesel fuel on performance and emission of dual fuel diesel engine powered by diesel/E85 fuels. *Transport* **2018**, *33*, 633–646. [CrossRef]

4. Chen, Z.; Yao, C.; Wang, Q.; Han, G.; Dou, Z.; Wei, H.; Wang, B.; Liu, M.; Wu, T. Study of cylinder-to-cylinder variation in a diesel engine fueled with diesel/methanol dual fuel. *Fuel* **2016**, *170*, 67–76. [CrossRef]

5. Directive 2009/28/EC of the European Parliament and of the Council, 23 April 2009. 2017. Available online: http://eur-lex.europa.eu/legal-content/PL/ALL/?uri=celex%3A32009L0028/ (accessed on 12 January 2019).

6. Atmanli, A.; Yilmaz, N. A comparative analysis of n-butanol/diesel and 1-pentanol/diesel blends in a compression ignition engine. *Fuel* **2018**, *234*, 161–169. [CrossRef]

7. Yilmaz, N.; Atmanli, A.; Vigil, F.M. Quaternary blends of diesel, biodiesel, higher alcohols and vegetable oil in a compression ignition engine. *Fuel* **2018**, *212*, 462–469. [CrossRef]

8. Ge, J.C.; Kim, M.S.; Yoon, S.K.; Choi, N.J. Effects of pilot injection timing and EGR on combustion, performance and exhaust emissions in a common rail diesel engine fueled with a canola oil biodiesel-diesel blend. *Energies* **2015**, *8*, 7312–7325. [CrossRef]

9. Tutak, W.; Jamrozik, A.; Gnatowska, R. Combustion of different reactivity fuel mixture in a dual fuel engine. *Therm. Sci.* **2018**, *22/3*, 1285–1297. [CrossRef]

10. Jamrozik, A.; Tutak, W.; Gruca, M.; Pyrc, M. Performance, emission and combustion characteristics of CI dual fuel engine powered by diesel/ethanol and diesel/gasoline fuels. *J. Mech. Sci. Technol.* **2018**, *32*, 2947–2957. [CrossRef]

11. Zahos-Siagos, I.; Karathanassis, V.; Karonis, D. Exhaust Emissions and Physicochemical Properties of n-Butanol/Diesel Blends with 2-Ethylhexyl Nitrate (EHN) or Hydrotreated Used Cooking Oil (HUCO) as Cetane Improvers. *Energies* **2018**, *11*, 3413. [CrossRef]

12. Kamaraj, R.K.; Raghuvaran, J.G.T.; Panimayam, A.F.; Allasi, H.L. Performance and Exhaust Emission Optimization of a Dual Fuel Engine by Response Surface Methodology. *Energies* **2018**, *11*, 3508. [CrossRef]

13. Tutak, W.; Jamrozik, A.; Pyrc, M.; Sobiepański, M. Investigation on combustion process and emissions characteristic in direct injection diesel engine powered by wet ethanol using blend mode. *Fuel Process. Technol.* **2016**, *149*, 86–95. [CrossRef]

14. Tutak, W.; Jamrozik, A.; Pyrc, M.; Sobiepański, M. A comparative study of co-combustion process of diesel-ethanol and biodiesel-ethanol blends in the direct injection diesel engine. *Appl. Therm. Eng.* **2017**, *117*, 155–163. [CrossRef]

15. Tutak, W.; Lukács, K.; Szwaja, S.; Bereczky, Á. Alcohol–diesel fuel combustion in the compression ignition engine. *Fuel* **2015**, *154*, 196–206. [CrossRef]

16. Tutak, W. Bioethanol E85 as a fuel for dual fuel diesel engine. *Energy Convers. Manag.* **2014**, *86*, 39–48. [CrossRef]

17. Kuszewski, H. Experimental investigation of the effect of ambient gas temperature on the auto-ignition properties of ethanol–diesel fuel blends. *Fuel* **2018**, *214*, 26–38. [CrossRef]

18. Huang, H.; Liu, Q.; Teng, W.; Pan, M.; Liu, Ch.; Wang, Q. Improvement of combustion performance and emissions in diesel engines by fueling n-butanol/diesel/PODE3–4 mixtures. *Appl. Energy* **2018**, *227*, 38–48. [CrossRef]

19. Paul, A.; Panua, R.; Debroy, D. An experimental study of combustion, performance, exergy and emission characteristics of a CI engine fueled by diesel-ethanol-biodiesel blends. *Energy* **2017**, *141*, 839–852. [CrossRef]

20. Alptekin, E. Evaluation of ethanol and isopropanol as additives with diesel fuel in a CRDI diesel engine. *Fuel* **2017**, *205*, 161–172. [CrossRef]

21. Wei, L.; Yao, C.; Han, G.; Pan, W. Effects of methanol to diesel ratio and diesel injection timing on combustion, performance and emissions of a methanol port premixed diesel engine. *Energy* **2016**, *95*, 223–232. [CrossRef]

22. Al-Hassan, M.; Mujafet, H.; Al-Shannag, M. An experimental study on the solubility of a diesel-ethanol blend and on the performance of a diesel engine fueled with diesel-biodiesel-ethanol blends. *Jordan J. Mech. Ind. Eng.* **2012**, *6*, 147–153.

23. Jamrozik, A. The effect of the alcohol content in the fuel mixture on the performance and emissions of a direct injection diesel engine fueled with diesel-methanol and diesel-ethanol blends. *Energy Convers. Manag.* **2017**, *148*, 461–476. [CrossRef]

24. Singh, P.; Chauhan, S.R.; Goel, V. Assessment of diesel engine combustion, performance and emission characteristics fuelled with dual fuel blends. *Renew. Energy* **2018**, *125*, 501–510. [CrossRef]

25. Jamrozik, A.; Tutak, W.; Pyrc, M.; Sobiepański, M. Experimental investigations on combustion, performance and emission characteristics of stationary CI engine fuelled with diesel-methanol and biodiesel-methanol blends. *Environ. Prog. Sustain. Energy* **2017**, *36*, 1151–1163. [CrossRef]

26. Li, T.; Zhang, X.Q.; Wang, B.; Guo, T.; Shi, Q.; Zheng, M. Characteristics of non-evaporating, evaporating and burning sprays of hydrous ethanol diesel emulsified fuels. *Fuel* **2017**, *191*, 251–265. [CrossRef]
27. Liu, H.; Ma, G.; Hu, B.; Zheng, Z.; Yao, M. Effects of port injection of hydrous ethanol on combustion and emission characteristics in dual-fuel reactivity controlled compression ignition (RCCI) mode. *Energy* **2018**, *145*, 592–602. [CrossRef]
28. Kyrtatos, P.; Brückner, C.; Boulouchos, K. Cycle-to-cycle variations in diesel engines. *Appl. Energy* **2016**, *171*, 120–132. [CrossRef]
29. Yasin, M.H.; Mamat, R.; Abdullah, A.A.; Abdullah, N.R.; Wyszynski, M.L. Cycle-to-Cycle Variations of a Diesel Engine Operating with Palm Biodiesel. *J. Kones* **2013**, *20/3*, 443–450.
30. Kyrtatos, P.; Zivolic, A.; Brückner, C.; Boulouchos, K. Cycle-to-cycle variations of NO emissions in diesel engines under long ignition delay conditions. *Combust. Flame* **2017**, *178*, 82–96. [CrossRef]
31. Pedrozo, V.B.; May, I.; Guan, W.; Zhao, H. High efficiency ethanol-diesel dual-fuel combustion: A comparison against conventional diesel combustion from low to full engine load. *Fuel* **2018**, *230*, 440–451. [CrossRef]
32. Taylor, J.R. *Introduction to Error Analysis: The Study of Uncertainties in Physical Measurements*; University Science Books: Sausalito, CA, USA, 1997.
33. Fayyazbakhsh, A.; Pirouzfar, V. Comprehensive overview on diesel additives to reduce emissions, enhance fuel properties and improve engine performance. *Renew. Sustain. Energy Rev.* **2017**, *74*, 891–901. [CrossRef]
34. Yusri, I.M.; Mamat, R.; Najafi, G.; Razman, A.; Awad, O.I.; Azmi, W.H.; Ishak, W.F.W.; Shaiful, A.I.M. Alcohol based automotive fuels from first four alcohol family in compression and spark ignition engine: A review on engine performance and exhaust emissions. *Renew. Sustain. Energy Rev.* **2017**, *77*, 169–181. [CrossRef]
35. Heywood, J.B. *Internal Combustion Engine Fundamentals*, 2nd ed.; McGraw-Hill Education: New York, NY, USA, 2018.
36. Kumar, M.S.; Nataraj, G.; Arulselvan, S. A comprehensive assessment on the effect of high octane fuels induction on engine's combustion behaviour of a Mahua oil based dual fuel engine. *Fuel* **2017**, *199*, 176–184. [CrossRef]
37. Wei, L.; Yao, C.; Wang, Q.; Pan, W.; Han, G. Combustion and emission characteristics of a turbocharged diesel engine using high premixed ratio of methanol and diesel fuel. *Fuel* **2015**, *140*, 156–163. [CrossRef]
38. Liu, J.; Yao, A.; Yao, C. Effects of injection timing on performance and emissions of a HD diesel engine with DMCC. *Fuel* **2014**, *134*, 107–113. [CrossRef]
39. Jamuwa, D.K.; Sharma, D.; Soni, S.L. Experimental investigation of performance, exhaust emission and combustion parameters of stationary compression ignition engine using ethanol fumigation in dual fuel mode. *Energy Convers. Manag.* **2016**, *115*, 221–231. [CrossRef]
40. Li, Y.; Zhang, C.; Yu, W.; Wu, H. Effects of rapid burning characteristics on the vibration of a common-rail diesel engine fueled with diesel-methanol dual-fuel. *Fuel* **2016**, *170*, 176–184. [CrossRef]
41. Chen, Z.; Yao, C.; Yao, A.; Dou, Z.; Wang, B.; Wei, H.; Liu, M.; Chen, C.; Shi, J. The impact of methanol injecting position on cylinder-to-cylinder variation in a diesel methanol dual fuel engine. *Fuel* **2017**, *191*, 150–163. [CrossRef]
42. Reyes, M.; Tinaut, F.V.; Giménez, B.; Pérez, A. Characterization of cycle-to-cycle variations in a natural gas spark ignition engine. *Fuel* **2015**, *140*, 752–761. [CrossRef]
43. Li, G.; Zhang, C.; Li, Y. Effects of diesel injection parameters on the rapid combustion and emissions of an HD common-rail diesel engine fueled with diesel-methanol dual-fuel. *Appl. Therm. Eng.* **2016**, *108*, 1214–1225. [CrossRef]
44. Liu, J.; Yao, A.; Yao, C. Effects of diesel injection pressure on the performance and emissions of a HD common-rail diesel engine fueled with diesel/methanol dual fuel. *Fuel* **2015**, *140*, 192–200. [CrossRef]

Article

An Experimental Study on the Performance and Emission of the diesel/CNG Dual-Fuel Combustion Mode in a Stationary CI Engine

Arkadiusz Jamrozik *, Wojciech Tutak and Karol Grab-Rogaliński

Faculty of Mechanical Engineering and Computer Science, Czestochowa University of Technology, 42-201 Czestochowa, Poland; tutak@imc.pcz.pl (W.T.); grab@imc.pcz.pl (K.G.-R.)
* Correspondence: jamrozik@imc.pcz.pl

Received: 23 September 2019; Accepted: 8 October 2019; Published: 12 October 2019

Abstract: One of the possibilities to reduce diesel fuel consumption and at the same time reduce the emission of diesel engines, is the use of alternative gaseous fuels, so far most commonly used to power spark ignition engines. The presented work concerns experimental research of a dual-fuel compression-ignition (CI) engine in which diesel fuel was co-combusted with CNG (compressed natural gas). The energy share of CNG gas was varied from 0% to 95%. The study showed that increasing the share of CNG co-combusted with diesel in the CI engine increases the ignition delay of the combustible mixture and shortens the overall duration of combustion. For CNG gas shares from 0% to 45%, due to the intensification of the combustion process, it causes an increase in the maximum pressure in the cylinder, an increase in the rate of heat release and an increase in pressure rise rate. The most stable operation, similar to a conventional engine, was characterized by a diesel co-combustion engine with 30% and 45% shares of CNG gas. Increasing the CNG share from 0% to 90% increases the nitric oxide emissions of a dual-fuel engine. Compared to diesel fuel supply, co-combustion of this fuel with 30% and 45% CNG energy shares contributes to the reduction of hydrocarbon (HC) emissions, which increases after exceeding these values. Increasing the share of CNG gas co-combusted with diesel fuel, compared to the combustion of diesel fuel, reduces carbon dioxide emissions, and almost completely reduces carbon monoxide in the exhaust gas of a dual-fuel engine.

Keywords: CNG; diesel fuel; dual fuel engine; rate of heat release; ignition delay; burn duration; exhaust gas emission

---

1. Introduction

The piston-based internal combustion engine is a heat machine, which is still the most common device for driving motor vehicles and various types of working machines [1]. Currently, internal combustion engines are increasingly used in stationary solutions, in small distributed energy, as a source of propulsion for cogeneration units, generating electricity and heat (CHP) for the needs of the local market [2,3]. For stationary applications, there is a wide range of internal combustion engines, both spark ignition and compression ignition. Stationary internal combustion engines are very often built as compression-ignition diesel engines, which have a number of advantages compared to spark-ignition engines. The most important of them are: higher reliability, fuel economy, higher power range, longer life, faster response to power demand and higher torque. Diesel fuel obtained from crude oil is a typical fuel supplying compression-ignition (CI) engines. The combustion of this type of fuel is accompanied by the emission of harmful compounds, such as carbon monoxide, carbon dioxide, hydrocarbons, nitrogen oxides and soot [4,5]. The vision of the depletion of oil resources, and the impact of CO_2, as a greenhouse gas, on the rise in the average temperature on Earth are widely known. To counteract this, restrictive emission standards, bans on vehicles with combustion engines entering city centers, or EU directives on the use of alternative and renewable energy sources have been introduced [6,7]. One of the possibilities to reduce the combustion of diesel fuel and at the same time reduce the emission of diesel engines is the use of alternative fuels, so far most commonly used to power spark ignition engines [8,9]. Most of that type of alternative fuels, however, due to their properties, cannot be burned independently in a compression-ignition engine; therefore, in recent years research has been ongoing on the technology of the co-combustion of alternative fuels with diesel in a dual-fuel engine [10]. Alternative fuels include hydrocarbon gas fuels [11,12]. The increase in interest in supplying combustion engines with hydrocarbon gas fuels observed in the last several years is caused by two important features of these fuels. The first is the low cost of energy obtained from them. The prices of gaseous fuels on European markets allow obtaining a cost of energy unit at the level of 30–50% in the case of natural gas and 40–60% in the case of liquefied gas, in relation to the cost of the energy unit contained in liquid fossil fuels. The second feature is the favorable ecological properties of gaseous fuels. Natural gas, unlike other fossil fuels such as coal and oil, due to its advantages, is one of the gaseous fuels that are increasingly used in the automotive industry, large energy and local distributed energy based on internal combustion engines [13,14]. Most often, natural gas is used in compressed form as CNG (compressed natural gas) [15]. CNG is a fuel increasingly used in dual-fuel compression-ignition engines, both naturally aspirated and supercharged [16,17]. Meng et al. [18] studied the co-combustion of a mixture of diesel fuel and n-butanol with CNG gas in a dual-fuel diesel engine. The use of a mixture of diesel fuel and n-butanol as a dose of pilot fuel was intended to improve engine performance and emissions. Three types of pilot-doses were used in the studies, including B0 (pure diesel), B10 (90% diesel and 10% n-butanol) and B20 (80% diesel and 20% n-butanol). Experiments were carried out for two loads, at different pilot-dose injection angles. Different CNG gas shares were analyzed for each load. For the first load (5 bar IMEP), the share of CNG gas was 60 and 80%. For B10CNG40, the highest thermal efficiency (ITE) and the lowest THC (total hydrocarbons) emissions were obtained with a slight increase in NO_x. For B20CNG70, a decrease in NO_x emission was obtained due to better homogeneity of the combustible mixture and higher latent heat of vaporization of n-butanol. In the case of higher loads (7.5 bar IMEP), the shares of CNG gas were 60% and 80%. For B10CNG60 and B20CNG60 compared to B10CNG80 and B20CNG80, ITE improvement and reduction of THC emissions were achieved, while the level of NO_x emissions remained constant. Ryu [19,20] conducted research on a single-cylinder, dual-fuel diesel engine fueled with biodiesel and CNG. In that experiment, he used a biodiesel pilot-dose injection to ignite the main charge consist of air and compressed natural gas (CNG). In paper [19], he studied the impact of pilot-dose injection angle on combustion characteristics, engine performance and emissions at a constant injection pressure of about 120 MPa, and a change in injection angle in the range of 11° to 23° before TDC. The results showed that engine performance can be improved by optimizing the start angle of biodiesel and CNG co-combusting. Improvement in performance, including a reduction in specific fuel consumption for low loads, was achieved due to the earlier angle of pilot-dose injection, while for high loads, it was beneficial to delay the injection angle of pilot-dose. The use of biodiesel-CNG dual fuel combustion mode compared to diesel single combustion mode caused a delay in ignition of the charge in the engine cylinder. The ignition delay decreased as the engine load increased. For the dual fuel engine compared to a conventional diesel engine, there was reduced smoke opacity, and NO_x and CO_2 emissions. In addition, relatively high CO and hydrocarbon (HC) emissions were obtained, especially under low load conditions, due to the low combustion temperature of CNG. The article [20] presents the results of research on the impact of biodiesel pilot injection pressure on performance and exhaust emissions during co-combustion of biodiesel and CNG. The results show that the indicated mean effective pressure (IMEP) for biodiesel-CNG dual fuel combustion mode is lower compared to Diesel single combustion mode, with the injection pressure increased above 30 MPa. Increasing the injection pressure resulted in a reduction in ignition delay and a reduction of combustion time, as well as a reduction of exhaust gas smoke opacity and an increase in NO_x emissions. The combustion stability of a dual-fuel engine increased with increasing injection pressure. Bari and Hossain [21] conducted experiments to examine the efficiency of a diesel engine fueled with CNG and diesel in dual-fuel mode, with various proportions of diesel, ranging from 10% to 100%. The tests were carried out for several selected engine loads. The results showed that the co-combustion of CNG and diesel, compared to the combustion of pure diesel, leads to a reduction in engine efficiency (BTE—brake thermal efficiency) and an increase in specific fuel consumption (BSFC). The largest decrease in efficiency and increase in fuel consumption was recorded for the smallest load—1.1 kW. At that load, BSFC growth was 68%. Analyses of exhaust gas composition have shown that the use of CNG to power a dual-fuel engine leads to a reduction in smoke and CO_2 emissions and an increase in CO emissions. The performances and emissions of dual-fuel compression-ignition engines are significantly influenced by the moment of starting the combustion process, which depends on the injection time of the pilot fuel dose and the amount of fuel. Liu et al. [22] studied the emission characteristics of a dual-fuel engine powered by CNG and diesel with the optimization of injection angle and various amounts of a diesel fuel pilot-dose. The results of the experiments showed that the CO emission in the dual-fuel engine was much higher and the NO_x emission was on average 30% lower compared to a conventional diesel engine. HC emissions in dual-fuel combustion mode are higher, especially at low and medium loads. In the partial load range, natural gas used in dual-fuel compression-ignition engines very often leads to reduced performance and reduced emissions. In order to eliminate some disadvantages of CNG gas, Karabektas et al. [23] proposed in a diesel-CNG dual-fuel engine, modification of the pilot fuel dose composition by adding diethyl ether (DEE) to diesel fuel. The research concerned the combustion of diesel alone, the co-combusting of diesel with 40% CNG and the co-combusting of diesel and DEE with 40% CNG. The pilot-doses of DEE were 5% and 10%. It was found that the co-combustion of diesel fuel with CNG gas compared to the combustion of diesel fuel alone causes deterioration of engine performance, especially at low and medium loads. In addition, it has higher CO and HC emissions at all loads and lower NO_x emissions at high loads. The use of DEE as an additive to a pilot-dose leads to an improvement in thermal efficiency and a reduction in specific energy consumption, resulting in lower CO and NO_x emissions.

The present work concerns tests of a stationary compression-ignition engine, adapted for dual-fuel operation, in which diesel fuel was combined with CNG (compressed natural gas). As part of the work, experimental studies were carried out on the impact of the CNG gas's energy share on selected engine operating parameters; i.e., cylinder pressure, pressure increase rate, heat release rate, autoignition delay or combustion time, and an analysis of harmful compounds' emissions in this engine exhaust gas was performed. As part of the presented research, an assessment was also performed, of the stability of operation of a dual-fuel diesel and CNG engine. In the available literature is a lack of sufficient information on that type of analysis. The engine stability, manifested in the uniqueness of subsequent cycles, was analyzed. This was determined on the basis of changes in the value of the uniqueness coefficient of the maximum pressure—COV_{pmax} and the probability density of the occurrence of the maximum pressure—$\Phi(p_{max})$. The tests were carried out for a wide range of CNG energy shares, from 0% to 95%.

2. Experimental Methodology

2.1. Experimental Setup

The research carried out as part of the study was experimental, in which the object of the study was a single-cylinder, naturally aspirated four-stroke diesel engine, Andoria 1CA90. This engine is a stationary, two-valve unit, with a vertical cylinder arrangement, in which an air-cooling system with an axial fan is used. The engine was designed to work with a constantly maximal load, at a constant rotation speed of 1500 rpm. The engine was adapted for dual-fuel operation by equipping it with an additional CNG gas supply system, injected under a pressure of approximately 3 bar into the intake manifold. The Servojet SP051S1 gas injector was operated by an external control system that made it possible to synchronize the injector with the engine and precisely control its opening time. This allowed for correct and even gas injecting during the suction stroke. Based on earlier tests of this engine, it was determined that the optimal diesel injection angle is 17° before TDC [24]. Figure 1 shows a diagram of a test bench with a test engine, while Table 1 shows the technical data of the 1CA90 engine. The stand was equipped with a system for measuring and recording combustion pressure in the engine cylinder, and systems for measuring the consumption of liquid fuel and gaseous fuel, air consumption, and the contents of engine exhaust components. Table 2 presents the parameters of the exhaust gas analyzer used during the tests. The pressure measurement system consisted of a piezoelectric pressure sensor, load amplifier, crankshaft angle sensor and data acquisition system with an analog-to-digital converter.

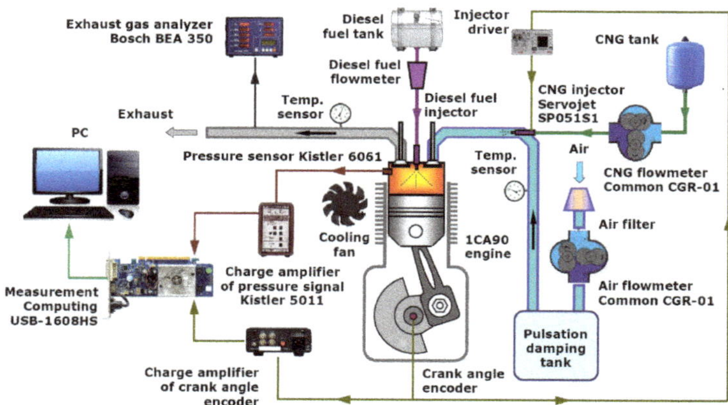

Figure 1. Schematic diagram of the experimental setup.

Table 1. Main engine parameters.

Parameter	Value
Engine	1CA90 Andoria
Type of engine	four stroke compression-ignition
Number of cylinders	1
Bore	90 mm
Stroke	90 mm
Displaced volume	573 cm^3
Number of valves	2
Compression ratio	17
Engine speed	1500 rpm
Diesel injection	direct injection
CNG injection	port injection
Diesel injection pressure	210 bar
CNG injection pressure	3 bar
Diesel injection timing	343°

The study used the following measurement apparatus:

(i)   Pressure sensor Kistler 6061—range 0–250 bar, linearity < ±0.5% FS;

(ii)   Charge amplifier Kistler 5011—range ±10–$\pm999{,}000$ pC for 10V FS, error $< \pm3\%$, linearity $<\pm$ 0.05% FS;

(iii)   Data acquisition module, Measurement Computing USB-1608HS—16 bits resolution, sampling frequency 20 kHz;

(iv)   Air rotor flowmeter Common CGR-01 G40 DN50—measuring range 0.65–65 m^3/h, accuracy class 1;

(v)   CNG rotor flowmeter Common CGR-01 G10 DN50—measuring range 0.25 . . . 25 m^3/h, accuracy class 1;

(vi)   Bosch BEA 350 analyzer.

Table 2. Parameters of the Bosch BEA 350 analyzer [25].

Apparatus	Measuring Range	Resolution	Accuracy from Measured Value	Absolute Accuracy
CO	0.000–10.000% vol. 0.000–5.000% vol.	0.001% vol. 0.001% vol.	. . . $\pm5\%$	. . . $\pm0.06\%$ vol.
HC	0–9999 ppm vol. 0–2000 ppm vol.	1 ppm vol. 1 ppm vol.	. . . $\pm5\%$	. . . ±12 ppm vol.
CO_2	0.00–18.00% vol. 0.00–16.00% vol.	0.01% vol. 0.01% vol.	. . . $\pm5\%$	. . . $\pm0.5\%$ vol.
O_2	0.00–22.00% vol. 0.00–21.00% vol.	0.01% vol. 0.01% vol.	. . . $\pm4\%$	. . . $\pm0.1\%$ vol.
NO	0–5000 ppm vol. 0–4000 ppm vol.	1 ppm vol. 1 ppm vol.	$\pm4\%$ $\pm8\%$	±25 ppm vol. ±50 ppm vol.
λ	0.500–9.999 0.700–1.300	0.001 0.001	. . . $\pm4\%$	. . . . . .

2.2. Methodology

During the tests, the engine was operated at full load and rotational speed dedicated to the serial engine. The tests were carried out after thermal stabilization of the engine. The tests were carried out at the factory settings of the serial engine, for which there was no knocking problem. Increasing the energy share of CNG gas consisted of a gradual increase in the gas injector opening time, while reducing the amount of diesel fuel fed. The energy dose of the two fuels supplied to the engine cylinder was kept approximately constant, corresponding to 1500 J per cycle. The main scope of research included recording the value of variable pressure in the cylinder of the research engine, every 1 degree of crank angle for three measurements containing 200 subsequent engine cycles. The program used for recording and analyzing in real time the pressure signal in the cylinder during the engine running used an experimental pressure recording system as a function of the crankshaft rotation angle, based on a data acquisition module with an A/D converter, a piezoelectric pressure sensor and a crankshaft rotation angle marker. In addition, engine speed, air consumption, diesel fuel consumption, CNG gas consumption, air temperature, exhaust gas temperature and ambient pressure and temperature were measured during the tests. The first stage of the experiment concerned the testing of an engine running on diesel alone as a reference fuel. The next, primary part of the work included tests of the engine working in a dual-fuel system, with the engine fed with diesel fuel and compressed natural gas. These studies were conducted for a wide range of CNG energy shares: from 30% to 95%.

2.3. Test Fuels

Diesel fuel, offered by the Polish refinery, was used as the reference fuel in the analysis, which is used in combination with liquid hydrocarbons separated from oil lubrication in the distillation processes. The fuel used includes the quality standards defined in the PN-EN 590 standard, applicable on the markets of the European Union and are available for motor vehicles equipped with compression-ignition engines. The key parameter of diesel fuel is the cetane number determining the ability of a fuel for auto-ignition under the influence of high temperature. The minimum value of the cetane number of diesel fuel offered on the Polish market, guaranteeing meeting the requirements related to the operation of the engine in various conditions of use, is 51 and is determined by regulations. Another important parameter of motor fuel is the calorific value, which determines the amount of heat released when burning a mass unit or a unit of fuel volume. The calorific value of diesel fuel is about 42.5 MJ/kg. The alternative fuel co-cobusted with diesel was CNG compressed natural gas. In practical applications, CNG is used in both spark-ignition engines and compression-ignition engines. In the CI engine, due to the low cetane number, close to 0, and high auto-ignition temperature (650 °C), a small amount of diesel fuel injection is used to ignite the air and natural gas mixture, whose task is to create self-ignition regions to initiate combustion [26]. The pro-ecological properties of natural gas are due to the fact that its main combustible component is methane—CH_4, the content of which ranges from 90% to 99% depending on the gas source. Methane is the simplest hydrocarbon, with one carbon atom, widely regarded as a non-toxic component. Due to the fact that in methane has up to four hydrogen atoms per one carbon atom, as a result of its combustion, about 24.5% less carbon dioxide is created compared to traditional liquid fuels [27]. Natural gas also includes hydrocarbons with a lower hydrogen to carbon ratio, such as ethane, propane, butane and pentane. The calorific value of CNG gas, compared to diesel, is higher and amounts to approximately 49 MJ/kg. Table 3 presents the basic properties of diesel fuel and CNG gas, while Table 4 presents the composition of natural gas.

Table 3. Fuel specifications.

Parameter	Diesel	CNG
Cetane number	51	0
Methane number	-	82
Research octane number	15–25	110–130
Density at 1 atm and 15 °C (kg/m^3)	840	0.72–0.76
Lower heating value (MJ/kg)	42.5	49.15
Heat of evaporation (kJ/kg)	243	510
Auto-ignition temperature (°C)	180–230	650
Stoichiometric air-fuel ratio	14.6	17.05
Viscosity at 20 °C (Pa·s)	2.8×10^{-3}	$1–1.4 \times 10^{-6}$
Boiling point (°C)	180–360	−162
Carbon content (%)	85	75

In the presented research, the process of co-combustion diesel fuel with CNG gas in a dual-fuel diesel engine was analyzed for CNG energy shares varying from 0% to 95%. The amount of energy in diesel fuel and CNG gas that was supplied to the engine cylinder for one work cycle was approximately constant and close to 1500 J/cycle (Figure 2a). With the change in the share of both co-combusted fuels, the total mass of fuel, due to different calorific values of diesel and CNG, was different. With the increase in the share of CNG gas, the total fuel mass decreased from 34.88 to 26.61 mg/cycle (Figure 2b).

Table 4. Compressed natural gas (CNG) composition.

Component	v/v (%)
Methane CH_4	96.1
Ethane C_2H_6	2.50
Propane C_3H_8	0.40
Butane C_4H_{10}	0.14
Pentane C_5H_{12}	0.01
Nitrogen N_2	0.59
Carbon dioxide CO_2	0.15

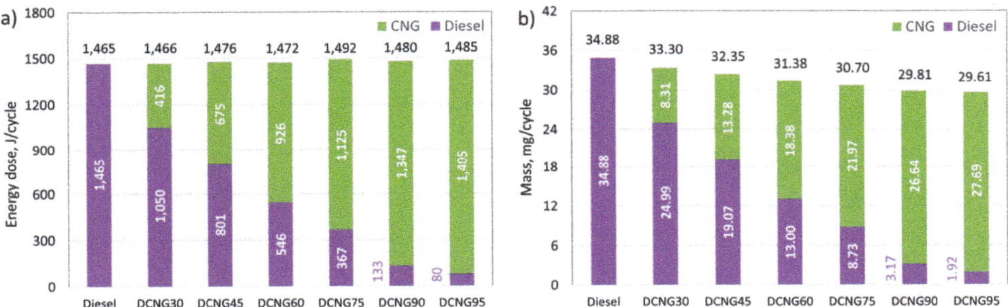

Figure 2. Energy (**a**) and mass (**b**) of the fuel delivered to the engine per one work cycle.

3. Test Method and Conditions

The study examined the impact of the energy share of compressed natural gas in the air-fuel mixture during co-combustion of this gas with diesel fuel on selected parameters of the combustion process. The energy share of CNG was calculated based on the equation:

$$CNG\% = \frac{\dot{m}_{CNG}\,LHV_{CNG}}{\dot{m}_{CNG}\,LHV_{CNG} + \dot{m}_D\,LHV_D}\,100\%, \tag{1}$$

where $\dot{m}_{CNG}$ and $\dot{m}_D$ correspond to CNG gas and diesel fuel consumption per engine cycle, LHV_{CNG} and LHV_D, respectively, represent the calorific values of CNG gas and diesel.

The heat release rate (HRR) is one of the indicators characterizing the combustion process in an internal combustion engine cylinder. HRR can be determined on the basis of registered changes in pressure in the cylinder, from the cylinder pressure graph, by calculating the changes in internal energy and the indicated work factor. Heat release rate:

$$HRR = \frac{1}{\chi - 1}\left[\chi p\frac{dV}{df} + V\frac{dp}{df}\right], \tag{2}$$

where χ is the ratio of specific heats, V is cylinder volume and p is in the cylinder pressure.

The nature of the work of a diesel engine is significantly influenced by the value of the pressure rise rate. The rate of pressure rise $dp/d\varphi$ was determined:

$$\frac{dp}{df} = \frac{p_k - p_{k-1}}{f_k - f_{k-1}}, \tag{3}$$

where p is in cylinder pressure, φ is crank angle and k is a current angle of crankshaft rotation for engine cycle.

Based on the combustion pressure charts, characteristic values representative of many individual engine cycles, such as average maximum combustion pressure, can be determined. Changes in these quantities can be calculated using statistical analysis methods and presented as coefficients of variation for the maximum pressure (COV_{pmax}) [28].

The average value of maximum pressure p_{max}, determined with the set of pressures, is:

$$\overline{P}_{max} = \frac{1}{N}\sum_{i=1}^{N} P_{maxi},\tag{4}$$

where N is the cycle index, $p_{max\,i}$ is the maximum pressure in individual cycles and *i* is the cycle number.

One of the most commonly used criteria for assessing the correct operation of an internal combustion engine is its cycle-by-cycle variation. As a measure of the cycle-by-cycle variation of the engine, the coefficient of variation for the maximum pressure COV_{pmax} can be taken, expressed as a percentage, and calculated as the ratio of the maximum standard pressure deviation to its average value over many recorded engine cycles.

Coefficient of variation of maximum pressure (COV_{pmax}) is defined as:

$$COV_{pmax} = \frac{STD_{pmax}}{\overline{P}_{max}}\,100\%,\tag{5}$$

The standard deviation of maximum pressure:

$$STD_{pmax} = \sqrt{\frac{1}{N}\sum_{i=1}^{N}\left(P_{maxi} - \overline{P}_{max}\right)^2},\tag{6}$$

where N is the cycle index and $\overline{P}_{max}$ is the mean value of maximum pressure of N cycle p_{maxi}.

One of the indicators enabling the assessment of the stability of operation of an internal combustion engine may also be the probability density of selected parameters of its operation [29]. The function that allows us to express the probability of obtaining or occurrence of a specific value of the analyzed parameter is the probability density function. The maximum pressure probability density distribution is an indicator of the repeatability (probability of occurrence) of individual p_{max} values obtained for many analyzed test engine work cycles. This function can also be used as an indicator for assessing the stability of an internal combustion engine. It indicates, inter alia, the frequency of occurrence of the most frequently repeated p_{max} value, close to the average value, showing the repeatability of subsequent engine cycles. The p_{max} probability density is a normal (Gaussian) distribution in which the density function is symmetrical in relation to the mean value of the distribution [30].

Maximum pressure probability density:

$$\Phi\left(P_{max}\right) = \frac{1}{STD_{pmax}\sqrt{2\pi}}\exp\left(\frac{-\left(P_{maxi} - \overline{P}_{max}\right)^2}{2\,STD_{pmax}^2}\right),\tag{7}$$

4. Results and Discussion

4.1. Combustion Characteristics

The basic source of information about the combustion process in the cylinder of the research engine was the results of pressure measurement expressed in the function of crankshaft rotation. Figure 3a,b presents the trace of pressure and heat release rates as well as the pressure rise rate in the cylinder of the engine co-combusting CNG with diesel fuel, for the entire range of gas shares, from 0%

to 95%. It can be seen that when the CNG energy share increases to 45%, the pressure and heat release rate increase, and the combustion pressure in the cylinder of the research engine increases. Increasing the gas share above 45% causes a decrease in p_{max}, HRRmax and dp/dφ due to ignition delay and prolonged combustion. The maximum value of dp/dφ = 0.76 MPa/degree obtained for DCNG45 is lower than the value obtained for diesel by 0.27 MPa/degree, and is lower than the permissible value for internal combustion engines of 1 MPa/degree [31]. This proves the lack of so-called hard work, which can be harmful to the engine structure, especially during a long operation. DCNG90 and DCNG95 show a rapid slowdown in combustion and a decrease in the rate of heat release. It can be seen that, for gas shares up to 75%, the dominant phase of the combustion process is the kinetic phase; combustion becomes too slow for larger CNG shares. Figure 3a shows a decrease in pressure in the compression stroke, along with an increase in the share of CNG gas. It results from the decrease in charge temperature in the intake manifold due to expansion of CNG gas from an injection pressure of 2 bar to atmospheric pressure. In addition, the pressure in the engine cylinder during compression is exponentially dependent on the specific heat ratio (which is the ratio of specific heat at constant pressure and constant volume), which for CNG gas is almost twice lower than for clean air.

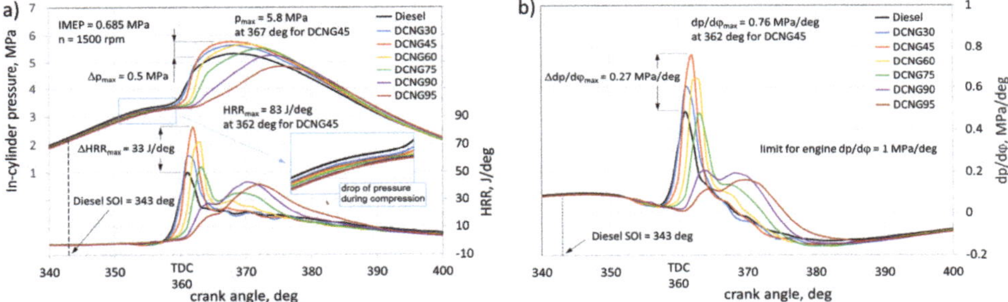

Figure 3. In cylinder combustion pressure, heat release rate (HRR) (**a**) and the rate of pressure rise (dp/dφ) (**b**) for co-combustion of CNG and diesel fuel.

Figure 4 shows the course of the normalized heat release rate function in the combustion process, as well as the ignition delay and duration of combustion for the analyzed shares of CNG gas co-combusted with diesel fuel. The normalized heat released rate can be used to determine two very important indicators characterizing the combustion process in a piston engine. The first is the auto-ignition delay, which was defined as the time expressed in crankshaft rotation angles, between the injection of diesel fuel pilot-dose until 10% of the total heat was released. The second parameter is the duration of combustion, defined as the time from the release of 10% of the heat until the release of 90% of the heat. The phenomenon of spontaneous combustion is characterized by the initial period in which chemical reactions before ignition play a key role. The so-called chemical delay is counted from the beginning of these reactions. The physical phenomena that cause a delay in the ignition of a diesel engine are as follows: the breakdown of fuel into separate drops, the heating and evaporation of drops, and finally, the diffusion of fuel vapors into the air. The rate of combustion of liquid fuel is determined by the rate of its evaporation and mixing of the atomized fuel with air. The dynamics of these physical and chemical phenomena that occur at the start of the ignition process depend on the temperature. Shorter ignition delay in compression-ignition engines improves engine efficiency (less fuel consumption), improves engine adjustment (more effective control of injection timing), improves engine starting and reduces the pressure rise rate in the combustion process in the cylinder—which reduces operating noise and reduces loads on the crankshaft and piston pin. Based on the graphs prepared, it can be concluded that the addition of CNG gas increases the ignition delay of the combustible mixture in the cylinder of a dual-fuel engine and reduces the total duration of combustion. During the combustion of the mixture with the largest 95% share of CNG gas, compared to the combustion of diesel fuel, a

31.5% increase in ignition delay (ID) and a 55% reduction in the burn duration (BD) was obtained. The increase in ignition delay was caused by the low self-ignition tendency of CNG gas expressed by high auto-ignition temperature and the low, close to 0, cetane number. The reduction in the duration of combustion for significant CNG shares was caused, among other things, by the improvement in the homogeneity of the combustible mixture, which, being homogeneous throughout the cylinder volume, burned more intensively and faster.

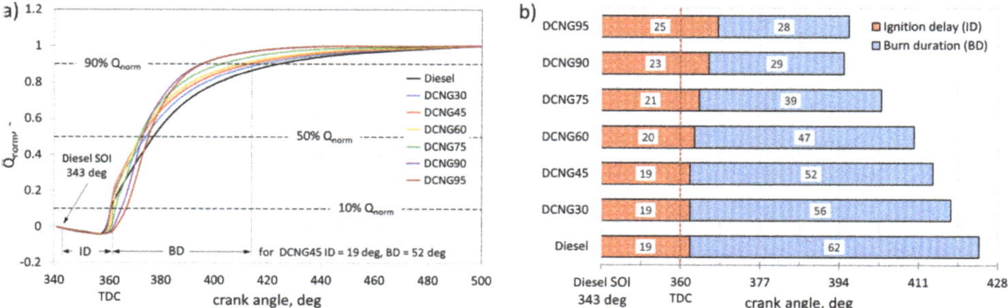

Figure 4. Normalized heat release (**a**) and ignition delay with burn duration (**b**) for analyzed share of CNG gas co-combusted with diesel fuel.

4.2. Engine Stability

The combustion engine's stability related to the cycle-by-cycle variation and the misfiring can be assessed on the basis of changes in the value of the coefficient of variation for maximum pressure—COV_{pmax}, and the probability density of the occurrence of the maximum pressure—$\Phi(p_{max})$. Figure 5a shows the average maximum pressure and the COV_{pmax} coefficient determined for the analyzed shares of CNG gas co-combusted with diesel fuel. Taking the COV_{pmax} factor as an indicator of the stability of dual-fuel engine operation, it can be seen that the most stable operation similar to diesel fuel combustion was ensured by diesel co-combusting with a 30% and 45% energy share of CNG gas ($COV_{pmax} \approx 1.2\%$). In the case of DCNG45, the highest value of the maximum combustion pressure $p_{max} = 5.75$ MPa was obtained. Figure 5b presents the probability density of p_{max} for a dual-fuel engine co-combusted diesel fuel with CNG gas. The maximum pressure probability density distribution ($\Phi(p_{max})$), based on the normal distribution, is an indicator of the repeatability of individual p_{max} values obtained for the 200 analyzed test engine work cycles. Changes in the maximum pressure probability density indicate, among others, the repeatability of the average p_{max} value, which may be an indicator of stable operation of the test engine. In the case of a dual-fuel engine, the highest values of p_{max} probability density were obtained for DCNG30 and DCNG45. They were similar to those obtained during the combustion of diesel alone. The individual p_{max} values obtained, in 200 subsequent engine cycles and in the largest number of cycles, approached the mean value of p_{max}, which proved the best repeatability of subsequent test engine work cycles. Those results confirmed earlier results of the analysis of the stability of dual-fuel engine operation based on changes in the COV_{pmax} coefficient.

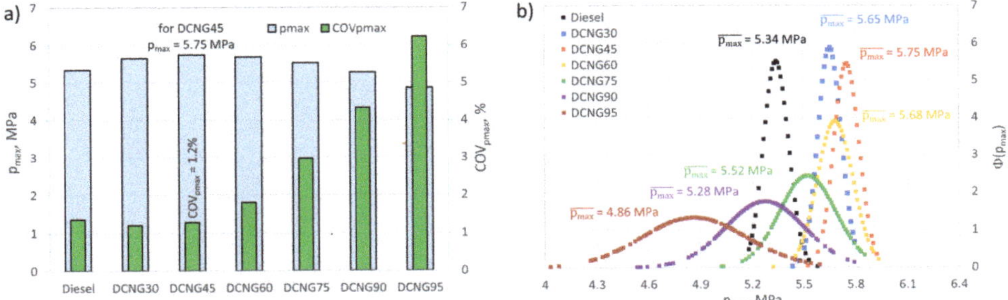

Figure 5. Mean maximum pressure, COV_{pmax} (**a**), and pressure probability density, $\Phi(p_{max})$ (**b**) determined for the CNG shares co-combusted with diesel.

4.3. Exhaust Emissions

The test stand built as part of the work in this study enabled the measurement of the content of components in the dual-fuel engine exhaust gas. Concentrations of toxic exhaust components were measured: nitrogen oxide—NO, HC—hydrocarbons and carbon monoxide CO, as well as CO_2 emissions. Figure 6a presents the impact of CNG gas on nitrogen oxide emissions. The main types of nitrogen oxides emitted by a piston engine are nitrogen oxide (NO) and nitrogen dioxide (NO_2). The formation of nitric oxide in a piston engine is a direct result of the reaction between nitrogen (N_2) and oxygen (O_2), under the favorable conditions prevailing in the engine cylinder during the combustion process. The combustion processes mainly produce NO, while NO_2 is formed by the oxidation of nitric oxide in atmospheric air. The NO_2 concentration in the exhaust gas achieves much lower concentrations compared to NO [32,33]. Increasing the share of CNG co-combusted with diesel fuel causes an increase in nitric oxide emissions. The highest NO emission value of 242 ppm was obtained for DCNG90. The conditions favoring the formation of NO during co-combustion of diesel fuel with CNG gas, whose share varied from 0% to 45%, resulted from the intensification of the combustion process in the kinetic phase and the increase in the rate of heat release. However, after exceeding 45%, the increase in nitrogen oxide emissions was the result of an increase in the availability of oxygen in the engine cylinder, unused in the prolonged combustion process. In addition, nitrogen-containing CNG gas increased its concentration in the cylinder and in the engine's exhaust gas. For DCNG95, there was a sudden decrease in NO emissions caused by a decrease in heat release in the disappearing kinetic stage of the combustion process. The intensification of the combustion process, presented by an increase in the heat release rate for DCNG30 and DCNG45, caused, compared to the combustion of diesel fuel alone, a reduction in HC emissions, despite an increase in ignition delay and a shorter combustion process (Figure 6a). After exceeding the 45% share of CNG gas, there was an increase in hydrocarbon emissions caused by a significant ignition delay and reduced burning time. The increase in THC emissions could be associated with the so-called crevice effect. Under these conditions, the fuel in the engine cylinder has no time to burn the fuel accumulated in the crevices of the combustion chamber (e.g., piston ring gaps). Figure 6b presents the impact of the share of natural gas on CO_2 and CO emissions of a dual-fuel engine powered by CNG and diesel. Research showed that increasing the share of CNG gas co-combusted with diesel fuel, compared to the combustion of diesel fuel, reduces carbon dioxide emissions by about 26% and almost completely reduces carbon monoxide in the exhaust gases of the engine. The reduction in CO_2 and CO emissions was primarily due to the reduction in the total mass of fuel (containing carbon) supplied to the engine per one cycle of operation, due to the higher calorific value of CNG gas compared to diesel. In addition, the natural gas molecule contains less carbon compared to diesel (Table 3). Carbon monoxide (CO) is a toxic gas formed in a piston engine cylinder as a result of incomplete combustion of fuel, which may be caused by the lack of homogeneity of the air and fuel mixture [34]. In addition, CO reduction may result from an improvement in the homogeneity of the air and fuel mixture, together with the supply of gaseous fuel. This means that the combustion process in the engine tends to complete combustion, during which carbon monoxide is oxidized to carbon dioxide.

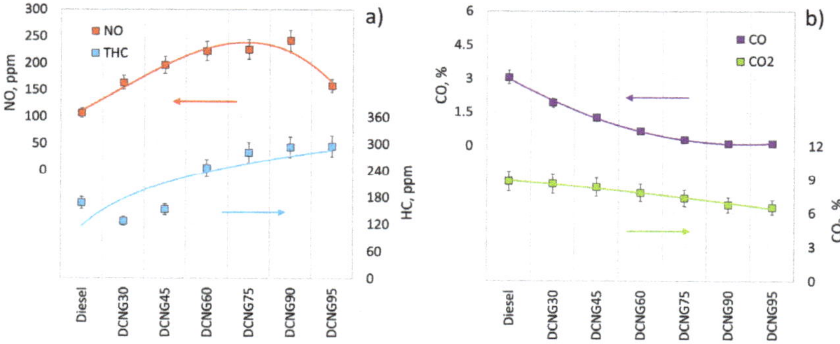

Figure 6. Emissions of NO and hydrocarbons (HC) (**a**), and CO and CO_2 (**b**) for the dual fuel engine tested.

5. Conclusions

The use of natural gas to power a compression-ignition engine is one possible way to reduce the combustion of diesel fuel, which is considered one of the main causes of the greenhouse effect. This paper presents experimental research on a CI dual-fuel engine—powered by diesel fuel injected directly into the cylinder, and CNG gas injected into the intake manifold. The energy share of CNG gas co-fired with diesel fuel ranged from 0% to 95%. Based on the analysis of test results, the following conclusions can be made:

(i) Increasing the energy share of CNG co-combusted with diesel in the CI engine from 0% to 45% increases the maximum combustion pressure, increases the rate of heat release and increases the combustion pressure. Increasing the CNG gas content above 45% causes a decrease in p_{max}, HRR_{max} and $dp/d\varphi$, due to the significant ignition delay and prolonged combustion.

(ii) Co-combustion of increasing amounts of CNG gas with diesel fuel increases the ignition delay of the combustible mixture in the cylinder of a dual-fuel engine and reduces the total duration of combustion. During the combustion of the mixture with the largest, 95% share of CNG gas, compared to the combustion of diesel alone, a 31.5% increase in ignition delay (ID) and a 55% reduction in the burn duration (BD) was obtained.

(iii) When assessing the stability of a dual-fuel engine co-combusted diesel with natural gas based on the COV_{pmax} coefficient and pmax probability density, it can be seen that the most stable operation similar to diesel fuel combustion is provided by diesel co-combustion with the 30% and 45% energy shares of CNG.

(iv) Increasing the share of CNG co-combusted with diesel fuel causes an increase in nitric oxide emissions of a dual-fuel engine. The highest NO emission value of 242 ppm was obtained for DCNG90.

(v) Compared to diesel fuel supply, co-combustion of this fuel with 30% and 45% CNG energy shares, despite the increase in ignition delay and shortening the combustion process, contributes to the reduction of HC emissions. After exceeding the 45% share of CNG, there is an increase in hydrocarbon emissions.

(vi) Increasing the share of CNG gas co-combusted with diesel causes, compared to the combustion of diesel fuel, a decrease in carbon dioxide emissions by about 26% and an almost complete reduction of carbon monoxide in the exhaust gas of a dual-fuel engine.

Author Contributions: Conceptualization, A.J., W.T. and K.G.-R.; Data curation, A.J., W.T. and K.G.-R.; Formal analysis, A.J.; Investigation, A.J., W.T. and K.G.-R.; Methodology, A.J. and W.T.; Writing—original draft, A.J.; Writing—review & editing, W.T. and K.G.-R.

Funding: This research was funded by the Ministry of Science and Higher Education of Poland from the funds dedicated to scientific research No. BS/PB 1-100-3010/2019/P. Publication supported financially under Contract No. 944/P-DUN/2019 from funds of MNiSW intended for dissemination of science (DUN).

Conflicts of Interest: The authors declare no conflict of interest.

Abbreviations

IMEP	indicated mean effective pressure, MPa
BTE	brake thermal efficiency, %
BSFC	brake specific fuel consumption, g/kWh
HRR	heat release rate, J/degree
COV_{pmax}	coefficient of variation of maximum pressure, %
STD_{pmax}	standard deviation of maximum pressure, MPa
$\dot{m}_{CNG}$	CNG consumption per cycle, mg/cyc
$\dot{m}_D$	diesel fuel consumption per cycle, mg/cyc
LHV_{CNG}	lower heating value of CNG, MJ/kg
LHV_D	lower heating value of diesel fuel, MJ/kg
ID	ignition delay, degrees
BD	burn duration, degrees
V_d	displaced cylinder volume, cm^3
P	pressure, bar
V	volume, cm^3
N	engine speed, rpm
$\Phi(p_{max})$	maximum pressure probability density
CI	compression ignition engine
CNG	compressed natural gas
TDC	top dead centre
SOI	start of injection
NO_x	nitrogen oxides
NO	nitrogen monoxide
HC	hydrocarbons
CO	carbon monoxide
CO_2	carbon dioxide
O_2	oxygenGreek letters
χ	ratio of specific heats
φ	crank angle, degrees

References

1. Kalghatgi, G. Development of Fuel/Engine Systems—The Way Forward to Sustainable Transport. *Engineering* **2019**, *5*, 510–518. [CrossRef]
2. Tutak, W.; Jamrozik, A. Validation and optimization of the thermal cycle for a diesel engine by computational fluid dynamics modeling. *Appl. Math. Model.* **2016**, *40*, 6293–6309. [CrossRef]
3. Tutak, W.; Jamrozik, A. Generator gas as a fuel to power a diesel engine. *Therm. Sci.* **2014**, *18*, 206–216. [CrossRef]
4. Park, H.; Shim, E.; Bae, C. Improvement of combustion and emissions with exhaust gas recirculation in a natural gas-diesel dual-fuel premixed charge compression ignition engine at low load operations. *Fuel* **2019**, *235*, 763–774. [CrossRef]
5. Manigandan, S.; Gunasekar, P.; Devipriya, J.; Nithya, S. Emission and injection characteristics of corn biodiesel blends in diesel engine. *Fuel* **2019**, *235*, 723–735. [CrossRef]
6. Directive 2009/28/EC of the European Parliament and of the Council. 23 April 2009. Available online: http://eur-lex.europa.eu/legal-content/PL/ALL/?uri=celex%3A32009L0028/ (accessed on 10 July 2019).

7. Braungardt, S.; Bürger, V.; Ziege, J.; Bosselaar, L. How to include cooling in the EU Renewable Energy Directive? Strategies and policy implications. *Energy Policy* **2019**, *129*, 260–267. [CrossRef]

8. Kuszewski, H. Experimental investigation of the autoignition properties of ethanol–biodiesel fuel blends. *Fuel* **2019**, *235*, 1301–1308. [CrossRef]

9. Jamrozik, A.; Tutak, W.; Gruca, M.; Pyrc, M. Performance, emission and combustion characteristics of CI dual fuel engine powered by diesel/ethanol and diesel/gasoline fuels. *J. Mech. Sci. Technol.* **2018**, *32*, 2947–2957. [CrossRef]

10. Jamrozik, A.; Tutak, W.; Pyrc, M.; Gruca, M.; Kocisko, M. Study on co-combustion of diesel fuel with oxygenated alcohols in a compression ignition dual-fuel engine. *Fuel* **2018**, *21*, 329–345. [CrossRef]

11. Lounici, M.S.; Boussadi, A.; Loubar, K.; Tazerout, M. Experimental investigation on NG dual fuel engine improvement by hydrogen enrichment. *Int. J. Hydrog. Energy* **2014**, *39*, 21297–21306. [CrossRef]

12. Verma, S.; Das, L.M.; Bhatti, S.S.; Kaushik, S.C. A comparative exergetic performance and emission analysis of pilot diesel dual-fuel engine with biogas, CNG and hydrogen as main fuels. *Energy Convers. Manag.* **2017**, *151*, 764–777. [CrossRef]

13. Wang, Z.; Du, G.; Wang, D.; Xu, Y.; Shao, M. Impact of pilot diesel ignition mode on combustion and emissions characteristics of a diesel/natural gas dual fuel heavy-duty engine. *Fuel* **2016**, *167*, 248–256. [CrossRef]

14. Nithyanandan, K.; Lin, Y.; Donahue, R.; Meng, X.; Zhang, J.; Lee, C.F. Characterization of soot from diesel-CNG dual-fuel combustion in a CI engine. *Fuel* **2016**, *184*, 145–152. [CrossRef]

15. Stelmasiak, Z.; Larisch, J.; Pietras, D. Selected problems of adaptation car diesel engine for dual fuel supplying. *Combust. Engines* **2015**, *162*, 1021–1029.

16. Papagiannakis, R.G.; Rakopoulos, C.D.; Hountalas, D.T.; Rakopoulos, D.C. Emission characteristics of high speed, dual fuel, compression ignition engine operating in a wide range of natural gas/diesel fuel proportions. *Fuel* **2010**, *89*, 1397–1406. [CrossRef]

17. Papagiannakis, R.G.; Krishnan, S.R.; Rakopoulos, D.C.; Srinivasan, K.K.; Rakopoulos, C.D. A combined experimental and theoretical study of diesel fuel injection timing and gaseous fuel/diesel mass ratio effects on the performance and emissions of natural gas-diesel HDDI engine operating at various loads. *Fuel* **2017**, *202*, 675–687. [CrossRef]

18. Meng, X.; Tian, H.; Long, W.; Zhou, Y.; Bi, M.; Tian, J.; Lee, C.F. Experimental study of using additive in the pilot fuel on the performance and emission trade-offs in the diesel/CNG (methane emulated) dual-fuel combustion mode. *Appl. Therm. Eng.* **2019**, *157*, 113718. [CrossRef]

19. Ryu, K. Effects of pilot injection timing on the combustion and emissions characteristics in a diesel engine using biodiesel–CNG dual fuel. *Appl. Energy* **2013**, *111*, 721–730. [CrossRef]

20. Ryu, K. Effects of pilot injection pressure on the combustion and emissions characteristics in a diesel engine using biodiesel–CNG dual fuel. *Energy Convers. Manag.* **2013**, *76*, 506–516. [CrossRef]

21. Bari, S.; Hossain, S.N. Performance of a diesel engine run on diesel and natural gas in dual-fuel mode of operation. *Energy Procedia* **2019**, *160*, 215–222. [CrossRef]

22. Liu, J.; Yang, F.; Wang, H.; Ouyang, M.; Hao, S. Effects of pilot fuel quantity on the emissions characteristics of a CNG/diesel dual fuel engine with optimized pilot injection timing. *Appl. Energy* **2013**, *110*, 201–206. [CrossRef]

23. Karabektas, M.; Ergen, G.; Hosoz, M. The effects of using diethylether as additive on the performance and emissions of a diesel engine fuelled with CNG. *Fuel* **2016**, *115*, 855–860. [CrossRef]

24. Jamrozik, A. The effect of the alcohol content in the fuel mixture on the performance and emissions of a direct injection diesel engine fueled with diesel-methanol and diesel-ethanol blends. *Energy Convers. Manag.* **2017**, *148*, 461–476. [CrossRef]

25. Repair Instructions. Bosch Emissions Analysis BEA 150, BEA 250, BEA 350. Available online: https://equiposbosch.es/ (accessed on 14 August 2019).

26. Gilowski, T.; Stelmasiak, Z. Impact of symmetrical division of initial dosage diesel oil on the selected thermodynamic parameters of the working medium and operation parameters dual fuel engine fuelled CNG. *Combust. Engines* **2013**, *154*, 879–886.

27. Stelmasiak, Z. Selected problems of application of natural gas to CI engine. *Arch. Automot. Eng.* **2006**, *1*, 13–30.

28. Chen, Z.; Yao, C.; Yao, A.; Dou, Z.; Wang, B.; Wei, H.; Liu, M.; Chen, C.; Shi, J. The impact of methanol injecting position on cylinder-to-cylinder variation in a diesel methanol dual fuel engine. *Fuel* **2017**, *191*, 150–163. [CrossRef]

29. Jamrozik, A.; Tutak, W.; Gnatowska, R.; Nowak, Ł. Comparative analysis of the combustion stability of diesel-methanol and diesel-ethanol in a dual fuel engine. *Energies* **2019**, *12*, 971. [CrossRef]

30. Taylor, J.R. *Introduction to Error Analysis: The Study of Uncertainties in Physical Measurements*; University Science Books: Sausalito, CA, USA, 1997.

31. Heywood, J.B. *Internal Combustion Engine Fundamentals*; McGraw-Hill Book Company: New York, NY, USA, 1988.

32. Żółtowski, A. Influence of after-treatment systems on NO_2 emissions in diesel engines. *Combust. Engines* **2017**, *170*, 24–29.

33. Olsen, D.B.; Kohls, M.; Arney, G. Impact of oxidation catalysts on exhaust NO_2/NO_x ratio from lean-burn natural gas engines. *J. Air Waste Manag. Assoc.* **2010**, *60*, 867–874. [CrossRef]

34. Darade, P.M.; Dalu, R.S. Performance and emissions of internal combustion engine fuelled with CNG-A Review. *Int. J. Eng. Innov. Res.* **2012**, *1*, 473–477.

Article

Thermal Decomposition of a Single AdBlue® Droplet Including Wall–Film Formation in Turbulent Cross-Flow in an SCR System

Kaushal Nishad [1,*], Marcus Stein [2], Florian Ries [1], Viatcheslav Bykov [2], Ulrich Maas [2], Olaf Deutschmann [3], Johannes Janicka [1] and Amsini Sadiki [1]

[1] Institute of Energy and Power Plant Technology, Technische Universität Darmstadt, 64287 Darmstadt, Germany
[2] Institut für Technische Thermodynamik, Karlsruher Institut für Technologie, 76131 Karlsruhe, Germany
[3] Institut für Technische Chemie und Polymerchemie, Karlsruher Institut für Technologie, 76131 Karlsruhe, Germany
* Correspondence: nishad@ekt.tu-darmstadt.de; Tel.: +49-6151-16-28756

Received: 23 May 2019; Accepted: 1 July 2019; Published: 6 July 2019

Abstract: The selective catalytic reduction (SCR) methodology is notably recognized as the widely applied strategy for NO_X control in exhaust after-treatment technologies. In real SCR systems, complex unsteady turbulent multi-phase flow phenomena including poly-dispersed AdBlue® spray evolve with a wide ranging relative velocity between the droplet phase and carrier gas phase. This results from an AdBlue® spray that is injected into a mixing pipe which is cross-flowing by a hot exhaust gas. To reduce the complexity while gaining early information on the injected droplet size and velocity needed for a minimum deposition and optimal conversion, a single droplet with a specified diameter is addressed to mimic a spray featuring the same Sauter Mean Diameter. For that purpose, effects of turbulent hot cross-flow on thermal decomposition processes of a single AdBlue® droplet are numerically investigated. Thereby, a single AdBlue® droplet is injected into a hot cross-flowing stream within a mixing pipe in which it may experience phase change processes including interaction with the pipe wall along with liquid wall–film and possible solid deposit formation. First of all, the prediction capability of the multi-component evaporation model and thermal decomposition is evaluated against the detailed simulation results for standing droplet case for which experimental data is not available. Next, exploiting Large Eddy Simulation features the effect of hot turbulent co- and cross-flowing streams on the dynamic droplet characteristics and on the droplet/wall interaction is analyzed for various droplet diameters and operating conditions. This impact is highlighted in terms of droplet evaporation time, decomposition efficiency, droplet trajectories and wall–film formation. It turns out that smaller AdBlue® droplet diameter, higher gas temperature and relative velocity lead to shorter droplet life time as the droplet evaporates faster. Under such conditions, possible droplet/wall interaction processes on the pipe wall or at the entrance front of the monolith may be avoided. Since the ammonia (NH_3) gas generated by urea decomposition is intended to reduce NO_X emission in the SCR system, it is apparent for the prediction of high NO_X removal performance that UWS injector system which allows to realize such operating conditions is favorable to support high conversion efficiency of urea into NH_3.

Keywords: AdBlue® injection; large eddy simulation; Eulerian–Lagrangian approach; thermal decomposition; wall–film formation; conversion efficiency

---

1. Introduction

Emission regulations in the automotive industry have become increasingly stringent, especially for diesel engines for both heavy-duty trucks, passenger cars and off-highway applications for which the limits on NO_X content in exhaust gas will be even further tightened [1–3]. A viable way to meet these requirements is in combining in-cylinder innovative combustion technologies and accurate exhaust after-treatment strategies [2,4]. Focusing on the exhaust after-treatment, the selective catalytic reduction (SCR) methodology is notably recognized as the widely applied technique for NO_X control [3,5,6].

To achieve the suitable high NO_X reduction rates, optimal atomization and evaporation of UWS droplets (UWS: urea–water solution), as well as the thermal decomposition of urea and the subsequent mixture formation of the resulting reducing agent ammonia in the exhaust system are crucial [5,7–9]. At the same time, the formation of undesirable deposit products on the exhaust duct wall must be avoided to ensure reliable operation. Both processes, NO_X reduction and the formation of deposits, are largely determined by the process evolving in the mixing section upstream to the SCR monolith and the positioning and type of UWS injector [10]. Many experimental and numerical investigations have been carried out in order to describe the evaporation process of UWS single droplet along with the thermal decomposition under various operating conditions [7–11].

Focusing on numerical simulations, an accurate way to account for all processes might consist in the use of a detailed simulation of both processes close to the droplet surface and within the droplet for the liquid phase. This art of simulations is unfortunately computationally very costly for turbulent poly-disperse sprays and only suitable for single droplet without complexities related to engineering applications [11,12]. With regards to the UWS binary system, the affect of convective flow around the AdBlue® droplet is neglected to reduce computational costs with over-saturation and crystallization of urea are not considered (e.g., [11]).

The state-of-the art for engineering predictions consists in applying droplet evaporation models mostly in the context of RANS technique in which the droplet phase is tracked by a Lagrange particle approach while the Eulerian framework is adopted to describe surrounding gas phase. In particular, instead of resolving the interface between gas-droplet phase, rather so-called film-based models including the 1/3 or 2/3 rule are used. A review of the various existing evaporation models can be found in [12–15]. The process within the droplet is captured by so-called droplet models (e.g., Rapid Mixing model [5], conduction limit model [7], effective conductivity model, diffusion model [11], vortex model, etc.). Thermal decomposition mechanisms are diversely implemented and various conversion rates achievements discussed (e.g., [5,7–9,16]). However, the complexity of evaporating urea sprays as they are encountered in large-scale exhaust systems does not facilitate the task once urea is sprayed within a mixing pipe configuration and evolves in an unsteady environment. From all these studies, it could be observed that under unfavorable operating conditions the UWS at the entrance of the SCR catalyst is not always fully evaporated or converted. As a result, droplets of UWS can get into the fine channels of the catalyst and interact with them leads to an inefficient NO_X reduction and in the worst case to failure of the SCR system due to deposit formation within the fine monolith channels [3,17,18]. This shows that the main issue in SCR system to find out the optimal velocity of urea injection and the UWS spray droplet size remains still unsatisfactory addressed.

Although diverse operating conditions have been considered in the literature as reviewed in [7–9,19], it is noteworthy that the effect of turbulence modulation on the evaporation and thermal decomposition process has not yet been really investigated in SCR systems. In [20] only the effect of turbulence on the droplet trajectory by means of a dispersion model within the RANS framework has been reported. It is therefore of great interest to investigate how the turbulent cross-flowing stream conditions may influence the modification of transport in the vicinity of the droplet interface. For single component droplet sprays, the effect of turbulence on vaporization and mixing of liquid-fuel sprays has been experimentally investigated in [14,21,22] and numerically in [14] resulting in the consideration of a so-called evaporation Damkoehler number defined as the ratio of the turbulence time scale to the vaporization time scale [14,21].

In SCR systems, contributions towards the analysis of mixing and evaporation have been reviewed in [5,7,8,20]. However, the most of these works focused solely on retrieving the evaporation and thermal decomposition of UWS along with the ammonia conversion rate in mono-disperse UWS sprays using

RANS models without particular consideration of turbulent flow modulation process. Only recently, hybrid LES/RANS (HLR) [19] and full LES [23] approaches have been implied to accurately describe the gas phase processes far around the droplet.

Kaario et al. [19] undertook simulation of single droplet in uniform ambient flow prior to their 3D (3 dimensional) simulations of mono-disperse spray. To get insight into the UWS injection, they used the hybrid LES/RANS (HLR) approach to create phase diagrams that can allow to predict the optimum operation regions for the SCR systems. Thereby, the injection process was primarily characterized as function of the gas velocity (gas velocity $U_g = 0, U_{bulk}, U_{inj}$) and droplet size (assuming a constant droplet temperature (T_p = 323 K) and gas temperature (T_g = 523 K), as well as as function of the initial droplet diameter (D_p = 10, 20, 30 40 µm) and the slip velocity. In their study, the UWS is considered as a single component for which the mass transfer during the droplet evaporation is described according to Bird et al. (1966) [24] where the mass transfer coefficient follows the Ranz and Marshall [25] ansatz. Gan et al. [16] pointed out the relation between evaporation time, droplet initial diameter and relative velocity, and found out that droplet with smaller diameter, higher gas temperature and relative velocity likely lead to shorter evaporation time. Operating temperature plays a key role as a determining factor for deposit component while temperature and reaction time strongly affect the deposit yield. However, this investigation was conducted within the RANS context completely missing the effect of turbulence modulation. Nishad et al. [7,23] developed an LES-based Eulerian–Lagrangian framework to study evaporation and thermal decomposition and applied it toAdBlue® single droplet as experimentally investigated in [26] without cross-flowing effect [7] and to AdBlue® spray in hot cross-flowing stream without consideration of turbulence modulation effect and wall–film formation process [23].

With respect to droplet trajectories, most of the studies in the literature have been carried out in the context of gas turbine combustion chamber applications for single component droplets under ambient conditions. Thereby, various semi-empirical correlations could be derived to describe the droplet trajectory [27]. Only few studies have been conducted at elevated pressure and temperature as pointed out by Prakash et al. [28] who suggested a new analytical correlation. This depends not only on the effective magnitude of the liquid-to-air ratio as in ambient conditions, but also on the density ratio of both the carrier phase and the single component liquid fluid. The case of multicomponent droplet with varying compositions during evaporation process as it occurs in SCR systems is not yet addressed.

In real SCR systems, complex unsteady turbulent multi-phase flow phenomena including poly-dispersed AdBlue® spray evolve with a wide ranging relative velocity between droplet phase and carrier gas phase. This results from an AdBlue® spray which is injected into a mixing pipe which is cross-flowing by a hot exhaust gas. To reduce the complexity while gaining early information on the injected droplet size and velocity needed for a minimum deposition and optimal conversion, a single droplet with a specified diameter is investigated in the present paper to mimic a spray featuring the same Sauter Mean Diameter. The LES-based Eulerian–Lagrangian model including a multi-component evaporation model along with a thermal decomposition mechanism as developed by Nishad et al. [7] is extended to include wall–film formation model and subsequent possible deposit formation in order to study the effect of turbulent cross-flowing stream conditions on the dynamic droplet characteristics. The effect is monitored in terms of evaporation time, decomposition efficiency, droplet trajectory and liquid wall–film formation. The objective is to provide early information on the injected droplet size and velocity required for a minimum deposition and optimal conversion.

The paper is organized as follows. In the next section (Section 2), the droplet configurations to be investigated are introduced. In Section 3, the modeling description of the evaporation, thermal decomposition and liquid wall–film formation is shortly introduced, followed by an outline of a detailed simulation model which will provide first reliable comparison data for the evaporation and thermal decomposition processes. The achievements of the LES-based Eulerian–Lagrangian model and the comparison with those obtained by the detailed numerical simulations of a standing droplet case (S1), for which no experimental data are available, are discussed in Section 4. The validated model is then used to retrieve the effects of the turbulent co- and cross-flowing streams (S2–S4) on the evaporating droplet dynamics and on the droplet/wall interaction process in Sections 5 and 6, respectively. The last section is devoted to concluding remarks.

2. Droplet Configurations under Study

For the specific purpose of this study, four droplet configurations are designed as depicted in Figure 1 mimicking the operating flow conditions in SCR systems. Since SCR systems are in reality streamed by complex unsteady turbulent multi-phase flows that include poly-dispersed AdBlue® droplets and carrier gas phase with non-vanishing relative velocity, the simulation of stagnant droplets with vanishing relative velocity (see Figure 1, case S1) will serve only for validation purposes with detailed numerical simulation data for which experimental data are not available. Note that the characteristics of the depletion process of a single UWS droplet have been reported in [23] based on various experimental temperatures operating conditions investigated by Wang et al. [26] with gravity effect.

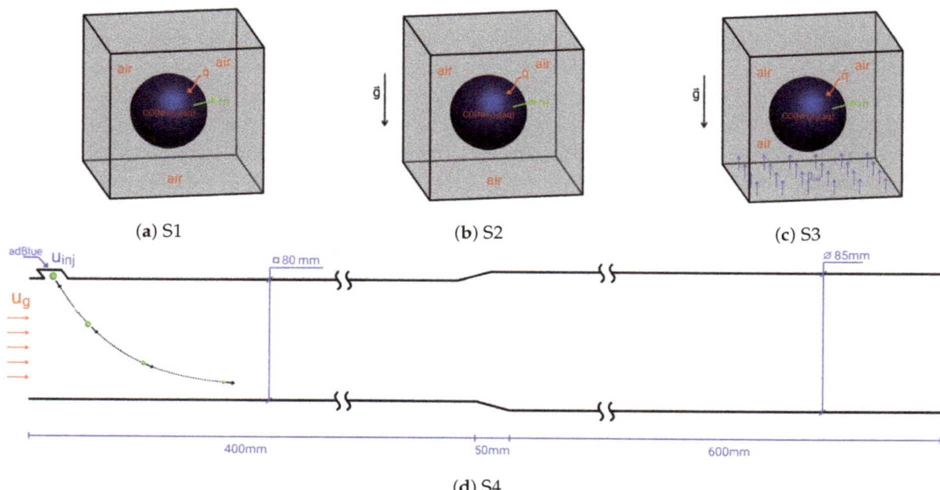

Figure 1. Schematic of simulation cases: (**a**) stagnant droplet no gravity (S1); (**b**) stagnant droplet with gravity (S2); (**c**) with relative velocity u_{rel}, m/s (S3); (**d**) hot cross-flow case (S4).

In the present paper, the influence of relative velocity on both the evaporation and thermal decomposition will be investigated for various droplet diameters and gas phase temperatures (see Figure 1, cases S2–S4) taking advantage of the LES capabilities of the LES-based Eulerian–Lagrangian module. Case S2 considers a standing droplet, while case S3 depicts a droplet into a convective environment with co-flow characterized by various relative velocities between the carrier phase and the droplet phase. S4 represents an injected droplet into a hot cross-flowing stream within a mixing pipe in which the droplet may experience phase change processes including interaction with the pipe wall along with liquid wall–film and possible solid deposit formation processes. A hex-hedral mesh with 216 grid points is used to investigate the standing droplet cases (case S1-no gravity, and case S2-with gravity). For the case with droplet relative velocity (case S3), the domain with $30 \times 30 \times 50$ mm^3 is discretized by a total of 45,000 control volumes (CVs). The case S4 is selected from a generic experimental configuration as reported in [10]. A hex-hedral mesh with $\approx$3.55 millions CVs is used to spatially discretize the complete domain (see Figure 1d). Since, the present simulations are near wall resolved, the mesh size varies from near wall to the duct core (from 120 μm at wall to 0.1 mm at flow core). Proper initial and inlet boundary conditions are prescribed by following our previous study [23].

3. Model Formulation

In this section, the LES-based Eulerian–Lagrangian numerical module is concisely described. Then, the mathematical modeling of the detailed simulation tool is outlined.

3.1. LES-Based Eulerian–Lagrangian Numerical Module

This module consists of an Eulerian description of the gas phase in the LES context and a Lagrangian framework for tracking the UWS droplet. It thus includes a multicomponent evaporation model, a thermal decomposition mechanism and a description of liquid wall–film formation.

3.1.1. Description of Carrier Phase and LES Model

In this paper, the flow turbulence is described by the a classical LES model. In this, large flow structures are fully resolved, while sub-grid scale (SGS) models are applied to closed the small scale non-filtered terms . The filtered transport equations read;

$$\frac{\partial \tilde{\Psi}}{\partial t} + \frac{\partial \tilde{u}_j \tilde{\Psi}}{\partial x_j} = \frac{\partial \tilde{\mathbf{F}}}{\partial \mathbf{x}} - \frac{\partial \,\mathbf{F}^{SGS}}{\partial \mathbf{x}} + \tilde{\Sigma}_{\tilde{\Psi}} + \tilde{S}_{\tilde{\Psi}} \tag{1}$$

Thereby the quantity $\tilde{\Psi}$ is defined as $\tilde{\Psi} = \frac{\overline{\rho\Psi}}{\bar{\rho}}$, t expresses the time, $\mathbf{x}$ the coordinate vector, $x_i, (i = 1, 2, 3)$ the coordinate component and u_i the velocity component. Overbars and tildes express spatially filtered with a filter width Δ_g and density-weighted (Favre-filtered) quantity, respectively. The remaining quantities in Equation (1) are provided in Table 1 for a Newtonian reacting fluid flow under investigation here.

Table 1. Specification of quantities appearing in the filtered transport equations.

Variables	Ψ	F	F^{SGS}	$\tilde{\Sigma}_{\tilde{\Psi}}$	$S_{\tilde{\Psi}}$
mass	$\bar{\rho}$	0	0	0	$S_{\bar{\rho}}$
momentum	$\bar{\rho}\tilde{u}_i$	$2\mu\left(\bar{D}_{ij} - \frac{1}{3}\bar{D}_{kk}\delta_{ij}\right) - \bar{p}\delta_{ij}$	τ_{ij}^{SGS}	$\bar{\rho}g_i$	$S_{\tilde{u}_i}$
species mass fraction	$\bar{\rho}\tilde{\phi}$	$\bar{\rho}\left(D_{\tilde{\phi}}\frac{\partial \tilde{\phi}}{\partial x_j}\right)$	$J_j^{\phi(SGS)}$	0	$S_{\tilde{\phi}}$
enthalpy	$\bar{\rho}\tilde{h}$	$\bar{\rho}\left(D_{\tilde{h}}\frac{\partial \tilde{h}}{\partial x_j}\right)$	$J_j^{h(SGS)}$	$\tilde{W}_{\tilde{h}}$	$S_{\tilde{h}}$

In Table 1 the quantities $\bar{\rho}$ and $\bar{p}$ are the Favre averaged density and pressure, respectively $\tilde{u}_i$ the gas phase filtered velocity, $\tilde{\mathbf{F}}$ the filtered molecular diffusion term, $\mathbf{F}^{SGS}$ the sub grid diffusion part. The quantity $\tilde{\Sigma}_{\tilde{\Psi}}$ represents a supply contribution, and $\tilde{W}_{\tilde{h}}$ is an extra source term of enthalpy that may be due to radiation. $S_{\tilde{\phi}}$, the so-called phase interaction/source term due to spray droplet ([29]). $D_{\tilde{\phi}}$ denotes the scalar diffusion coefficient.

Since the turbulent co- and cross-flow features a wall-bounded flow, a Wall-Adapting Local Eddy-Viscosity (WALE) Model [30] is considered appropriate with reasonable computational cost as pointed out in [30–32], among others. In which, the term $\tau_{ij}^{SGS} = \widetilde{u_i u_j} - \tilde{u}_i \tilde{u}_j$ also called SGS stress tensor is closed as:

$$\tau_{ij}^{SGS} - \tfrac{1}{3}\tau_{kk}^{SGS}\delta_{ij} = -2\mu_t \tilde{S}_{ij} - \tfrac{2}{3}\delta_{ij}\tilde{S}_{kk} \quad \text{with} \quad \mu_t = \rho\left(C_w \Delta_g\right)^2 \frac{\left(S_{ij}^d S_{ij}^d\right)^{2/3}}{\left(\bar{D}_{ij}\bar{D}_{ij}\right)^{5/2}\left(S_{ij}^d S_{ij}^d\right)^{5/4}}$$

$$S_{ij}^d = \tfrac{1}{2}\left(\left(\frac{\partial \bar{u}_i}{\partial x_j}\right)^2 + \left(\frac{\partial \bar{u}_i}{\partial x_j}\right)^2\right) - \tfrac{1}{3}\delta_{ij}\left(\frac{\partial \bar{u}_k}{\partial x_k}\right)^2 \quad \text{and} \quad D_{ij} = \tfrac{1}{2}\left(\frac{\partial \bar{u}_i}{\partial x_j} + \frac{\partial \bar{u}_i}{\partial x_j}\right) + \tfrac{1}{3}\delta_{ij}\frac{\partial \bar{u}_k}{\partial x_k}. \tag{2}$$

In these equations, S_{ij}^d is the deviatoric part of the square of velocity gradient, D_{ij} the symmetric part of the velocity gradient, Δ_g the filter width and $C_w = 0.5$, a model coefficient expressed in terms of Smagorinsky coefficient as $C_w^2 = 10.6 \cdot C_s^2$ [32] (for $C_s = 0.17$).

Note that in the present study the size of wall mesh is kept close to "$Y^+ \leq 1.0$" in order to carry out a wall resolved LES according to [33].

Further, for the filtered scalar (mass fraction, and energy) field, the quantity $J_j^{SGS} = \overline{u_j \Psi} - \tilde{u}_j \tilde{\Psi}$ is closed as:

$$J_j^{SGS} = \widetilde{u_j \Psi} - \tilde{u}_j \tilde{\Psi} = D_{ed} \frac{\partial \tilde{\Psi}}{\partial x_j} \quad \text{with} \quad D_{ed} = \frac{\nu_t}{Sc_t(Pr_t)}. \tag{3}$$

where Sc_t (Pr_t) expresses the turbulent Schmidt (Prandtl) number.

It is worth noting that the two-way coupling for drag, evaporations is realized between the gas and droplet particle phase. Their expressions are provided in [34]. Thereby, both carrier and dispersed phases in this formulation are treated separately and in a non-conservative form following [7,34]. Further, as the LES considered do at least resolve 80% of the instantaneous carrier phase turbulence, the droplet particle dispersion and the fluctuations on the scalars are fairly considered. In LES methodology the state of turbulence at inlet and initial conditions is crucial as reported in many research contributions [23,30,32,33]. Therefore, we follow the method of Nishad et al. [23] to provide realistic initial and inlet conditions for LES in the present paper.

3.1.2. Lagrange Particle Tracking

In this section, a brief overview of the governing equations for the liquid phase is provided. It is based on Lagrangian formulation and provided by a set of ordinary differential equations for droplet position X_i and velocity u_i as:

$$\frac{dX_i}{dt} = u_i, \quad \frac{du_i}{dt} = \frac{C_D}{\tau_d} \frac{Re_P}{24} (U_i - u_i) + g_i \tag{4}$$

Thereby, the drag and buoyancy forces are only considered as the density ratio of dispersed and gas phase is in the order of 10^3. Where, $Re_P = |U_i - u_i| d_p \rho_g / \mu_g$ is the droplet Reynold's Number, $\tau_d = \rho_p d_p^2 / (18 \mu_g)$ the droplet relaxation time, and the droplet drag coefficient C_D is given as (see in [7,35]);

$$C_D = \begin{cases} \frac{24}{Re_P} \left(1 + \frac{1}{6} Re_P^{1/3} \right) & \text{if } Re_P < 1000 \\ 0.424 & \text{if } Re_P \geq 1000 \end{cases} \tag{5}$$

Along with the droplet motion, the heating and evaporation processes within the droplet are described by solving a 1D (1 dimensional) heat and mass transport equation to track the species and thermal evolution along the droplet radius according to [7]. In particular, equidistant 11 CVs along the radial coordinate are enough to properly resolve these evolutions. Heat and mass exchanges with the gaseous phase are computed according to the evaporation film-based model as reported in [15]. The evaporation rate is, thus, described as:

$$\dot{m} = \frac{dm_d}{dt} = \sum_i^n \dot{m}_i = \sum_i^n \left[\pi d_p (\bar{\rho} \bar{D})_{i,g} Sh_i^* ln \left(1 + B_{M,i} \right) \right], \tag{6}$$

while the heat balance is given as:

$$\frac{dT_d}{dt} = -\frac{\dot{m} c_{p,vap,ref} \left(T_g - T_P \right) / B_T - H_{vap}}{m_d C_{p,d}}. \tag{7}$$

where, d_p is the droplet diameter, T_d the droplet temperature, $\dot{m}$ and $\dot{m}_i$ represents the total evaporation rate of droplet and individual species i, respectively. $D_{i,g}$ the binary vapor/gas diffusion of component

i in the gas, H_{vap} the latent heat, $c_{p,d}$ and $c_{p,vap,ref}$ are the specific heat capacity of liquid and vapor phase, respectively . The quantities B_T and B_M are the Spalding mass and heat transfer numbers, respectively. $Sh_i^\star$ denotes the modified Sherwood number given by:

$$Sh_i^\star = 2 + \frac{Sh_{0i} - 2}{F(B_{M,i})} \quad \text{with} \quad B_{M,i} = \frac{Y_{i,s} - Y_{i,\infty}}{1 - Y_{i,s}}, \quad F(B_{M,i}) = (1 + B_{M,i})^{0.7} \frac{\ln(1 + B_{M,i})}{B_{M,i}} \tag{8}$$

$$\text{and} \quad Sh_{0i} = 2.0 + 0.6 Re^{1/2} Sc_i^{1/3} \quad \text{where} \quad Sc_i = \frac{\mu_g}{\rho_g D_{ig}} \tag{9}$$

define the Reynolds and Schmidt numbers, respectively. In particular $Y_{i,\infty}$ and $Y_{i,s}$ are the mass fraction of species i far field and at droplet surface, respectively. Equations (6) and (7) can be combined to deliver the heat transfer correlation as [12];

$$\dot{m} = \pi d_p \frac{\lambda_g}{C_{pg}} Nu^\star \ln(1 + B_T) \tag{10}$$

where, the updated Nusselt number $Nu^\star$ is defined as [12];

$$Nu^\star = 2 + \frac{Nu_0 - 2}{F(B_T)} \quad \text{with} \quad B_T = \frac{C_{pg}(T_\infty - T_d)}{L}, \quad Nu_0 = 2.0 + 0.6 Re^{1/2} Pr^{1/3}, \quad Pr = \frac{C_{pg}\mu_g}{\lambda_g}. \tag{11}$$

thereby, the Prandtl number Pr is expressed as function of specific heat C_{pg}, viscosity μ_g, and thermal conductivity λ_g of carrier gas, respectively. It should be noted that B_T is related to B_M by

$$B_T = (1 - B_{M,i})^\phi \quad \text{where} \quad \phi = \frac{C_{pl} Sh^\star}{C_{pg} Nu^\star} \frac{1}{Le}. \tag{12}$$

In this study, the Lewis number Le is set to value 1. In order consider natural convection, the expressions above have been extended as reported in [23].

3.1.3. Thermal Decomposition

The AdBlue® droplet is a binary mixture of 32.5 wt % urea and 67.5 wt % water which is injected upstream of the SCR monolith into the exhaust duct, where the evaporation process first takes place by utilizing the available exhaust gas heat:

$$CO(NH_2)_2(aq) \rightarrow CO(NH_2)_2(s,l) + H_2O(g). \tag{13}$$

The resulting urea first melts (melting point of 407 K) if it is in solid state and, then starts to decompose thermally. In addition to ammonia production, the urea decomposition of also leads to the formation of ammonium isocyanate, biuret, and triuret. The cyanuric acid and other higher compounds are produced above 453 K. For very fast urea heating, the thermal decomposition processes which include both the thermolysis and hydrolysis follow after most of the water is evaporated. During thermolysis, urea decomposes into ammonia (NH_3) and isocyanic acid (HNCO) :

$$CO(NH_2)_2(s,l) \rightarrow NH_3(g) + HNCO(g). \tag{14}$$

The produced gaseous NH_3 thus takes part in the SCR reactions to reduce the engine NO_X, while the resulting HNCO will further react with water vapor to produce NH_3 through hydrolysis reaction:

$$HNCO(g) + H_2O \rightarrow NH_3(g) + CO_2(g). \tag{15}$$

Following [23] where only the thermolysis has been considered, in the present paper in which both the thermolysis and the hydrolysis are accounted for, the hydrolysis given by the reaction rate of Equation (15) is also described by an Arrhenius-type equation according to [5,20].

As already mentioned above, the SCR performance is greatly hindered by the incomplete thermolysis of urea ahead the SCR catalyst, among others. Which can be attributed to both the incomplete evaporation of water or/and to urea thermolysis process. This may lead to undesirable deposition solid by-products on the SCR duct walls and substrates inlets.

3.1.4. Liquid Wall Film and Deposit Formation

The droplet impingement over the wall may result in a liquid wall–film formation. Since the liquid film may flow along the wall and evaporate, this may evolve to form deposition after a certain characteristic time depending on the temperature and the flow rate in presence. The length scale of wall–film and solid deposit formation can vary from micro to macro scale depending upon the operational stage of SCR system. The rate of chemical reactions for deposit by-product formation can also vary extensively with certain reactions taking minutes to evolve completely. This way temporal and spatial evolution of the deposition features a highly complex phenomenon which is not very well understood. Using both simulations and experiments, Munnannur et al. [36] provided a regime map for droplet-wall interaction with respect to fundamentals of thermal and fluid dynamic properties. Birkhold et al. [5,37] utilized a drop impingement model based on the work by Kuhnke [38]. Obviously, the spray wall-impingement and subsequent deposit formation depends on individual injection system and spray properties and decomposition, configuration of SCR mixing chamber, conditions on the exhaust pipe surface, etc. Given a specific SCR system, the chemistry models must consider some reaction pathways also resulting from incomplete decomposition (e.g., [5,7,8,16]). Nevertheless, the single step global mechanism for urea decomposition into NH_3 and HNCO as provided in Equations (14) and (15) is used in the present paper.

Regardless of this limitation, high-fidelity sub-models are generally used to investigate the key processes such as cross-flow turbulence and droplet-wall interaction to allow useful inferences to be drawn on urea deposit formation, especially the likelihood of wall–film development and impingement induced breakup which can influence the overall evaporation and conversion dynamics. The wall/droplet interaction description in the present paper relies on an advanced extension of the approach reported by O'Rourke et al. [39]. This extension includes droplet deposition, spreading of the film owing to impingement pressure/film inertia, rebound, and splashing (see in [40]). In the case of droplet deposition, impacting droplets are treated as so-called wall particles. For these conditions, 2D (2 dimensional) equations for film mass, energy and momentum are solved. The resulting film thickness, which in this case results from the total volume of the deposited droplets, is described by an evolution equation. The various regimes of the drop-wall interaction (breakup/splash, deposition, rebound) are differentiated or determined by means of an effective impact parameter, K, and a suitable dimensionless temperature, T*, given as [37,40].

$$K = We^{1/2} Re^{1/4}; \quad T^* = \frac{T_w}{T_{sat}} \quad \text{with} \quad We = \frac{\rho u_n^2 d_p}{\sigma}; \quad Re = \frac{\rho u_n d_p}{\mu} \tag{16}$$

with u_n is the normal velocity to the impinging wall. T_w and T_{sat} are the wall and droplet saturation temperature, respectively.

3.1.5. Numerical Procedure

In this study, a KIVA-4mpi CFD code which is open source is used (LANL (Los Alamos National Laboratory), Los Alamos, NM, USA) for its appropriate features recently integrated for LES investigations in [40–42]. The code is based on on finite volume formulation, and is also suitable for compressible flow simulations. An arbitrary Eulerian–Lagrangian (ALE) method is applied to track the droplets within an LES framework. More details about the governing equations and respective formulations for the dispersed and gas phase solver is provided in [35]. See also [7,40–42] for more details.

3.2. Detailed Mathematical Modelling

To determine the influence of the evaporation model, the chemical kinetics and the ambient conditions on the evaporation and decomposition of UWS, it is numerically suitable to use a detailed numerical simulation. To reduce the computational efforts, the droplet is considered spherical so that a one dimensional models can be formulated [43]. In this subsection we rely on Stein et al. [11]. To avoid repetition, only the essential steps are recalled here.

3.2.1. Governing Equations

The required equations consist of mass, momentum and energy evolution equations as described in [44]. Integrating the convective terms often causes difficulties so the equations are transformed into modified Lagrangian coordinates [43]. For the gas-phase, these are [11]

$$\left(\frac{\partial r}{\partial \psi}\right)_t = \frac{1}{\rho r^2},\tag{17}$$

$$\frac{\partial w_i}{\partial t} + z\frac{\partial w_i}{\partial \psi} + \frac{\partial}{\partial \psi}(r^2 j_i) = \frac{\dot{\omega}_i M_i}{\rho},\tag{18}$$

$$
\begin{aligned}
0 = &\frac{\partial T}{\partial t} - \frac{1}{\rho c_p}\frac{\partial p}{\partial t} + z\frac{\partial T}{\partial \psi} - \frac{1}{c_p}\frac{\partial}{\partial \psi}\left(\rho r^4 \lambda \frac{\partial T}{\partial \psi}\right)\\
&- \frac{r^2}{c_p}\sum_{i=1}^{n_S} j_i c_{p,i}\frac{\partial T}{\partial \psi} + \frac{1}{\rho c_p}\sum_{i=1}^{n_S}\dot{\omega}_i h_i M_i,
\end{aligned}\tag{19}
$$

$$\rho = \frac{p\bar{M}}{RT},\tag{20}$$

$$z = v(\psi_0)\rho(\psi_0)r_D^2.\tag{21}$$

The equations of the liquid phase are transformed in the same way and read [11]

$$\left(\frac{\partial r}{\partial \eta}\right)_t = \frac{\psi_D^0}{\rho r^2},\tag{22}$$

$$\frac{\partial w_i}{\partial t} - \frac{1}{\psi_D^0}(\eta \vartheta^0)\frac{\partial w_i}{\partial \eta} + \frac{1}{\psi_D^0}\frac{\partial}{\partial \eta}(r^2 j_i) = 0,\tag{23}$$

$$
\begin{aligned}
0 = &\frac{\partial T}{\partial t} - \frac{1}{\psi_D^0}(\eta \vartheta^0)\frac{\partial T}{\partial \eta} - \frac{1}{c_p \psi_D^0}\frac{\partial}{\partial \eta}\left(\rho r^4 \lambda \frac{1}{\psi_D^0}\frac{\partial T}{\partial \eta}\right)\\
&- \frac{r^2}{c_p}\sum_{i=1}^{n_S} j_i c_{p,i}\frac{1}{\psi_D^0}\frac{\partial T}{\partial \eta},
\end{aligned}\tag{24}
$$

$$\rho = \rho(w_1, \ldots, w_{n_S}, p, T),\tag{25}$$

$$\vartheta^0(t) = \frac{d\psi_D^0}{dt} = -\phi_{vap}r_D^2.\tag{26}$$

In these equations, the quantities r, and ψ are the spatial radial coordinate, and Lagrangian coordinate, respectively. The superscript "$()^0$" represent gas phase and "$()_D^0$" the liquid phase at droplet surface before second transformation with η the Lagrangian coordinate transformed for the liquid phase. The variables w and x are the mass and molar fraction, respectively. The thermo physical properties c_p, p, R, λ, and h represent the constant pressure specific heat capacity, pressure, gas constant, heat conductivity, and enthalpy, respectively. r_D the droplet radius, j_i the diffusion flux density of species i, $j_{q,c}$ the heat flux density due to conduction, z the mass flux at droplet surface, ϕ the mass flux density, n_S the number of species, r_{dec} the reaction rate of urea decomposition, and $\dot{\omega}$ the source term due to chemical reaction.

In the gas phase detailed models are applied to describe the transport processes. An approximation by Hirschfelder and Curtiss [24] is used to determine the diffusion coefficients and heat fluxes are calculated using Fourier's law. As the droplet size is small convection can be neglected as it would only have a limited effect on heat and mass transfer. According to earlier studies, UWS shows only slight deviations from an ideal mixture [45]. It was also found that crystallization and over-saturation of urea show no strong influence on the evaporation process of UWS [46]. Furthermore, after most water has evaporated from an UWS droplet the temperature increases above the urea melting point in a very short time due to the small droplet sizes and thus, crystallized, urea is not expected to be present for a relevant amount of time [5,47,48]. For these reasons, urea is assumed to be liquid and can be modeled as an ideal mixture of liquids with property data taken from Yaws et al. [49]. During the water evaporation stage, the diffusion of urea within UWS is approximated by the diffusion coefficient of solid urea in water. When only a little amount of water is left in the droplet the influence of diffusion in the liquid phase becomes negligible. To ensure consistency, correlations from Reid et al. [50] are used to determine a diffusion coefficient of water in liquid urea during this phase.

3.2.2. Numerical Solution

Finite differences are employed to discretize the spatial coordinates for the gas phase (ψ) and for the liquid phase (η) on a non-equidistant adaptive grid. In this detailed numerical simulation, a total number of 100 grid points is sufficient to resolve the temperature and the species evolution in the liquid, gas phase and at the droplet interface. In the gas phase near the droplet transient steep gradients are expected. These gradients are resolved by a meshing procedure based on a grid function [51], which is applied during the evaporation process. In the liquid phase the steepest gradients are expected close to the droplet surface and the grid is refined towards this position [52]. At the center of the droplet, symmetry boundary conditions are applied. Neumann boundary conditions with zero gradients are used at the outer boundary of the gas phase. They approximate an adiabatic vessel without mass flow through the boundary, which makes it easier to determine the overall mass and reaction rates of all species in the gas phase after the droplet is evaporated completely. The system of partial differential and algebraic equations for both phases is then integrated using the linearly implicit extrapolation method LIMEX [53].

3.2.3. Interface

The multi-component evaporation model used in the detailed simulations is based on a model from Stauch [43]. It was modified in order to properly handle conditions with a strong difference in vapor pressure of the evaporating species. During the evaporation of one species, the second species is always near the vapor-liquid equilibrium. Such conditions are often found during UWS evaporation, especially when there is a high water content in the exhaust gas. In the resulting model, each species can independently condensate or evaporate irrespective of the overall mass flow. The assumptions of a continuous temperature profile $T^g = T^l$ as well as a local phase equilibrium at the surface of the droplet were retained from the original model. To calculate the molar fraction of the evaporating species directly above the surface Raoult's law is invoked. The resulting equation for all evaporating species, transformed for mass fractions, is

$$w_i^g = \frac{M_i}{\bar{M}_v^g} \frac{p_{vap,i}}{p_{amb}} x_i^l,$$
(27)

with

$$x_i^l = w_i^l \frac{\bar{M}_v^l}{M_i},$$
(28)

The mass flux at the surface for each species is [11]

$$\phi_{vap,i} = w_i^g \phi_{vap} + j_i^g = w_i^l \phi_{vap} + j_i^l. \tag{29}$$

The overall vaporization rate is the sum of Equation (29) over all evaporating species. The boundary condition for other non-evaporating species reads

$$w_i^g = -\frac{j_i^g}{\phi_{vap}}. \tag{30}$$

The system Equation (29) is thus closed by applying it to the liquid side. The equations for the liquid phase and for the gas phase are connected by the interface conditions and solved in a fully coupled way. The energy conservation equation for the interface is given in simplified form as [43]

$$0 = \sum_i \phi_{vap,i} \Delta h_{vap,i} + j_{q,c}^g - j_{q,c}^l. \tag{31}$$

The Equations (17)–(21) thus represents the resulting system of partial differential algebraic equations for the gas phase and Equations (22)–(26) for the liquid phase. They are closed by Equation (27), and Equations (29)–(31) and the boundary conditions given in Section 3.2.2.

4. Assessment of the Multicomponent Evaporation Model: Evaporation Characteristics

In this section, the numerical data for the evaporation and thermal decomposition processes obtained by means of the detailed numerical simulations are used to validate the LES-based Eulerian–Lagrangian model applied to a generic standing droplet case (see Figure 1, case S1), for which experimental data are not available.

Using the boundary conditions as outlined above for a detailed description and considering the associated computational costs, it is reasonable to analyze the overall mass and reaction rate in the gas phase after the droplet evaporation without convection around the droplet. Thus, a standing AdBlue® droplet similar to that investigated in [23] is considered and relative velocity for which no experimental data is available. The numerical domain is extended to 100 droplet diameters or more so that possible interactions between the evaporation process and the outer boundary are inhibited. The obtained results are employed to assess the prediction capability of the multi-component evaporation model and thermal decomposition description. The properties used for comparisons between detailed simulations and the LES-based Eulerian–Lagrangian model are listed in Table 2. The respective comparisons have been carried out under consideration of chemistry (Equations (15) and (14)) and without chemistry, respectively. Considering both dry air and moist air environmental conditions, the results exhibit a very good agreement, demonstrating the suitability of the designed LES-based Eulerian–Lagrangian model when chemistry is not included. This is shown in Figure 2 for various temperatures (423 K, 573 K, 773 K).

Table 2. Comparisons properties between detailed simulations and the LES-based Eulerian–Lagrangian.

Properties	Detailed Simulation [B07]	Reduced Model [C05]
Droplet phase	Resolved 1D	Resolved 1D
Gas phase close to droplet	Resolved 1D	Not resolved, 0D (Film model)
Ambient conditions	Given as boundary condition of gas-phase	Taken from LES
Interface	Local evaporation balance Connects gas-phases and droplet	Film model connects ambient onditions and droplet
Values at interface	Vapor pressure and partial pressure	Weighted average of droplet data and ambient conditions
Gradients at interface	Local gradients (first element of gas phase)	Interface value to ambient condition
Diffusion	Mixture average	Only evaporating species

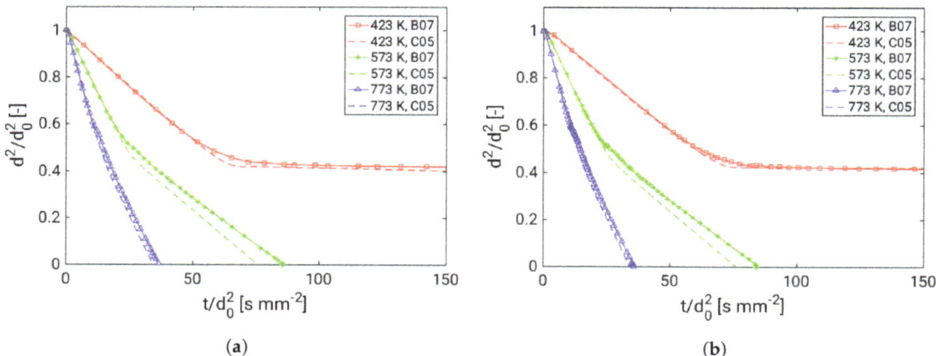

Figure 2. Comparison between detailed simulations (B07) and reduced model (LES based, C05) without chemistry (only evaporation). (**a**) dry air; (**b**) moist air.

The results from both the simulations clearly capture the two typical slopes for the mass transfer of the AdBlue® droplet. The first slope represents an initial stage of water evaporation while the second slope shows the urea decomposition and mass transfer process. Regarding the temperature evolution within the droplets, this corresponds to a small temperature gradient during the initial stage of evaporation followed by four stages (heating of UWS, constant temperature value until water fully evaporates, heating up of urea, constant temperature value until urea fully evaporates/decomposes) of thermal conditions inside the droplet as pointed out in [7,23]. Subsequently, the predictive capability of the multi-component evaporation model is then appraised against the experimental data reported in [26]. Since the experiment is carried out for a static single droplet, the simulation results only from reduced model are validated (see Figure 3). The comparison shows a good agreement with experiment for two operating gas phase temperatures. The well-known evolution of four stages of the droplet surface temperature (T_s) can be clearly seen as dashed line in Figure 3. More details about experimental configuration and adopted measurement techniques can be found in [26].

After the successful validation of the LES-based Eulerian–Lagrangian module, focus is now put on the impact of relative velocity taking advantage of the LES features on both the evaporation and thermal decomposition examined for various droplet diameters and gas phase temperatures (see Figure 1, cases S2-4).

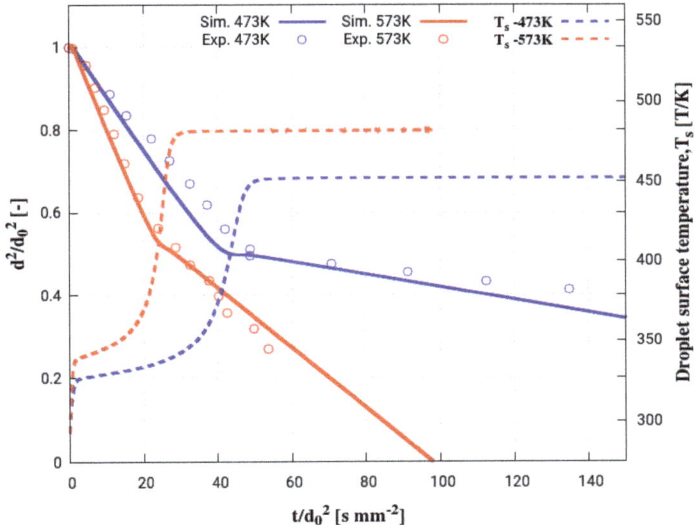

Figure 3. Validation of evaporation model (C05) for AdBlue® droplet evaporation under gravitational influence (experiment [26]).

5. Evaporation and Thermal Decomposition Characteristics: Influence of Droplet Relative Velocity

In order to carry out early design analysis, the droplet residence and life time together with the possibilities of droplet-wall impingement are of great relevance. At first, Figure 4a shows the comparison of droplet life time for a droplet with $d_p = 50$ μm under stagnant and convective carrier gas conditions at various temperatures. It can be clearly seen that the convective gas at higher temperature enhances the mass transfer rate to many folds. This strongly influences the conversion rate as seen in Figure 4b. The effect of gas velocity will be further analyzed next. In fact, the usual operating conditions in automotive applications for the SCR exhaust after-treatment lie in the range of 5–100 m/s for the exhaust gas velocity, 400–1000 K for the exhaust gas temperature and 350–900 K for the wall temperature. The UWS is usually injected with the injection velocity ranging within 5–25 m/s and the injection temperature of about 300–350 K to atomize sprays characterized by Sauter Mean Diameters between 20–150 μm, (see in [5]. To gain further information, the phase change dynamics of AdBlue® droplets of various diameter (25, 50, 100, 150, and 200 μm) typical for SCR systems are simulated. All the numerical simulations carried out are summarized in Table 3. The respective results are shown in Figure 5 in which the noticeable influence of droplet relative velocity can be seen for each combination of droplet diameter and carrier gas temperature. The mass transfer is greatly enhanced when the gas temperature is increased from 473 K to 573 K, while it is more gradual between temperatures of 573 K to 673 K and 673 K to 773 K. With increasing gas temperature, the influence of droplet relative velocity becomes very minimal. This observation also holds for the other droplet diameters. However, as expected, the larger droplets last for considerably longer times than smaller droplet.

Table 3. Operating parameters for droplet life parametric study.

Initial Diameter, (μm)	T_g, (K)	Relative Velocity, u_{rel} (m/s)
25	473	0, 5, 10, 20, 40
	573	0, 5, 10, 20, 40
	673	0, 5, 10, 20, 40
	773	0, 5, 10, 20, 40
50	473	0, 5, 10, 20, 40
	573	0, 5, 10, 20, 40
	673	0, 5, 10, 20, 40
	773	0, 5, 10, 20, 40
100	473	0, 5, 10, 20, 40
	573	0, 5, 10, 20, 40
	673	0, 5, 10, 20, 40
	773	0, 5, 10, 20, 40
150	473	0, 5, 10, 20, 40
	573	0, 5, 10, 20, 40
	673	0, 5, 10, 20, 40
	773	0, 5, 10, 20, 40
200	473	0, 5, 10, 20, 40
	573	0, 5, 10, 20, 40
	673	0, 5, 10, 20, 40
	773	0, 5, 10, 20, 40

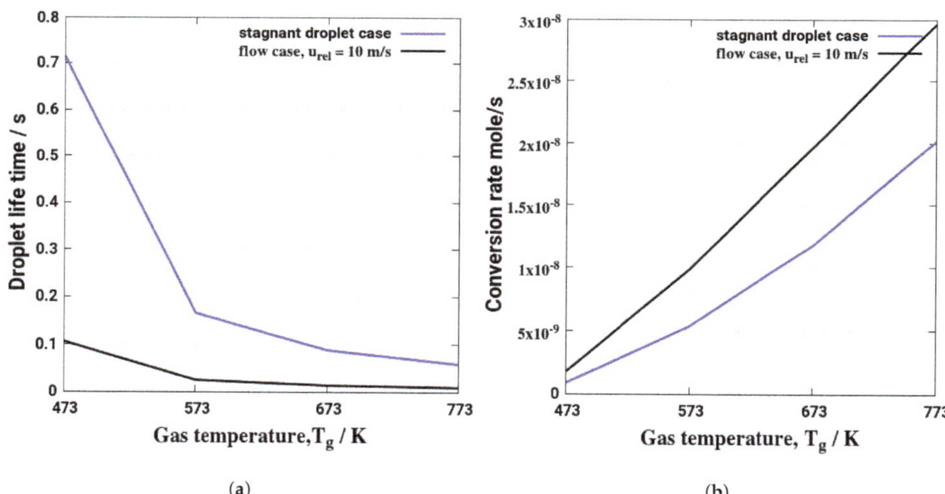

Figure 4. Comparison of evaporation/conversion dynamics of d_p = 50μm for stagnant droplet u_{rel} = 0, and with relative velocity u_{rel} = 10 m/s: (**a**) droplet lifetime; (**b**) average NH_3 conversion rate.

This behavior is clearly displayed in Figure 6 for two temperatures, namely, 573 K and 773 K. For the lowest small relative velocity (5 m/s) droplet life time for biggest droplet is almost 20 times longer than for the small droplets (see Figure 6a) and around ten times longer for higher relative velocity (40 m/s). These differences are smaller for higher gas phase temperature as seen in Figure 6b.

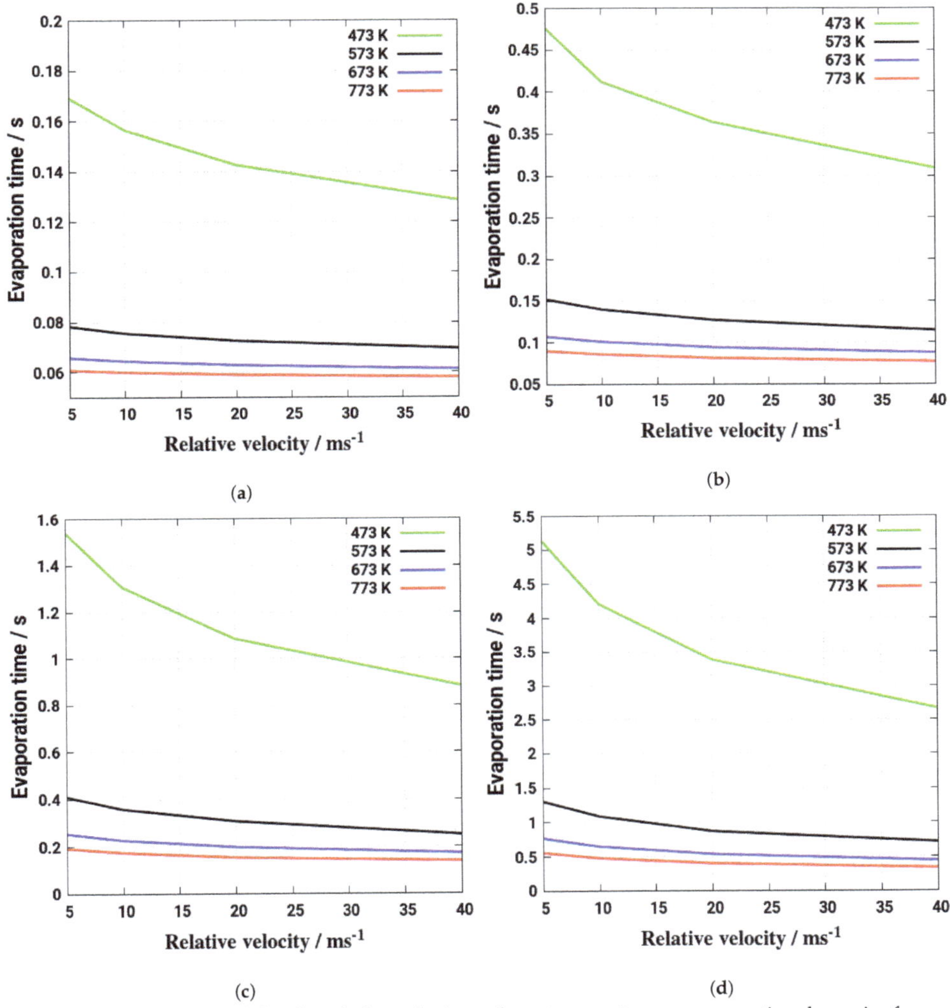

Figure 5. Influence of droplet relative velocity and gas temperature on evaporation dynamics for droplet diameters of: (**a**) 25 μm; (**b**) 50 μm; (**c**) 100 μm; (**d**) 200 μm.

Focusing on the droplet life time inside the SCR system, it appears that this parameter is a vital indicator of SCR system performance, once the AdBlue® evaporation and subsequent mixing and NH_3 conversion efficiency are of interest. For a given SCR duct size (diameter and length), the droplet life time can then be utilized to draw inferences about liquid wall–film and subsequent deposit formation on the mixing duct wall and on the entrance cross-section of the monolith allowing at the same time to select the optimum injection characteristics. Since droplet may travel within the mixing pipe while thermally decomposing, the time and length scale of processes of interest are of utmost importance. In this respect, Figure 7 depicts the percentage of evaporated mass for various droplets at two gas temperatures, 573 K and 673 K, as a function of the droplet life time. Since the associated time scale is considerably larger for the bigger droplet at both temperature levels, it can be deduced that big droplets may have enough time to impact the pipe wall or the entrance front of the monolith.

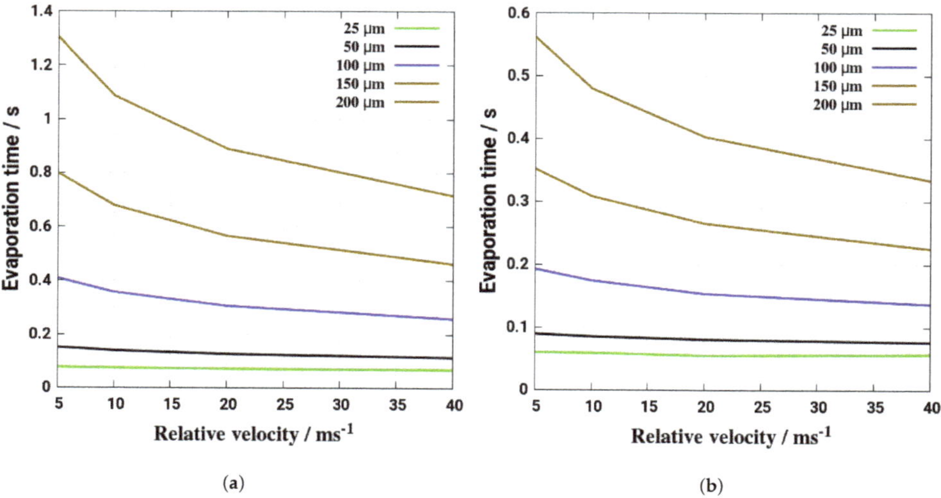

(a)

(b)

Figure 6. Influence of droplet relative velocity and diameter on evaporation dynamics for gas temperature. (**a**) 573 K; (**b**) 773 K.

Considering all the results accounting for various relative velocities, Figure 8 provides a clear view of how the two parameters, relative velocity and initial diameter, impact the prediction of the droplet life time, exemplary at two different gas phase temperatures, T_g = 573 K and T_g = 773 K. It can be partially concluded that for a given droplet diameter, both high relative velocity and high gas temperature enhance the droplet evaporation rate while reducing the droplet life time.

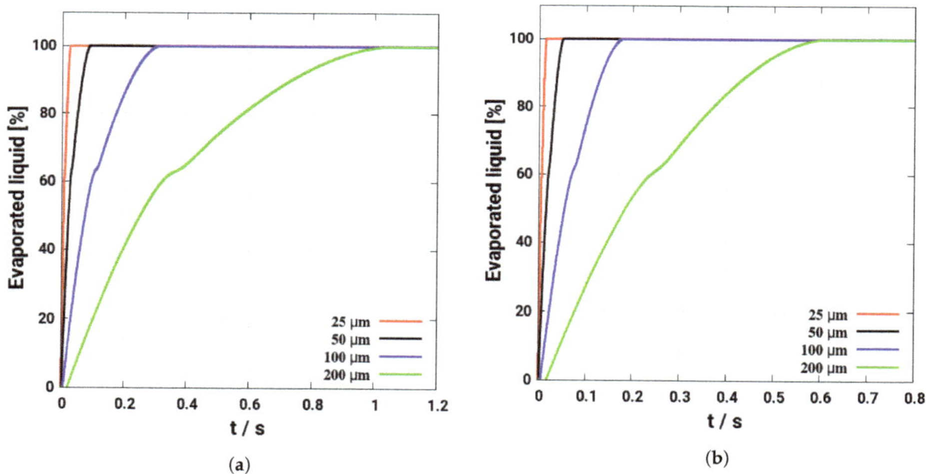

(a)

(b)

Figure 7. *Cont.*

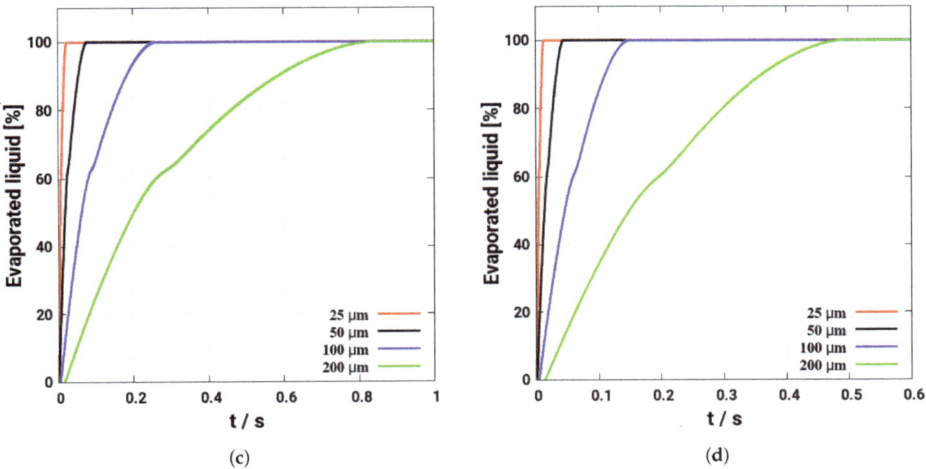

Figure 7. Evaporated droplet mass with respect to various diameter for: (**a**) u_{rel} = 10 m/s, T_g = 573 K; (**b**) u_{rel} = 10 m/s, T_g = 673 K; (**c**) u_{rel} = 20 m/s, T_g = 573 K; (**d**) u_{rel} = 20 m/s, T_g = 673 K.

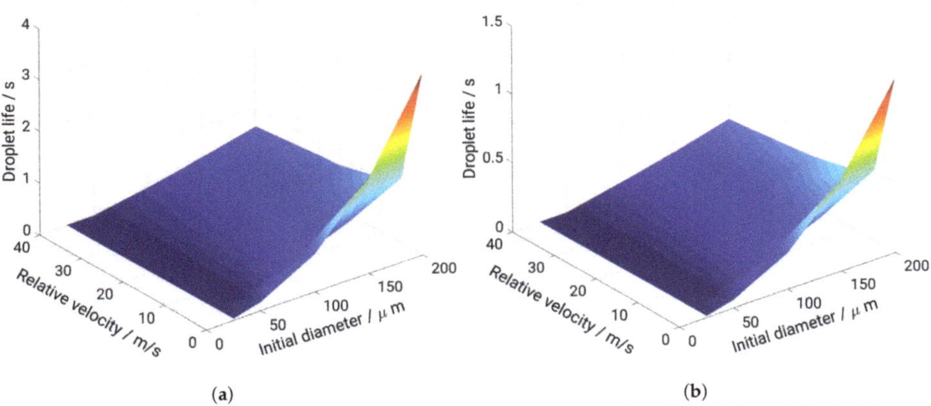

Figure 8. Droplet life with respect to various diameter and relative velocity for: (**a**) T_g = 573 K; (**b**) T_g = 773 K.

How the resulting urea is decomposed to provide the NH_3 gas needed to reduce NO_X emission in the SCR system is depicted in Figure 9. Thereby, the urea conversion efficiency which is defined as the ratio of decomposed mass of pure urea to the initial mass of urea is displayed as a function of droplet life time and imposed relative velocity for various droplet diameters at two different temperatures. It is apparent that for a given droplet diameter, higher relative velocity results in faster conversion. Further, higher relative velocity and temperature lead to higher conversion efficiency. With respect to the droplet size, it can be observed that larger droplets need more time to first evaporate water and then slowly decompose. The decomposition is faster with higher relative velocity for small droplets. By reducing the droplet diameter exemplary from 200 μm to 100 μm (case, T_g = 573 K, relative velocity u_{rel} = 20 m/s), it can be quantitatively seen that the time needed only for water evaporation process in the case of d = 200 μm (t_{evap} = 0.35245 s) is larger than the total time for the overall decomposition (evaporation and thermal conversion) of the droplet with d = 100 μm (t_{evap} + $t_{decomposition}$ = 0.2575 s).

This implies that small droplets and higher relative velocity are to be favored for optimal conversion of NH$_3$ in addition to higher temperature.

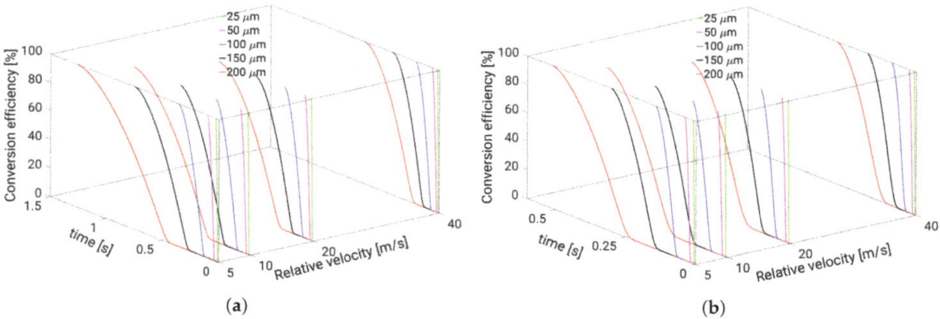

(a)   (b)

Figure 9. Conversion efficiency with respect to various diameter and relative velocity for: (**a**) T_g = 573 K; (**b**) T_g = 773 K.

6. Effect of Turbulent Cross-Flowing on Droplet Dynamics Characteristics

In real SCR systems, complex unsteady turbulent multi-phase flow phenomena including polydispersed AdBlue® spray evolve with a wide range of relative velocity between droplet phase and carrier gas phase. This results from AdBlue® droplets which are sprayed into a mixing pipe which is flowing by a hot exhaust gas. To reduce the complexity while gaining early information on the injected droplet size needed for a minimum deposition and optimal conversion, only a single AdBlue® is considered. It is injected into a hot turbulent cross-flowing stream within a mixing pipe under consideration of the operating parameters listed in Table 4. The droplet may experience phase change processes including interaction with the pipe and monolith front wall; see Figure 1, case S4).

Table 4. Operating parameters for hot-cross flow cases (Wall temperature T_w is assumed equal to T_g).

Case	Gas Temperature, T_g (K)	Gas Flow Rate, $\dot{m}_g$ (kg/h)	Droplet Diameter (µm)
T1M1	573	100	25, 50, 55, 100, 200
T1M2	573	200	25, 50, 55, 100, 200
T2M1	673	100	25, 50, 55, 100, 200
T2M2	673	200	25, 50, 55, 100, 200
3TM1	573, 523, 473	100	55

The injection location and the dimension of SCR duct are shown in Figure 1d. The droplet is injected with an injection velocity of 28.7 m/s at from an injector inclined at an angle of 50° with respect to the horizontal plane (see Figure 1, (S4)). Especially, the droplet residence time along with the droplet path length are investigated for various droplet diameters under (1) variation of the gas flow rate (T1M1 and T1M2, T2M1 and T2M2), and (2) variation of the gas temperature (T1M1 and T2M1, T1M2 and T2M2), respectively. In addition the case 3TM1 is designed to especially meet the condition for liquid wall–film formation during the droplet impingement. Figure 10 shows a regime map based on the impact parameters "K" and "T*" (see Equation (10)) for droplet-wall impingement of AdBlue® as reported in [37]. It provides a clear demarcation among various impingement outcomes. The wall deposition is expected with droplet having very little momentum and lower wall temperature, while droplet can rebound when temperature is sufficiently high. By further increasing the droplet momentum, the droplet can splash and break into smaller droplets as suggested in [38], which obviously can enhance the evaporation of AdBlue® droplet and conversion rate (see Figure 11).

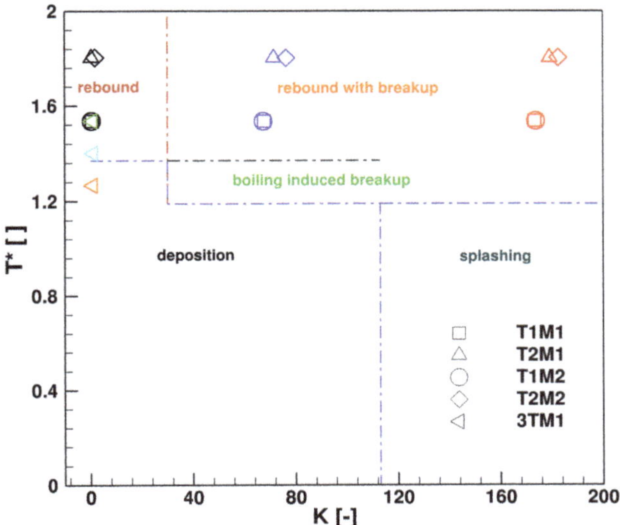

Figure 10. Droplet -wall impingement regime map [5,37] with operating points for single droplet study in cross flow (red: 200 μm; blue: 100 μm; black: 55 μm; green: 573 K; cyan: 523K; orange: 473K).

Focusing on the residence time of the droplet, let us monitor the trajectory of the droplet resulting from two essential motions experienced by the droplet: the transverse motion pushing the droplet away from the injection point and a stream-wise motion carrying the droplet downstream with the cross-flow.

The obtained droplet trajectories for all cases under consideration are plotted in Figure 11. To recall that both higher droplet relative velocity and higher gas temperature enhance considerably the droplet evaporation rate and, in turn, reduce the droplet life time (see previous section), indicating favorable SCR operating conditions. Apart from that, how further a droplet evolves and how long it takes to travel the SCR mixing duct may also give useful hint for possible wall impingement and subsequent liquid wall–film and deposit formation.

Figure 11a depicts the droplet trajectory for the case "T1M1" (T_g = 573 K, $\dot{m}_g$ = 100 kg/h). It can be clearly seen that the droplet (except d_0 = 100 μm) completely evaporates before reaching the end of the mixing section where the entrance cross-section of the monolith may be situated. In such a case, tendency to deposition formation on the SCR mixing duct wall or/and on the monolith entrance cross-section is avoided. Droplet/wall interaction is obvious for droplet diameters of 100 μm and 200 μm, only, for which splashing is observed. After splashing the droplet breaks into smaller droplets giving rise to an increase of evaporation. The degree of splashing is strong for droplet diameter of 200 μm while the longer droplet life time and longer droplet path length can be realized for diameter of 100 μm.

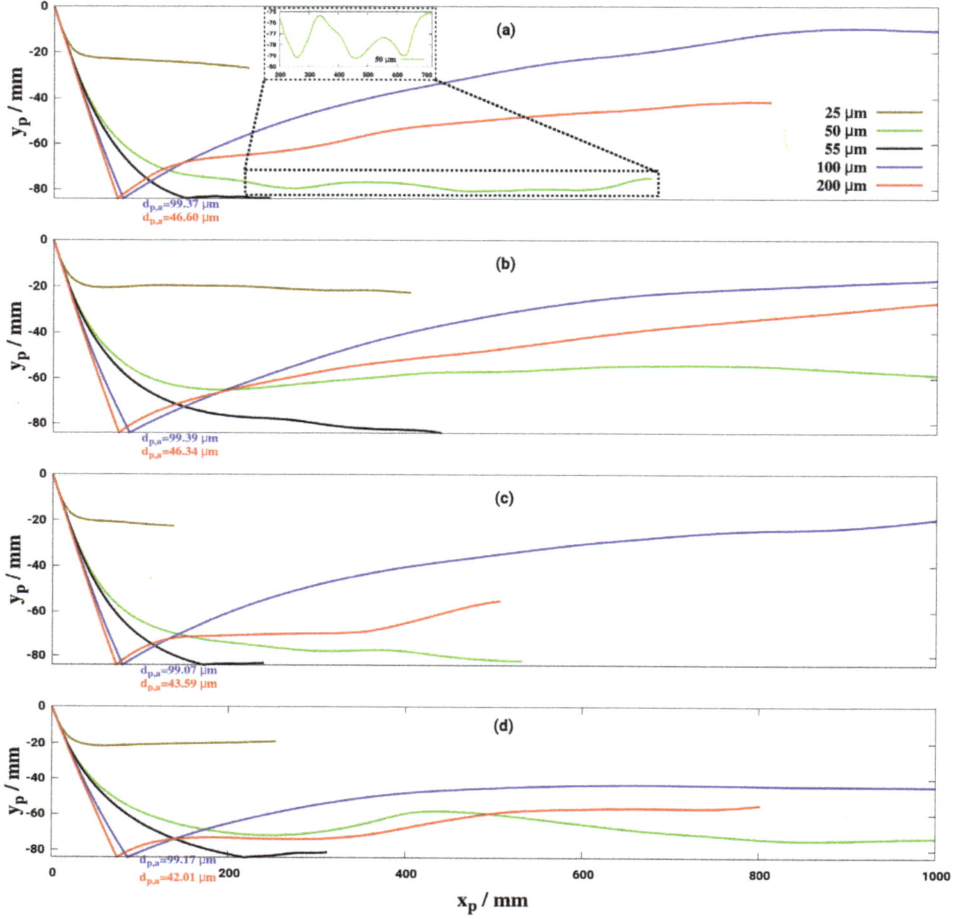

Figure 11. Droplet trajectory and life under various gas temperatures and mass flow rate cases: (**a**) T1M1; (**b**) T1M2; (**c**) T2M1; (**d**) T2M2. Initial post-impingement droplet diameters ($d_{p,a}$) are given in colors at respective impingement locations for droplets which undergo splashing/rebound.

Comparisons with case "T1M2" (T_g = 573 K, $\dot{m}_g$ = 200 kg/h) (see Figure 11a,b) allow to assess the influence of cross-flow on droplet dynamics. Compared to case "T1M1" all the droplets achieved longer droplet path length. Moreover, all the droplets, except d = 25 μm, can potentially cross the mixing duct and impinge the entrance cross-section of the monolith. Such a droplet can enter and further decompose in the fine channel of monolith or may lead to deposits formation across the SCR-monolith cross-section. To examine the influence of gas temperature, further comparison can be made especially between cases "T1M1" and "T2M1" (see Figure 11a,c). Compared to case "T1M1" all the droplets have shorter trajectory, except for d = 100 μm. This can be attributed to complex wall-impingement dynamics predominantly bouncing in nature for droplet with diameter of 100 μm. Note that longer droplet path length is observed for all droplets in "T2M2" of strong gas flow rate in contrast to "T2M1" under the same gas temperature of T_g = 673 K. Once different temperatures are considered, droplet trajectories in case of high temperature "T2M2" are shorter than in the case with low temperature "T1M2" under the same gas flow rate of $\dot{m}_g$ = 200 kg/h. The initial post-impingement droplet diameters are also provided at the respective impingement location for droplets which undergo splashing/rebound (see Figure 11). It can be noted that the droplet with d_0 = 100 µm undergoes bouncing without any noticeable breakup in all cases. However, droplet with d_0 = 200 µm solely experience intense splashing which give rise to smaller droplets after impingement. This in turn contributes to faster evaporation and reduced droplet life inside the SCR duct as shown in Figure 11. Further, in order to evidence the influence of gas phase turbulence on the droplet dispersion, the instantaneous velocity of the droplet is plotted in Figure 12 for the case of d_0 = 25/50 µm, $\dot{m}_g$ = 100 kg/h, and T_g = 573 K during its full lifetime.

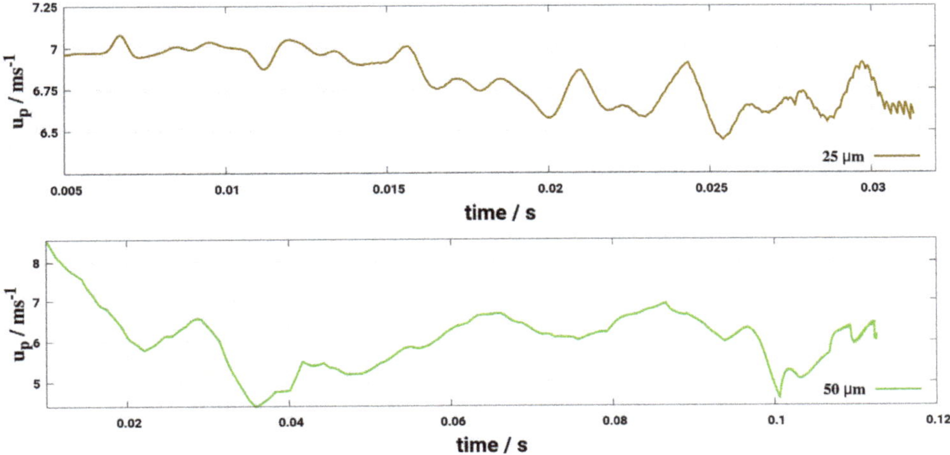

Figure 12. Evolution of the instantaneous velocity of the droplets during its lifetime: for mass flow rate $\dot{m}_g$ = 100 kg/h at gas temperatures of 573 K, examplarily d_0 = 25, 50 µm.

The turbulence impact on the droplet velocity is even higher at the last stage as the droplet diameter becomes smaller and smaller during its evaporation process. It can be further observed that for d_0 = 25 µm droplet, the dispersion is more significant as the droplet size is smaller, thus the droplet strongly follows the turbulent flow. In fact, by considering the two-way coupling between the gas phase and droplet, it turns out that the turbulent dispersion is adequately captured for the case of Stoke's number larger than one as it appears in this study without accounting for subgrid scale dispersion model [54]. According to [55], it is worth noting that for RANS (URANS) additional dispersion model must be included which may require more numerical care with respect to the phase coupling procedure.

Additional simulations are performed with d = 55 µm to assess the wall–film formation dynamics for respective cross-flow conditions. Due to very weak impact, the droplet forms a wall–film (see Figure 11a) and further convected along the duct wall until completely decomposed. Similar behavior is also observed for other two cases (T2M1, T1M2, see Figure 11b,c), while at higher wall temperature in the case T2M2 , the droplet is bounced off after impinging, and convected closely parallel to the duct wall due to a weak impact. Considering the case 3TM1, the droplet impinges the wall and sticks on it and forms a liquid wall–film. This is clearly depicted in Figure 13.

These characteristic behaviors can be imprinted within the diagram K-T* . This is done in Figure 10 for all cases under consideration in Table 4. It turns out that the case 3TM1 features the condition for liquid wall–film formation on the duct wall that may lead to deposit formation as observed in [37].

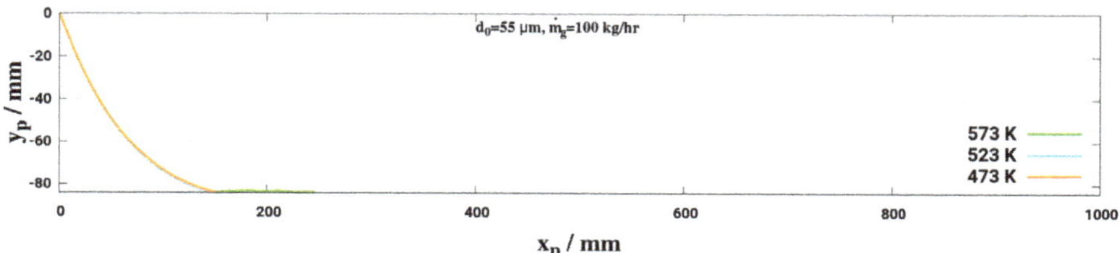

Figure 13. Droplet trajectory and life: deposition case for mass flow rate $\dot{m}_g$ = 100 kg/h at various gas temperatures of 573 K, 523 K, and 423 K.

7. Conclusions

A LES-based Eulerian–Lagrangian model tool including a multi-component evaporation model along with a thermal decomposition mechanism has been extended to include wall–film formation model in order to study the effect of turbulent cross-flowing stream conditions on the dynamic droplet characteristics. The impact has been recorded in terms of droplet trajectory and evaporation time, decomposition efficiency and liquid wall–film formation. The objective was to provide early information on the injected droplet size and velocity required for a minimum deposition and optimal conversion in terms of droplet trajectories, droplet evaporation time, decomposition efficiency and wall–film formation dynamics.

For that purpose, four droplet configurations have been designed. First, a standing droplet case without any relative velocity followed by a standing droplet case. Next, a droplet into convective environment with co-flow characterized by various relative velocities between the carrier phase and the droplet phase, and an injected droplet into a hot cross-flowing stream within a mixing pipe in which the droplet may experience phase change processes including interaction with the pipe wall along with liquid wall–film and possible solid deposit formation processes.

After a successful validation of the numerical tool by means of numerical experimental data from a detailed numerical model of a standing droplet case, the results from the investigations allow to draw following conclusions.

1. As the evaporation characteristics obviously depend on the AdBlue® droplet diameter, gas phase temperature and droplet relative velocity, it turns out that a smaller droplet diameter, higher temperature and relative velocity lead to shorter droplet life time as the droplet evaporates faster. Under such conditions, possible droplet/wall interaction processes on the pipe wall or/and at the entrance front of the monolith are avoided.

2. Since the gaseous NH_3 generated by urea decomposition is intended to reduce NO_X emission in the SCR system, it is apparent for the prediction of high NO_X removal performance that UWS injector system which allows to realize such operating conditions (droplet with smaller diameter, higher relative velocity and gas temperature) will support high conversion efficiency of urea into NH_3.

3. Lower mass flow rate and higher gas temperature reduce considerably the droplet trajectory allowing a droplet to fully decompose before interacting with the entrance section of the monolith. This may help to avoid impinging and deposition on the pipe wall for small droplet (d < 50 µm) for the investigated conditions.

4. The wall deposition consists essentially of two main processes: first, droplet impingement on the wall for which, based on K and T* values, the droplet can undergo various physical processes such as deposition, splashing rebound, breakup. Second, only deposited droplet may form solid deposition based on chemical kinetics and phase thermodynamics. In the present study only phase thermodynamics are considered to define the solid deposition, while chemical kinetics for solid deposit formation is scope of our ongoing research activities. On this basis, it has been

clearly shown that the droplet wall impingement can be favorable scenario as it can be responsible for breakup into smaller droplets and subsequently cause faster evaporation and conversion.

5. Considering the two-way coupling between the gas phase and droplet, it turns out that the turbulent dispersion is adequately captured by LES for a case of a Stoke's number that is larger than one as it appears in this study without accounting for a subgrid scale dispersion model. Note that for RANS (URANS) additional dispersion model must be included which may require more numerical care with respect to the phase coupling procedure.

Author Contributions: For this research, K.N., F.R. and A.S. conceived and designed the numerical setup. Detailed numerical simulation was perfomed by M.S. Reaction kinetics were reduced by M.S. and V.B. from the detail kinetics provided by O.D. K.N. carried out the numerical simulations using the LES-based Eulerian–Lagrangian tool and exploited the data. Together with A.S., analyzed, discussed and interpreted the overall numerical results. A.S., U.M., V.B. and J.J. contributed by providing materials and computing resources.

Funding: Deutsche Forschungsgemeinschaft (DFG): 237267381.

Acknowledgments: The authors gratefully acknowledge the financial support by the DFG (German Research Council)-Projektnummer 237267381—TRR 150, the support of the numerical simulations on the Lichtenberg High Performance Computer (HHLR) at the University of Darmstadt and the financial support by the Open Access Publishing Fund of Technische Universität Darmstadt.

Conflicts of Interest: The authors declare no conflict of interest.

References

1. *Regulation (EC) No 715/2007 of the European Parliament and of the Council of 20 June 2007 on Type Approval of Motor Vehicles with Respect to Emissions from Light Passenger and Commercial Vehicles (Euro 5 and Euro 6) and on Access to Vehicle Repair and Maintenance Information*; Technical Report; Official Journal of the European Union: Brussels, Belgium, 2007.

2. Nova, I.; Tronconi, E.; Eds. *Urea-SCR Technology for deNOx After Treatment of Diesel Exhausts*; Springer Press: Singapore, 2014.

3. Tripathi, G.; Dhar, A.; Sadiki, A. Recent Advancements in After-Treatment Technology for Internal Combustion Engines—An Overview. In *Advances in Internal Combustion Engine Research. Energy, Environment, and Sustainability*; Srivastava, D., Agarwal, A., Datta, A., Maurya, R., Eds.; Springer Press: Singapore, 2018; pp. 159–179.

4. Nishad, K.; Sadiki, A.; janicka, J. Numerical Modeling of adBlue Droplet Evaporation in the Context of SCR-DeNOx. In Proceedings of the 18th Annual Conference of Liquid Atomization and Spray Systems-Asia (ILASS-Asia), Chennai, India, 6–10 November 2016.

5. Birkhold, F.; Meingast, U.; Wassermann, P.; Deutschmann, O. Modeling and simulation of the injection of urea-water-solution for automotive SCR DeNOx-systems. *Appl. Catal. B Environ.* **2007**, *70*, 119–127. [CrossRef]

6. Praveena, V.; Martin, M.L.J. A review on various after treatment techniques to reduce NOx emissions in a CI engine. *J. Energy Inst.* **2017**. [CrossRef]

7. Nishad, K.; Sadiki, A.; Janicka, J. Numerical investigation of adBlue droplet evaporation and thermal decomposition in the context of NOx-SCR using a multi-component evaporation model. *Energies* **2018**, *11*, 222. [CrossRef]

8. Ebrahimian, V.; Nicolle, A.; Habchi, C. Detailed modeling of the evaporation and thermal decomposition of urea-water solution in SCR systems. *AIChE J.* **2012**, *58*, 1998–2009. [CrossRef]

9. Ström, H.; Sasic, S.; Andersson, B. Effects of the Turbulent-to-Laminar Transition in Monolithic Reactors for Automotive Pollution Control. *Ind. Eng. Chem. Res.* **2011**, *50*, 3194–3205. [CrossRef]

10. Spiteri, A. Experimental Investigation of the Injection Process in Urea-SCR deNOx Exhaust Gas After-Treatment Systems. Ph.D. Thesis, Department of Mechanical and Process Engineering, ETH Zürich, Switzerland, 2016.

11. Stein, M.; Bykov, V.; Mass, U. The Effect of Evaporation Models on Urea Decomposition from Urea-Water-Solution Droplets in SCR Conditions. *Emiss. Control Sci. Technol.* **2018**, *3*, 263–274. [CrossRef]

12. Abramzon, B.; Sirignano, W. Droplet vaporization model for spray combustion calculations. *Int. J. Heat Mass Transf.* **1989**, *32*, 1605–1618. [CrossRef]
13. Sazhin, S.S.; Elwardany, A.E.; Heikal, M.R. New approaches to the modelling of multi-component fuel droplet heating and evaporation. *J. Physics: Conf. Ser.* **2015**, *585*, 012014. [CrossRef]
14. Sadiki, A.; Chrigui, M.; Janicka, J.; Maneshkarimi, M. Modeling and Simulation of Effects of Turbulence on Vaporization, Mixing and Combustion of Liquid-Fuel Sprays. *Flow Turbul. Combust.* **2005**, *75*, 105–130. [CrossRef]
15. Miller, R.S.; Harstad, K.; Bellan, J. Evaluation of equilibrium and non-equilibrium evaporation models for many-droplet gas-liquid flow simulations. *Int. J. Multiph. Flow* **1998**, *24*, 1025–1055. [CrossRef]
16. Gan, X.; Yao, D.; Wu, F.; Dai, J.; Wei, L.; Li, X. Modeling and simulation of urea-water-solution droplet evaporation and thermolysis processes for SCR systems. *Chin. J. Chem. Eng.* **2016**, *24*, 1065–1073. [CrossRef]
17. Olsson, L.; Sjövall, H.; Blint, R.J. A kinetic model for ammonia selective catalytic reduction over Cu-ZSM-5. *Appl. Catal. B Environ.* **2008**, *81*, 203–217. [CrossRef]
18. Wurzenberger, J.C.; Wanker, R. *Multi-Scale SCR Modeling, 1D Kinetic Analysis and 3D System Simulation*; SAE Technical Paper; SAE International: Warrendale PA, USA, 2005.
19. Kaario, O.T.; Vuorinen, V.; Zhu, L.; Larmi, M.; Liu, R. Mixing and evaporation analysis of a high-pressure SCR system using a hybrid LES-RANS approach. *Energy* **2017**, *120*, 827–841. [CrossRef]
20. Ström, H.; Lundström, A.; Andersson, B. Choice of urea-spray models in CFD simulations of urea-SCR systems. *Chem. Eng. J.* **2009**, *150*, 69–82. [CrossRef]
21. Gökalp, I.; Chauveau, C.; Simon, O.; Chesneau, X. Mass transfer from liquid fuel droplets in turbulent flow. *Combust. Flame* **1992**, *89*, 286–298. [CrossRef]
22. Sornek, R.; Dobashi, R.; Hirano, T. Effect of turbulence on vaporization, mixing, and combustion of liquid fuel soray. *Combust. Flame* **2000**, *120*, 479–491. [CrossRef]
23. Nishad, K.; Ries, F.; Janicka, J.; Sadiki, S. Analysis of spray dynamics of urea-water-solution jets in a SCR-DeNOx system: An LES based study. *Int. J. Heat Fluid Flow* **2018**, *70*, 247–258. [CrossRef]
24. Hirschfelder, J.O.; Curtiss, C.F.; Bird, R.B. *Molecular Theory of Gases and Liquids*; Wiley: New York, NY, USA, 1964.
25. Ranz, W.; Marshall, W. Evaporation from drops: Part 2. *Chem. Eng. Prog.* **1952**, *18*, 173–180. [CrossRef]
26. Wang, T.J.; Baek, S.W.; Lee, S.Y.; Kang, D.H.; Yeo, G.K. Experimental Investigation on Evaporation of Urea-Water-Solution Droplet for SCR Applications. *AIChE J.* **2009**, *55*, 3267–3276. [CrossRef]
27. Wu, P.K.; Kirkendall, K.A.; Fuller, R.P.; Nejad, A.S. Spray Structures of Liquid Jets Atomized in Subsonic Cross-Flows. *J. Propuls. Power* **1998**, *14*, 173–182. [CrossRef]
28. Prakash, R.S.; Sinha, A.; Raghunandan, B.; Tomar, G.; Ravikrishna, R. Breakup of Volatile Liquid Jet in Hot Cross Flow. *Procedia IUTAM* **2015**, *15*, 18–25. [CrossRef]
29. Chrigui, M.; Sadiki, A.; Ahmadi, G. Study of interaction in spray between evaporating droplets and turbulence using second order turbulence RANS modelling and a Lagrangian approach. *Prog. Comput. Fluid Dyn.* **2004**, *4*, 162–174. [CrossRef]
30. Nicoud, F.; Ducros, F. Subgrid-scale stress modelling based on the square of the velocity gradient Tensor. *Flow Turbul. Combust.* **1999**, *62*, 183–200. [CrossRef]
31. Montorfano, A.; Piscaglia, F.; Onorati, A. Wall-adapting subgrid-scale models to apply to large eddy simulation of internal combustion engines. *Int. J. Comput. Math.* **2014**, *91*, 62–70. [CrossRef]
32. Ries, F.; Nishad, K.; Dressler, L.; Janicka, J.; Sadiki, A. Evaluating large eddy simulation results based on error analysis. *Theor. Comput. Fluid Dyn.* **2018**, 1–20. [CrossRef]
33. Pope, S.B. *Turbulence Flow*; Cambridge University Press: Cambridge, UK, 2000.
34. Chrigui, M.; Masri, A.R.; Sadiki, A.; Janicka, J. Large Eddy Simulation of a Polydisperse Ethanol Spray Flame. *Flow Turbul. Combust.* **2013**, *90*, 813–832. [CrossRef]
35. Torres, D.J.; O'rourke, P.J.; Amsden, A.A. Efficient multicomponent fuel algorithm. *Combust. Theory Model.* **2003**, *7*, 66–86. [CrossRef]
36. Munnannur, A.; Chiruta, M.; Liu, Z.G. Thermal and Fluid Dynamic Considerations in Aftertreatment System Design for SCR Solid Deposit Mitigation. In Proceedings of the SAE 2012 World Congress & Exhibition, Detroit, MI, USA, 24–26 April 2012; SAE International: Warrendale, PA, USA, 2012.
37. Börnhorst, M.; Deutschmann, O. Single droplet impingement of urea water solution on a heated substrate. *Int. J. Heat Fluid Flow* **2018**, *69*, 55–61. [CrossRef]

38. Kuhnke, D. Spray Wall Interaction Modelling by Dimensionless Data Analysis. Ph.D. Thesis, Department of Mathematics, Darmstadt University of Technology, Darmstadt, Germany, 2004.
39. O'Rourke, P.J.; Amsden, A.A. A Spray/Wall Interaction Submodel for the KIVA-3 Wall Film Model. In Proceedings of the SAE 2000 World Congress, Detroit, MI, USA, 6–9 March 2000; SAE International: Warrendale, PA, USA, 2000.
40. Nishad, K. Modeling and Unsteady Simulation of Turbulent Multi-Phase Flow Including Fuel Injection in IC-Engines. Ph.D. Thesis, Department of Mechanical Engineering, Technische Universität Darmstadt, Darmstadt, Germany, 2013.
41. Nishad, K.; Pischke, P.; Goryntsev, D.; Sadiki, A.; Kneer, R. LES Based Modeling and Simulation of Spray Dynamics including Gasoline Direct Injection (GDI) Processes using KIVA-4 Code. In Proceedings of the SAE 2012 World Congress & Exhibition, Detroit, MI, USA, 24–26 April 2012; SAE International: Warrendale, PA, USA, 2012.
42. Goryntsev, D.; Nishad, K.; Sadiki, A.; Janicka, J. Application of LES for Analysis of Unsteady Effects on Combustion Processes and Misfires in DISI Engine. *Oil Gas Sci. Technol. Rev. IFP Energies Nouvelles* **2014**, *69*, 129–140. [CrossRef]
43. Stauch, R.; Lipp, S.; Maas, U. Detailed numerical simulations of the autoignition of single n-heptane droplets in air. *Combust. Flame* **2006**, *145*, 533–542. [CrossRef]
44. Maas, U. Mathematische Modellierung instationärer Verbrennungsprozesse unter verwendung detaillierter chemischer Reaktionsmechanismen. Ph.D. Thesis, Universität Heidelberg, Heidelberg, Germany, 1988.
45. Smith, L.J.; Berendsen, H.J.C.; van Gunsteren, W.F. Computer simulation of urea-water mixtures: A test of force field parameters for use in biomolecular simulation. *J. Phys. Chem. B* **2004**, *108*, 1065–1071. [CrossRef]
46. Kontin, S.; Höfler, A.; Koch, R.; Bauer, H.J. Heat and Mass Transfer accompanied by Crystallisation of single Particles containing Urea-water-solution. In Proceedings of the 23rd Annual Conference on Liquid Atomization and Spray Systems, Brno, Czech Republic, 12–14 September 2010.
47. Ryddner, D.T.; Trujillo, M.F. Modeling Urea-Water Solution Droplet Evaporation. *Emiss. Control Sci. Technol.* **2015**, *1*, 80–97. [CrossRef]
48. Wei, L.; Youtong, Z.; Asif, M. Investigation on UWS Evaporation for Vehicle SCR Applications. *AIChE J.* **2016**, *62*, 880–890. [CrossRef]
49. Yaws, C.L. *Chemical Properties Handbook*; McGraw-Hill: New York, NY, USA, 1999.
50. Reid, R.; Prausnitz, J.; Poling, B. *The Properties of Gases and Liquids*, 4th ed.; McGraw–Hill: New York, NY, USA, 1989.
51. Maas, U.; Warnatz, J. Ignition processes in hydrogen-oxygen mixtures. *Combust. Flame* **1988**, *74*, 53 – 69. [CrossRef]
52. Stauch, R. Detaillierte Simulation von Verbrennungsprozessen in Mehrphasensystemen. Ph.D. Thesis, Universität Karlsruhe, Karlsruhe, Germany, 2007.
53. Deuflhard, P.; Hairer, E.; Zugck, J. One-step and Extrapolation Methods for Differential-Algebraic systems. *Numer. Math.* **2004**, *51*, 501–516. [CrossRef]
54. Fede, P.; Simonin, O.; Villedieu, P.; Squires, K. Stochastic Modeling of the Turbulent Subgrid Fluid Velocity Along Inertial Particle Trajectories. In Proceedings of the Summer Program Center for Turbulence Research Stanford University, Stanford, CA, USA, 4–9 July 2006.
55. Ahmadi, W.; Mehdizadeh, A.; Chrigui, M.; Sadiki, A. Numerical Evaluation of Unsteadiness in Particle Dispersion Modeling. *J. Fluids Eng.* **2014**, *137*, 034502. [CrossRef]

MDPI

St. Alban-Anlage 66

4052 Basel

Switzerland

Tel. +41 61 683 77 34

Fax +41 61 302 89 18

www.mdpi.com

Energies Editorial Office

E-mail: energies@mdpi.com

www.mdpi.com/journal/energies